# PMI-CPMAI Practice Exam Workbook



550+ Situational, Scenario-Based Questions & Detailed Answers Explanations

Ronald Clarence Squash

**First Edition: November, 2025**

**ISBN: 978-1-923646-15-5**

# Table of Contents

# Chapter 1.0: AI Fundamentals and Project Management

The PMI-CPMAI certification starts with a crucial foundation: understanding what AI actually is, when to use it, and how it differs from traditional automation. This chapter isn't about making you a data scientist—it's about equipping you to make smart decisions about AI projects and lead them effectively.

## 1.1 Understanding AI versus Automation

Here's where most AI projects go wrong from day one: teams confuse automation with artificial intelligence. Let me be clear about this. **Automation** means using machines to repeat deterministic, rule-based tasks. You can code the rules explicitly. **Artificial intelligence** handles probabilistic, learning-based tasks where you can't easily code all the rules (Cognilytics, 2024).

Think about processing insurance claims. If you're automating the extraction of data from forms that always have the same fields in the same locations, that's automation—robotic process automation (RPA), specifically. But if you need a system that can read handwritten doctor's notes, understand context, and make judgment calls about medical necessity? That requires AI.

The CPMAI methodology emphasizes this distinction because applying AI frameworks to simple automation projects wastes resources and creates unnecessary complexity (Project Management Institute, 2025). Consider this real-world scenario:

**Case Example: The Manufacturing Mix-Up**

Carlos, a project manager at a mid-sized manufacturing company, was asked to lead what his executives called an "AI initiative" to optimize inventory ordering. The company wanted to automatically reorder parts when stock fell below certain thresholds.

Carlos spent his first week digging into the requirements. What he found was straightforward: IF inventory drops below X units, THEN order Y units. The rules were deterministic and could be explicitly coded. This wasn't an AI problem—it was an automation problem.

By recognizing this early, Carlos saved his company from purchasing expensive ML platforms and hiring data scientists they didn't need. Instead, he implemented a simple RPA solution that cost 80% less and delivered results in weeks instead of months. The project succeeded because Carlos understood the fundamental question: Can you code the rules, or do you need the system to learn patterns from data?

## 1.2 The Seven Patterns of AI

The CPMAI framework identifies seven distinct patterns of AI applications (PMI, 2025). Understanding these patterns helps you match business problems to appropriate AI solutions. Let's break them down with practical examples.

**Conversational AI** enables natural language interactions between humans and machines. Think chatbots, virtual assistants, and voice interfaces. A hospital system might deploy conversational AI to handle appointment scheduling, answer basic medical questions, and triage patient concerns before they reach human staff.

**Recognition patterns** involve identifying objects, faces, text, speech, or other entities in unstructured data. A retail bank using computer vision to automatically process check deposits is applying recognition AI. The system must learn to read various handwriting styles, handle poor image quality, and distinguish legitimate checks from fraudulent ones.

**Predictive analytics** forecasts future outcomes based on historical data patterns. An e-commerce company predicting which customers are likely to churn within the next 90 days uses predictive AI. The model learns patterns from past customer behavior—purchase frequency, customer service interactions, browsing patterns—to identify at-risk accounts.

**Goal-driven systems** (also called autonomous systems) make sequential decisions to achieve defined objectives. A supply chain optimization system that determines the most efficient routing for hundreds of delivery trucks, accounting for traffic patterns, delivery windows, vehicle capacity, and fuel costs, exemplifies goal-driven AI.

**Anomaly detection** identifies unusual patterns that deviate from expected behavior. Credit card fraud detection systems use anomaly detection to flag suspicious transactions that don't match a cardholder's typical spending patterns.

**Hyper-personalization** delivers individualized experiences based on detailed user profiles and behaviors. Netflix's recommendation engine uses hyper-personalization AI to suggest content based not just on what you've watched, but when you watch, how long you watch, what you rewatch, and how your patterns compare to similar users (Project Management Institute, 2025).

**Autonomous systems** operate independently, making real-time decisions without human intervention. Self-driving vehicles represent the most complex form of autonomous systems, integrating recognition (seeing pedestrians and obstacles), prediction (anticipating other vehicles' movements), and goal-driven decision-making (reaching the destination safely).

## 1.3 Machine Learning Fundamentals

Machine learning enables computers to learn patterns from data without being explicitly programmed for every scenario (Russell & Norvig, 2020). The three main learning paradigms are supervised learning, unsupervised learning, and reinforcement learning.

**Supervised learning** trains models on labeled data—you show the algorithm examples with known correct answers. Imagine training an email spam filter. You provide thousands of emails labeled "spam" or "not spam," and the model learns patterns that distinguish spam (certain keywords, sender patterns, link structures) from legitimate emails. When new emails arrive, the model applies these learned patterns to classify them.

Common supervised learning tasks include classification (categorizing items into predefined groups) and regression (predicting continuous numerical values). A hospital predicting patient readmission risk within 30 days uses classification. A real estate company predicting home sale prices uses regression.

**Unsupervised learning** finds patterns in unlabeled data. You give the algorithm data without telling it what to look for—it discovers structure on its own. Customer segmentation often uses unsupervised learning. A retailer might analyze purchase histories, demographics, and browsing behavior to identify natural customer groups without predefined categories.

The algorithm might discover that one segment consists of "high-value, infrequent luxury shoppers," another of "frequent, price-sensitive everyday purchasers," and a third of "seasonal, gift-focused buyers." Nobody told the algorithm to look for these specific groups—it found meaningful patterns in the data.

**Reinforcement learning** trains agents through trial and error, rewarding desired behaviors and penalizing undesired ones. A robot learning to navigate a warehouse uses reinforcement learning. It tries different paths, receives positive rewards for efficient routes that avoid obstacles and negative rewards for collisions or inefficient paths. Over thousands of trials, it learns optimal navigation strategies (Sutton & Barto, 2018).

## 1.4 Neural Networks and Deep Learning

**Neural networks** are computing systems inspired by biological neural networks in animal brains. They consist of interconnected nodes (neurons) organized in layers: an input layer, one or more hidden layers, and an output layer. Information flows through the network, with each connection having a weight that adjusts as the network learns.

**Deep learning** uses neural networks with many hidden layers—hence "deep"—enabling the model to learn increasingly abstract representations of data. Early layers might detect simple features (edges in images), while deeper layers recognize complex patterns (faces, objects, scenes).

Deep learning excels at handling unstructured data like images, speech, and text. A medical imaging system using deep learning to detect cancerous tumors in CT scans might have a network with 50+ layers. Early layers identify basic visual features like edges and textures. Middle layers recognize anatomical structures. Deep layers learn subtle patterns that distinguish malignant from benign growths.

The power of deep learning comes with costs: these models require massive amounts of training data, substantial computing resources, and considerable expertise to develop and deploy effectively. They're also "black boxes"—it's often difficult to explain why they made specific decisions, which raises concerns in regulated industries (Goodfellow et al., 2016).

## 1.5 Generative AI and Large Language Models

**Generative AI** creates new content—text, images, code, music—rather than just analyzing or classifying existing content. This represents a fundamental shift in AI capabilities. Instead of asking "Is this image a cat or a dog?" you're asking "Generate an image of a cat wearing a business suit."

**Large Language Models (LLMs)** like GPT-4 and Claude are trained on vast amounts of text data to understand and generate human language. They learn statistical patterns in how words, sentences, and ideas relate to each other. An LLM doesn't "understand" language the way humans do, but it predicts probable next words based on context with remarkable accuracy (Brown et al., 2020).

Organizations are deploying LLMs for customer service automation, content generation, code assistance, document summarization, and knowledge extraction. A legal firm might use an LLM to review contracts, identify potential issues, and draft standard clauses—tasks that previously required hours of attorney time.

The game-changer with generative AI is its versatility. Traditional AI models are narrow—trained for one specific task. LLMs demonstrate broader capabilities across diverse tasks without task-specific training. This flexibility makes them powerful tools, but also introduces new challenges around accuracy, bias, and responsible use.

## 1.6 When to Choose AI

Not every problem needs an AI solution. The CPMAI methodology provides clear guidance: use AI when you can't easily code the rules, when you can't scale with people alone, or when patterns are probabilistic rather than deterministic (Cognilytics, 2024).

Ask these questions before committing to an AI approach:

**Can you write explicit rules?** If yes, use traditional automation. Processing expense reports with clear approval thresholds doesn't need machine learning—it needs good business logic.

**Is the pattern deterministic or probabilistic?** Deterministic patterns (every input produces one correct output) suit traditional programming. Probabilistic patterns (multiple possible correct outputs, context-dependent decisions) may need AI.

**Do you have sufficient quality data?** AI models learn from data. Without adequate, representative, labeled data, even the best algorithms fail. A company with six months of sales data probably can't build reliable demand forecasting models.

**Is the problem's value worth AI's cost?** Training, deploying, and maintaining AI systems requires significant resources. Make sure the business value justifies the investment.

**Case Example: The Insurance Intelligence Decision**

Priya leads technology projects at a regional insurance company. Her CEO asked her to explore AI for two initiatives: automated document classification and personalized policy recommendations.

For document classification—sorting incoming claims, medical records, and correspondence into categories—Priya found the problem perfectly suited for AI. Documents arrived in hundreds of formats with inconsistent layouts. Writing rules to handle every variation would be impractical. A recognition AI trained on historical documents could learn to classify new documents accurately.

For policy recommendations, Priya discovered the company already had explicit business rules: customer age, income, dependents, assets, and risk tolerance mapped directly to product recommendations. No learning was needed—the patterns were deterministic. She recommended a rules engine, not AI.

By correctly matching problems to solutions, Priya's projects delivered faster, cost less, and worked more reliably than if she'd forced AI into both scenarios.

## 1.7 The CPMAI Methodology

The CPMAI framework provides a structured, iterative approach to managing AI projects through six phases: Business Understanding, Data Understanding, Data Preparation, Model Development, Model Evaluation, and Model Operationalization (PMI, 2025).

Unlike traditional waterfall project management, CPMAI emphasizes iteration. You might reach Model Development, discover data quality issues, and return to Data Preparation. This flexibility acknowledges AI's inherent uncertainty—you often don't know if an approach will work until you try it.

Each phase includes go/no-go criteria to assess project viability before investing more resources. After Business Understanding, you evaluate whether the problem is appropriate for AI and whether expected ROI justifies costs. After Data Understanding, you determine whether adequate data exists and is accessible.

This iterative, gated approach reduces the catastrophic failures that plague AI projects. Rather than discovering nine months in that your data is inadequate, you identify that issue in week three and either fix it or pivot (Piatetsky, 2019).

The methodology is vendor-agnostic and tool-agnostic. Whether you're using Python, R, TensorFlow, PyTorch, cloud platforms, or on-premise infrastructure, CPMAI provides the framework for successful project delivery.

## 1.8 Core Concepts Recap

AI differs fundamentally from automation—it handles probabilistic patterns and learns from data rather than following explicit rules. The seven patterns of AI (conversational, recognition, predictive, goal-driven, anomaly detection, hyper-personalization, and autonomous systems) provide a taxonomy for matching business problems to AI approaches.

Machine learning comes in three main flavors: supervised learning with labeled data, unsupervised learning finding patterns in unlabeled data, and reinforcement learning through trial and error. Deep learning extends traditional neural networks with many layers, enabling breakthrough performance on unstructured data like images and text. Generative AI and LLMs represent the cutting edge, creating new content rather than just analyzing existing content.

Choosing AI requires careful analysis. Use it when rules can't be explicitly coded, patterns are probabilistic, and sufficient quality data exists. The CPMAI methodology provides six iterative phases with go/no-go gates to reduce project failure risk and increase delivery success.

## 1.9 Practice Questions

### Pattern Matching Scenarios

**Question 1:** A retail company wants to automatically categorize thousands of customer service emails into topics like "shipping issues," "product defects," "billing questions," and "returns." Which AI pattern is most appropriate?

A) Goal-driven systems B) Recognition (text classification) C) Anomaly detection D) Hyper-personalization

**Answer: B) Recognition (text classification)**

**Explanation:** Text classification is a recognition pattern where the system identifies and categorizes text into predefined groups. This is distinct from conversational AI (which would respond to emails) or hyper-personalization (which would tailor responses to individuals).

---

**Question 2:** A hospital emergency department receives hundreds of patients daily and needs a system that determines the optimal allocation of medical staff, equipment, and room assignments while considering patient severity, wait times, and resource constraints. Which pattern best fits?

A) Predictive analytics B) Conversational AI C) Goal-driven (autonomous) systems D) Recognition

**Answer: C) Goal-driven (autonomous) systems**

**Explanation:** Goal-driven systems make sequential decisions to achieve defined objectives while considering multiple constraints. The ED system must optimize resource allocation across multiple variables to achieve the goal of efficient, safe patient care.

---

**Question 3:** A manufacturing company wants to identify unusual vibration patterns in equipment that might indicate impending failure, even though they've never seen these specific patterns before. Which AI pattern applies?

A) Predictive analytics B) Anomaly detection C) Recognition D) Hyper-personalization

**Answer: B) Anomaly detection**

**Explanation:** Anomaly detection identifies unusual patterns that deviate from normal behavior without requiring labeled examples of failures. This differs from predictive analytics, which forecasts specific outcomes based on historical patterns.

---

**Question 4:** An e-commerce platform wants to show different homepage layouts, product recommendations, pricing, and marketing messages to each customer based on their browsing history, purchase patterns, demographics, and real-time behavior. Which pattern is this?

A) Recognition B) Predictive analytics C) Hyper-personalization D) Conversational AI

**Answer: C) Hyper-personalization**

**Explanation:** Hyper-personalization delivers individualized experiences based on detailed user profiles and behaviors. While this uses predictive analytics as a component, the defining characteristic is the individualized tailoring of the entire experience.

---

**Question 5:** A logistics company needs a chatbot that can answer questions about shipping status, update delivery addresses, and help customers file claims through natural language conversations. Which pattern is primary?

A) Conversational AI B) Recognition C) Goal-driven systems D) Hyper-personalization

**Answer: A) Conversational AI**

**Explanation:** Conversational AI enables natural language interactions between humans and machines. While the system might use recognition for understanding text and goal-driven logic for processing requests, the primary pattern is conversation.

---

## Technology Selection Dilemmas

**Question 6:** Your team needs to forecast monthly sales for 500 products across 100 stores. You have three years of historical sales data, promotional calendars, and weather data. Which approach should you start with?

A) Deep neural networks with LSTM layers B) Ensemble of gradient boosting machines C) Simple linear regression as baseline, then try tree-based methods D) Reinforcement learning agents

**Answer: C) Simple linear regression as baseline, then try tree-based methods**

**Explanation:** Always start with simple baselines to establish minimum performance expectations. Linear regression provides an interpretable starting point. Tree-based methods (random forests, gradient boosting) handle non-linear relationships well for tabular data without requiring the complexity of deep learning.

---

**Question 7:** A radiology department has 50,000 labeled chest X-rays and wants to detect pneumonia. Radiologists need to understand why the system flagged specific images. What approach balances performance and explainability?

A) Deep CNN trained from scratch B) Pretrained CNN (ResNet/EfficientNet) with fine-tuning plus SHAP/Grad-CAM for explanations C) Simple logistic regression on hand-crafted features D) Unsupervised clustering of image features

**Answer: B) Pretrained CNN with fine-tuning plus SHAP/Grad-CAM for explanations**

**Explanation:** Transfer learning with pretrained CNNs delivers strong performance on medical images. Grad-CAM highlights which image regions influenced predictions, and SHAP provides feature importance. This balances performance needs with explainability requirements for clinical use.

---

**Question 8:** You're building a real-time fraud detection system that must evaluate credit card transactions in under 50 milliseconds. You have 100 million historical transactions. Which model type is most appropriate?

A) Deep neural network with 10 hidden layers B) Gradient boosting with 1,000 trees C) Logistic regression or shallow neural network D) K-nearest neighbors with full dataset

**Answer: C) Logistic regression or shallow neural network**

**Explanation:** Latency requirements demand fast inference. Logistic regression and shallow networks provide sub-millisecond inference times. Deep networks and large ensembles, while potentially more accurate, can't meet strict latency requirements. K-NN is too slow for real-time scoring at scale.

---

**Question 9:** A startup with limited resources wants to build a document classification system. They have 5,000 labeled documents and need production deployment in four weeks. What's the best approach?

A) Train a transformer model from scratch B) Fine-tune a pretrained language model (BERT/RoBERTa) C) Develop custom neural architecture after extensive research D) Wait to collect 1 million labeled documents first

**Answer: B) Fine-tune a pretrained language model**

**Explanation:** Pretrained language models deliver strong performance with limited data and training time. Fine-tuning takes hours to days rather than weeks. Training from scratch requires massive data and compute. The 5,000 labeled documents are sufficient for fine-tuning but insufficient for training from scratch.

---

**Question 10:** Your model needs to process customer support calls in real-time to assist agents, but your organization prohibits sending call audio to external cloud services due to privacy concerns. What deployment strategy applies?

A) Real-time cloud API deployment B) Batch processing overnight C) Edge deployment on local servers D) Manual processing by human analysts

**Answer: C) Edge deployment on local servers**

**Explanation:** Privacy constraints requiring on-premises processing combined with real-time needs dictate edge deployment. Local servers can run models without sending data externally. Cloud deployment violates privacy requirements; batch processing doesn't meet real-time needs.

---

## AI vs. Non-AI Solution Decisions

**Question 11:** A company wants to automate approval of expense reports following clear rules: amounts under $500 approved automatically, amounts $500-$2,000 require manager approval, amounts over $2,000 require director approval. Should they use AI?

A) Yes, this is perfect for machine learning B) Yes, but only deep learning can handle this complexity C) No, use rule-based automation (RPA) D) No, keep manual processing

**Answer: C) No, use rule-based automation (RPA)**

**Explanation:** The decision logic is deterministic and explicitly codable. AI is unnecessary when rules are clear and don't require pattern learning. Rule-based automation is simpler, more reliable, more explainable, and less expensive than AI for this use case.

---

**Question 12:** A bank receives 10,000 loan applications monthly in various formats—PDFs, scanned documents, photos, handwritten forms. They want to automatically extract applicant information (name, income, employment). Is AI appropriate?

A) No, rules can extract all information reliably B) Yes, use recognition AI (OCR + NLP) to handle format variability C) No, the problem is too complex for current AI D) Yes, but only for handwritten forms

**Answer: B) Yes, use recognition AI (OCR + NLP) to handle format variability**

**Explanation:** The variability in formats makes rule-based extraction impractical. OCR handles different document formats, and NLP extracts structured information from unstructured text. This is a probabilistic problem where rules can't easily be coded for every variation.

---

**Question 13:** A manufacturing plant wants to monitor temperature sensors on equipment and trigger alerts when readings exceed defined thresholds (above 150°F or below 32°F). Is AI needed?

A) Yes, anomaly detection AI is required B) Yes, predictive maintenance models are necessary C) No, simple threshold-based monitoring suffices D) Yes, deep learning can detect subtle patterns

**Answer: C) No, simple threshold-based monitoring suffices**

**Explanation:** The monitoring logic is deterministic with explicit thresholds. Simple rule-based systems handle this reliably. AI would be appropriate if they wanted to predict failures before they occur or detect unusual patterns, but for threshold-based alerting, rules are sufficient.

---

**Question 14:** An insurance company reviews 1,000 claims daily, each requiring evaluation of medical records, policy terms, treatment appropriateness, and fraud indicators. Experienced adjusters take 45 minutes per claim. Should they use AI?

A) No, claims are too complex for AI B) Yes, use AI to fully automate all claim decisions C) Yes, use AI to assist adjusters by flagging potential issues and providing recommendations D) No, hire more adjusters instead

**Answer: C) Yes, use AI to assist adjusters by flagging potential issues**

**Explanation:** The complexity and volume justify AI, but full automation is inappropriate for high-stakes decisions. AI augmentation helps adjusters by highlighting relevant information, detecting patterns, and providing recommendations while keeping humans in the decision loop.

---

**Question 15:** A retail chain wants to predict which products will be popular next season. Historical sales patterns exist, but fashion trends, economic conditions, and competitor actions create uncertainty. Is AI appropriate?

A) No, prediction is impossible with so much uncertainty B) Yes, predictive analytics can forecast demand based on historical patterns and external signals C) No, use fixed ordering based on last year's sales D) Yes, but only if you have 50+ years of data

**Answer: B) Yes, predictive analytics can forecast demand**

**Explanation:** Uncertainty doesn't eliminate the value of prediction—it makes it more valuable. Predictive models learn from historical patterns and can incorporate external signals (economic indicators, trends). While predictions won't be perfect, they improve decision-making versus ignoring data.

---

## Foundational Concept Applications

**Question 16:** You're training a spam classifier with 100,000 emails, of which 1% are spam. Using standard accuracy metrics, a model predicting "not spam" for all emails achieves 99% accuracy. What's the problem?

A) 99% accuracy is excellent; deploy immediately B) The model has high accuracy but fails completely at the actual task C) The training data is too small D) The model is overfitting

**Answer: B) The model has high accuracy but fails completely**

**Explanation:** This illustrates the importance of choosing appropriate metrics for imbalanced datasets. The model achieves 99% accuracy by exploiting class imbalance but catches zero spam emails. Precision, recall, and F1-score reveal the failure that accuracy masks.

---

**Question 17:** A supervised learning model needs to classify customer reviews as positive, neutral, or negative. What type of machine learning problem is this?

A) Regression B) Binary classification C) Multi-class classification D) Clustering

**Answer: C) Multi-class classification**

**Explanation:** Classification assigns examples to discrete categories. Binary classification has two classes; multi-class has three or more. This problem has three sentiment classes (positive, neutral, negative), making it multi-class classification.

---

**Question 18:** You're building a recommendation system but don't have explicit ratings from users—only implicit data like clicks, views, and purchases. Which learning approach applies?

A) Supervised learning with labeled ratings B) Unsupervised learning to find patterns in behavior C) Reinforcement learning through trial and error D) Transfer learning from other domains

**Answer: B) Unsupervised learning to find patterns**

**Explanation:** Without explicit labels (ratings), you must find patterns in implicit behavior. Collaborative filtering, matrix factorization, and clustering techniques discover user preferences from behavioral signals without requiring labeled training data.

---

**Question 19:** Your neural network achieves 95% training accuracy but only 65% validation accuracy. What problem does this indicate?

A) Underfitting B) Overfitting C) Perfect performance D) Insufficient data

**Answer: B) Overfitting**

**Explanation:** Large gap between training and validation performance indicates overfitting—the model memorized training data rather than learning generalizable patterns. Solutions include regularization, dropout, more training data, or simpler model architecture.

---

**Question 20:** A robotics company is training a warehouse robot to navigate optimally by rewarding efficient paths and penalizing collisions. Which learning paradigm applies?

A) Supervised learning B) Unsupervised learning C) Reinforcement learning D) Transfer learning

**Answer: C) Reinforcement learning**

**Explanation:** Reinforcement learning trains agents through trial and error with rewards and penalties. The robot learns optimal navigation policies by receiving positive rewards for efficient paths and negative rewards for collisions, without explicit examples of correct actions for each situation.

---

**Question 21:** You have customer data with features like age, income, purchase history, and website behavior, but no labels indicating customer lifetime value. You want to group similar customers together. What type of learning applies?

A) Supervised classification B) Supervised regression C) Unsupervised clustering D) Reinforcement learning

**Answer: C) Unsupervised clustering**

**Explanation:** Without labels, unsupervised learning finds patterns. Clustering algorithms like K-means or hierarchical clustering group similar customers based on feature similarity. This discovers natural customer segments without predefined categories.

---

**Question 22:** A convolutional neural network (CNN) is most appropriate for which type of data?

A) Tabular data with 50 numerical features B) Time series sensor readings C) Images and visual data D) Graph-structured social networks

**Answer: C) Images and visual data**

**Explanation:** CNNs excel at image processing through convolutional layers that detect spatial patterns. While CNNs can sometimes work on other data types, they're specifically designed for and most appropriate for visual data where spatial relationships matter.

---

**Question 23:** Your project requires predicting house sale prices (continuous values) based on features like square footage, bedrooms, location, and age. What type of problem is this?

A) Binary classification B) Multi-class classification C) Regression D) Clustering

**Answer: C) Regression**

**Explanation:** Regression predicts continuous numerical values rather than discrete categories. House prices are continuous values that can take any number within a range, making this a regression problem.

---

**Question 24:** You're building a fraud detection model and discover that 99.5% of training examples are legitimate transactions. What challenge does this present?

A) Too much data; delete some examples B) Imbalanced dataset requiring special handling C) Perfect dataset; proceed normally D) Insufficient data; collect more fraud examples

**Answer: B) Imbalanced dataset requiring special handling**

**Explanation:** Severe class imbalance (0.5% fraud rate) requires special techniques: oversampling minority class, undersampling majority class, class weighting, or specialized algorithms. Standard training approaches fail on highly imbalanced data.

---

**Question 25:** A language model like GPT is an example of what type of AI?

A) Recognition AI B) Generative AI C) Anomaly detection D) Goal-driven system

**Answer: B) Generative AI**

**Explanation:** Generative AI creates new content (text, images, code) rather than classifying or analyzing existing content. Language models like GPT generate text by predicting probable next words, making them generative models.

---

## Advanced Scenario Questions

**Question 26:** A hospital has 500 CT scans showing a rare condition and 50,000 scans without it. They want to train a detection model. What's the biggest challenge?

A) Too much data to process B) Severe class imbalance (1% positive rate) C) Insufficient computing power D) Data privacy regulations

**Answer: B) Severe class imbalance**

**Explanation:** With only 1% positive examples, models can achieve 99% accuracy by always predicting negative. This requires resampling techniques, class weighting, careful metric selection (precision/recall instead of accuracy), and possibly data augmentation for the minority class.

---

**Question 27:** You're deciding between a random forest with 500 trees and logistic regression for a production credit scoring model. The random forest achieves 82% accuracy; logistic regression achieves 78%. Regulators require explainability. What should you choose?

A) Random forest for better accuracy B) Logistic regression for explainability C) Neither; wait for perfect model D) Deploy both and average predictions

**Answer: B) Logistic regression for explainability**

**Explanation:** Regulatory requirements for explainability trump small accuracy differences. Logistic regression provides clear feature coefficients showing how each variable affects decisions. Random forests, while more accurate, are harder to explain to regulators and applicants.

---

**Question 28:** Your team built an excellent model on data from January-October. It's now November, and performance has dropped 15%. What's the most likely cause?

A) The model stopped working B) Seasonal patterns not present in training data C) Hardware failure D) Too many users

**Answer: B) Seasonal patterns not in training data**

**Explanation:** Models trained on January-October data lack November-December holiday season patterns. This temporal distribution shift (concept drift) causes performance degradation. The solution is retraining with data spanning full seasonal cycles.

---

**Question 29:** A startup has 1,000 labeled examples and needs high accuracy. A senior data scientist suggests: "Let's build an ensemble of gradient boosting, random forest, and neural networks." What's the risk?

A) Ensembles always overfit B) With limited data, complex ensembles may overfit C) This approach is too simple D) Ensembles never work

**Answer: B) With limited data, complex ensembles may overfit**

**Explanation:** Complex ensembles have many parameters and require substantial data. With only 1,000 examples, simpler models (single gradient boosting or random forest) may generalize better. Start simple, increase complexity only if validation performance improves.

---

**Question 30:** You're choosing between batch predictions (overnight) and real-time predictions (API calls) for a marketing campaign system that selects offers to display. Campaign decisions are made daily. Which deployment approach fits?

A) Real-time API predictions for immediate response B) Batch predictions overnight for next day's campaigns C) Stream processing for continuous updates D) Manual decisions without AI

**Answer: B) Batch predictions overnight**

**Explanation:** Daily decision-making doesn't require real-time predictions. Batch processing overnight provides predictions for the next day's campaigns efficiently. Real-time API deployment adds complexity and cost without business benefit when decisions are made daily.

---

**Question 31:** Your model uses 200 input features. After training, you discover 150 features have near-zero importance. What should you do?

A) Keep all features; more is better B) Remove low-importance features and retrain C) Collect 1,000 more features D) Start over with different algorithms

**Answer: B) Remove low-importance features and retrain**

**Explanation:** Removing irrelevant features reduces overfitting risk, speeds training and inference, improves interpretability, and simplifies maintenance. Feature selection identifying truly important variables often improves model performance while reducing complexity.

---

**Question 32:** A financial services company wants to use AI for loan approvals but faces strict regulations requiring they explain rejection reasons to applicants. Which principle is most critical?

A) Model accuracy above 95% B) Model explainability and interpretability C) Training speed under 1 hour D) Using the latest deep learning architectures

**Answer: B) Model explainability and interpretability**

**Explanation:** Regulatory requirements for explanation trump other considerations. Models must provide clear, understandable reasons for decisions. This might mean sacrificing some accuracy for interpretability or using post-hoc explanation methods.

---

**Question 33:** You're evaluating three approaches: (1) rules-based system, (2) machine learning, (3) deep learning. The problem has well-defined rules covering 95% of cases but complex edge cases. What's the best approach?

A) Pure rules-based system B) Pure machine learning C) Hybrid: rules for common cases, ML for edge cases D) Pure deep learning

**Answer: C) Hybrid: rules for common cases, ML for edge cases**

**Explanation:** Hybrid approaches leverage strengths of each method. Well-defined rules handle the 95% reliably, simply, and explainably. Machine learning handles the 5% of complex cases where rules are impractical. This is often optimal for problems with clear rules for most cases.

---

**Question 34:** A model performs well on the test set but fails in production. Investigation reveals production users behave differently than test users. What went wrong?

A) Model overfitting B) Training-serving skew due to distribution mismatch C) Hardware limitations D) Insufficient training data

**Answer: B) Training-serving skew due to distribution mismatch**

**Explanation:** When training/test data doesn't represent production users, models fail despite good test metrics. This requires collecting representative data from production populations or weighting training data to match production distributions.

---

**Question 35:** Your company has proprietary business rules developed over 20 years. A vendor claims their "AI black box" will outperform your rules but can't explain how. What's the primary risk?

A) The AI might actually work too well B) Loss of accumulated business knowledge and inability to debug failures C) AI is always worse than rules D) Vendor solutions are always perfect

**Answer: B) Loss of accumulated business knowledge**

**Explanation:** Black box systems that can't be explained risk abandoning validated business logic without understanding what the AI learned. When failures occur (and they will), debugging is impossible. Critical business applications need explainability.

---

**Question 36:** You're building a computer vision model for quality inspection in manufacturing. You have 1,000 labeled images. A pretrained ImageNet model is available. What should you do?

A) Train from scratch; 1,000 images is plenty B) Use transfer learning with the pretrained model C) Give up; you need 10 million images D) Use traditional computer vision without ML

**Answer: B) Use transfer learning with pretrained model**

**Explanation:** Transfer learning leverages knowledge from ImageNet's millions of images. Fine-tuning pretrained models on 1,000 domain-specific images typically outperforms training from scratch, which would likely overfit on limited data.

---

**Question 37:** A retail forecasting model trained on pre-pandemic data (2017-2019) performs poorly in 2024. What's the issue?

A) The model is broken B) Concept drift; consumer behavior changed post-pandemic C) Insufficient training data D) Wrong algorithm choice

**Answer: B) Concept drift; consumer behavior changed**

**Explanation:** Major events like pandemics fundamentally change patterns. E-commerce adoption, work-from-home trends, and shopping behaviors shifted. Models trained on pre-pandemic data don't capture these changes. Retraining on recent data is essential.

---

**Question 38:** You need to detect fraudulent insurance claims but have only 200 labeled fraud examples versus 100,000 legitimate claims. What technique helps most?

A) Delete legitimate claims until classes balance B) Use oversampling or SMOTE for minority class C) Ignore the imbalance D) Wait until you have 100,000 fraud examples

**Answer: B) Use oversampling or SMOTE for minority class**

**Explanation:** Oversampling techniques create synthetic fraud examples, helping models learn fraud patterns despite scarcity. SMOTE generates synthetic examples by interpolating between existing fraud cases. This addresses imbalance without waiting for more fraud (which is unethical to desire) or discarding valuable legitimate claim data.

---

**Question 39:** Your neural network training loss decreases steadily, but validation loss increases after epoch 20. What's happening and what should you do?

A) Perfect training; deploy the final model B) Overfitting; use early stopping at epoch 20 C) Underfitting; train longer D) Normal behavior; ignore it

**Answer: B) Overfitting; use early stopping at epoch 20**

**Explanation:** Diverging training and validation loss signals overfitting—the model memorizes training data rather than learning generalizable patterns. Early stopping saves the model at epoch 20 (before overfitting begins) rather than continuing to train as performance degrades.

---

**Question 40:** A healthcare provider wants to use AI for diagnosis. They have 10,000 patient records with diagnoses but realize 80% were diagnosed by one physician who may have systematic biases. What's the risk?

A) Too little data B) Model will learn and perpetuate the physician's biases C) Perfect training data D) Insufficient computing power

**Answer: B) Model will learn and perpetuate physician's biases**

**Explanation:** Models learn patterns in training data, including biases. If one physician's diagnostic patterns dominate training data, the model learns those patterns (including any systematic errors or biases). This requires diverse training data from multiple physicians and bias testing.

---

**Question 41:** You're comparing supervised learning (you have 10,000 labeled examples) versus unsupervised learning (using all data but no labels) for customer segmentation. Which approach is better?

A) Supervised learning with the 10,000 examples B) Unsupervised learning on all data C) Neither can work D) Both simultaneously

**Answer: B) Unsupervised learning on all data**

**Explanation:** Customer segmentation is inherently an unsupervised task—discovering natural groupings in data without predefined categories. Unless you have a specific target to predict (like customer lifetime value), unsupervised clustering is more appropriate than supervised learning.

---

**Question 42:** Your sentiment analysis model trained on Twitter data performs poorly on customer service emails. Why?

A) The algorithm is wrong B) Domain shift; Twitter language differs from business emails C) Insufficient training data D) Twitter is better than email

**Answer: B) Domain shift; different language patterns**

**Explanation:** Twitter uses informal language, slang, emojis, and character limits. Business emails are formal and structured. Models trained on one domain don't generalize to others without adaptation. This requires domain-specific training data or domain adaptation techniques.

---

**Question 43:** A project manager asks, "Should we use supervised or unsupervised learning?" What's the most important clarifying question?

A) How much compute power do you have? B) Do you have labeled data with known correct answers? C) What's your budget? D) How many data scientists are available?

**Answer: B) Do you have labeled data with known correct answers?**

**Explanation:** The supervised versus unsupervised decision fundamentally depends on label availability. Supervised learning requires labeled examples; unsupervised doesn't. Resources, budget, and team size matter but don't determine which learning paradigm applies.

---

**Question 44:** You're building a recommendation system for a new product line with no historical data. Which approach could work?

A) Collaborative filtering based on user-item interactions B) Content-based filtering using product attributes C) Supervised learning on past purchases D) Wait until you have 10 years of data

**Answer: B) Content-based filtering using product attributes**

**Explanation:** Cold start problems (no historical data) prevent collaborative filtering. Content-based approaches recommend items similar to what users liked previously based on product features (genre, price, category) rather than requiring historical interaction data.

---

**Question 45:** A model uses 50 features but three features account for 90% of prediction power. What's the best practice?

A) Add 100 more features B) Remove the three important features C) Consider simplifying to use primarily those three features D) Keep everything unchanged

**Answer: C) Consider simplifying to use primarily those three features**

**Explanation:** When few features dominate prediction, simpler models using just those features often perform nearly as well as complex models while being faster, more interpretable, and less prone to overfitting. Test simplified models before committing to full complexity.

---

**Question 46:** Your classification model outputs probabilities: 0.51 for Class A, 0.49 for Class B. How should you interpret this?

A) Definitely Class A B) High uncertainty; consider thresholding or human review C) Model failure D) Definitely Class B

**Answer: B) High uncertainty; consider thresholding or human review**

**Explanation:** Probabilities near 50% indicate high uncertainty. For high-stakes decisions, flag uncertain predictions for human review rather than acting on borderline classifications. Proper calibration and uncertainty quantification enable appropriate confidence-based decision-making.

---

**Question 47:** A startup claims their AI can predict stock prices with 99% accuracy. What's the most skeptical question to ask?

A) What programming language did you use? B) How did you evaluate accuracy? On what data? Using what metrics? C) How many GPUs do you have? D) What neural architecture did you use?

**Answer: B) How did you evaluate accuracy? On what data? Using what metrics?**

**Explanation:** Extraordinary claims require extraordinary evidence. Many fraudulent or mistaken accuracy claims come from inappropriate evaluation—testing on training data, not accounting for time series structure, or using misleading metrics. Evaluation methodology matters more than architectural details.

---

**Question 48:** Your model performs poorly on recent data. You retrain monthly using the last six months of data. What problem might this cause?

A) Models will improve steadily B) Potential "catastrophic forgetting" of older patterns C) This approach is always optimal D) Computational waste

**Answer: B) Potential catastrophic forgetting of older patterns**

**Explanation:** Retraining on only recent data risks forgetting patterns that occur less frequently or seasonally. Balance is needed—include enough historical data to capture seasonal patterns while incorporating recent data to adapt to drift.

---

**Question 49:** A team debates: "Should we optimize for precision or recall?" What context is essential to answer?

A) Size of training dataset B) Relative costs of false positives versus false negatives C) Choice of algorithm D) Computing power available

**Answer: B) Relative costs of false positives versus false negatives**

**Explanation:** Precision-recall tradeoffs depend on error costs. Medical screening prioritizes recall (catching all diseases) over precision (some false alarms acceptable). Spam filtering prioritizes precision (don't block legitimate email) over recall (some spam reaching inboxes acceptable). Business context determines the right balance.

---

**Question 50:** You have strong performance on validation data but stakeholders don't trust the model. What's missing?

A) Better accuracy B) More data C) Explainability and transparency D) Faster inference

**Answer: C) Explainability and transparency**

**Explanation:** Trust requires understanding. Even accurate models face adoption barriers without explainability. Stakeholders need to understand how models make decisions, what factors matter, and why predictions make sense. Technical performance is necessary but insufficient for organizational adoption.

---

# Chapter 2.0: Phase I Business Understanding

Phase I of the CPMAI methodology is where AI projects succeed or fail—usually fail, if we're being honest. Research consistently shows that 70-85% of AI projects never make it to production, and the vast majority of those failures trace back to poor business understanding (Gartner, 2022). Teams rush to model development without clearly defining what success looks like, why AI is necessary, or whether the business problem is even appropriate for an AI solution.

## 2.1 Formulating AI Business Questions

The first critical task in Phase I is formulating specific, measurable business questions that AI can address. This sounds simple. It's not. Most teams start with vague aspirations like "use AI to improve customer experience" or "leverage machine learning for operational efficiency." Those aren't business questions—they're wishes.

An AI-specific business question must be concrete enough to guide solution design and enable success measurement. Transform "improve customer experience" into "Can we predict which customers will abandon their shopping carts and intervene with personalized offers to increase conversion rates by 15%?" Now you have something actionable.

The CPMAI framework emphasizes the DIKUW pyramid—Data, Information, Knowledge, Understanding, Wisdom—as a tool for refining business questions (PMI, 2025). Start at the bottom: What data exists? Move up: What information can we extract? What knowledge can we build? What understanding does that provide? What wise business decisions result?

### Case Example: The Retail Revelation

Marcus, a program manager at a national retail chain, was tasked with an "AI initiative to boost sales." His executive sponsor kept saying "just get AI working in our stores" without defining what that meant.

Marcus scheduled stakeholder interviews and started asking uncomfortable questions: What specific business outcome do we want? What decisions would we make differently with AI? How would we measure success? What would justify the investment?

Through these conversations, Marcus discovered the real pain point: the company struggled with markdowns. Store managers waited too long to discount seasonal merchandise, resulting in dead stock that sold for pennies on the dollar. Discounting too aggressively left money on the table.

Marcus reformulated the vague initiative into a precise business question: "Can we predict optimal markdown timing and depth for seasonal merchandise to maximize revenue while minimizing leftover inventory?" That question drove everything else—data requirements, model design, success metrics, and deployment strategy.

The project succeeded because Marcus refused to start coding until he understood exactly what business problem he was solving.

## 2.2 Aligning AI with Business Objectives

AI projects must tie directly to strategic business objectives, not exist as science experiments. This alignment goes beyond saying "this project supports our digital transformation strategy." You need explicit connections between AI capabilities and measurable business outcomes (Davenport & Ronanki, 2018).

Start by identifying which business objectives the AI initiative serves. Are you aiming to increase revenue, reduce costs, improve customer satisfaction, mitigate risk, enhance operational efficiency, or gain competitive advantage? Get specific. "Increase revenue" becomes "increase revenue from existing customers by 12% through personalized product recommendations."

Next, map how AI capabilities enable those objectives. Predictive models might identify high-value prospects for targeted marketing. Anomaly detection could reduce fraud losses. Computer vision might accelerate quality inspection on manufacturing lines.

The critical question: How much improvement is needed to justify the investment? AI projects aren't cheap. You're looking at data infrastructure, specialized talent, computing resources, ongoing maintenance, and opportunity costs. Small improvements rarely justify these costs.

## 2.3 Estimating ROI and Value

Every AI project needs a clear value proposition before you write a single line of code. The CPMAI methodology requires teams to estimate time-to-ROI for various AI project types and calculate expected returns based on realistic assumptions (PMI, 2025).

Start with baseline metrics. What's the current performance of whatever process you're trying to improve? If you can't measure current state, you can't measure improvement. A manufacturer pursuing predictive maintenance needs baseline data on equipment downtime, maintenance costs, and unplanned outage frequency.

Estimate realistic improvements AI could deliver. Don't assume 80% improvement because a vendor's case study claimed that. Factor in your data quality, organizational readiness, and implementation challenges. A 20-30% improvement on a well-defined problem is often realistic; anything beyond 50% should trigger skepticism (Fountaine et al., 2019).

Calculate costs comprehensively. Include data preparation (often 60-80% of effort), infrastructure, tools and platforms, specialized personnel, training and change management, and ongoing operational costs. Many organizations dramatically underestimate total cost of ownership.

**Case Example: The Healthcare Reality Check**

Dr. Elena Chen, a clinical informatics leader, was evaluating an AI solution to predict patient deterioration in hospital wards. The vendor promised 85% accuracy in identifying patients who would require ICU transfer within 24 hours.

Elena built a detailed ROI model. Benefits included: reduced ICU transfers through early intervention, decreased mortality from missed deterioration, lower liability exposure, and improved patient satisfaction. She estimated $2.8 million in annual benefits.

Then she calculated real costs: $800,000 for the software platform, $300,000 in integration with existing systems, $200,000 for retrospective data preparation, $150,000 for clinician training, and $250,000 in annual operational costs. Total five-year cost: $3.4 million.

Even with optimistic benefit assumptions, the ROI was marginal. More importantly, Elena recognized the project depended on clinicians actually using the alerts—and alert fatigue was already a major problem in her hospitals. She recommended a smaller pilot focused on specific high-risk units where clinical champions could drive adoption. That measured approach proved crucial. The pilot revealed alert accuracy was closer to 65% in real-world conditions, requiring significant refinement before broader deployment.

## 2.4 Stakeholder Identification and Engagement

AI projects affect multiple stakeholder groups, often in ways you don't anticipate at project outset. The CPMAI framework emphasizes early stakeholder identification and ongoing engagement throughout the project lifecycle (PMI, 2025).

Primary stakeholders typically include business sponsors providing funding and strategic direction, end users who will interact with the AI system, IT operations responsible for deployment and maintenance, data engineers managing data pipelines, data scientists building models, and compliance/legal/ethics teams ensuring regulatory adherence.

Secondary stakeholders might include customers affected by AI decisions, employees whose jobs change due to AI deployment, regulatory bodies overseeing your industry, and external partners integrating with your AI systems.

Map stakeholders' interests, influence, and concerns. Business sponsors care about ROI and strategic fit. End users worry about workflow disruption and job security. IT operations stress system stability and security. Data scientists want interesting problems and cutting-edge tools. These interests often conflict.

Develop engagement strategies for each stakeholder group. Executives need high-level progress updates tied to business metrics. End users need hands-on demos and opportunities to provide feedback. Technical teams need detailed architecture discussions and resource commitments.

## 2.5 Go No Go Decision Criteria

Phase I culminates in a go/no-go decision. Should you proceed to data assessment, or abandon the project before investing significant resources? The CPMAI methodology provides structured criteria for this decision (Cognilytics, 2024).

**Is the problem appropriate for AI?** Can you clearly articulate why traditional approaches are insufficient? Is the problem probabilistic rather than deterministic? Are patterns too complex to code explicitly?

**Does adequate data likely exist?** You don't need complete data assessment yet—that's Phase II—but you need reasonable confidence that relevant data exists and is accessible. If the answer is "we'll collect data as we go," that's usually a red flag.

**Is the business value sufficient?** Does expected ROI justify the investment? Are business stakeholders committed to adoption and change management? Will success be measured and rewarded?

**Do you have necessary capabilities?** Can you access required data science expertise, computing infrastructure, and operational support? Are you willing to invest in building missing capabilities?

**Are risks manageable?** Have you identified ethical concerns, regulatory requirements, and technical challenges? Do mitigation strategies exist?

A "no" answer to any of these questions should pause the project for remediation or termination. Proceeding despite red flags virtually guarantees failure.

## 2.6 Feasibility Assessment

Feasibility assessment in Phase I takes a high-level view of whether the project can succeed given current constraints and capabilities. You're not building detailed technical architectures yet—that comes later—but you are identifying obvious blockers.

**Technical feasibility** asks whether the problem is solvable with current AI techniques. Some problems remain beyond AI's capabilities. Others are theoretically solvable but require resources or expertise beyond your reach. A small regional bank probably can't build GPT-4 level language models, but they might deploy existing LLMs effectively.

**Operational feasibility** examines whether you can integrate AI into existing workflows and systems. An AI model that requires complete process redesign faces higher implementation risk than one that augments current workflows. Change management complexity scales with operational disruption.

**Economic feasibility** evaluates whether benefits justify costs at realistic estimates. Don't base feasibility on best-case scenarios. Use conservative estimates with reasonable contingency.

**Legal and ethical feasibility** becomes increasingly critical as AI regulation expands. Can you use the required data? Can you explain decisions to affected parties? Does the use case create unacceptable discrimination risk? The EU AI Act, for instance, bans certain high-risk AI applications and strictly regulates others (European Commission, 2023).

## 2.7 Matching Problems to AI Patterns

One of Phase I's practical deliverables is matching your business problem to one or more of the seven AI patterns: conversational, recognition, predictive analytics, goal-driven, anomaly detection, hyper-personalization, and autonomous systems (PMI, 2025).

This matching guides solution architecture, technology selection, and team composition. A conversational AI project needs NLP expertise and dialogue design skills. A computer vision project requires image processing specialists and massive computing resources for model training.

Some problems map to single patterns. Chatbot customer service is conversational AI. Fraud detection is anomaly detection. Sales forecasting is predictive analytics.

Complex problems often combine multiple patterns. An autonomous vehicle integrates recognition (seeing the environment), prediction (anticipating other vehicles' behavior), and goal-driven decision-making (planning routes). A sophisticated recommendation system might use predictive analytics (forecasting preferences), hyper-personalization (tailoring to individuals), and conversational AI (interactive refinement of suggestions).

**Case Example: The Supply Chain Solution**

Aisha, an operations manager, was asked to reduce logistics costs for a distribution network serving 500 retail locations. Leadership vaguely mentioned "maybe AI could help."

Aisha broke the business problem into components. Route optimization required goal-driven AI that could consider multiple constraints—delivery windows, vehicle capacity, traffic patterns, driver schedules—and find near-optimal solutions. Demand forecasting needed predictive analytics to anticipate product needs at each location. Anomaly detection could identify unusual demand patterns suggesting data errors or special events requiring attention.

By mapping to specific AI patterns, Aisha could evaluate build-versus-buy decisions for each component. Route optimization had mature commercial solutions; buying made sense. Demand forecasting needed industry-specific customization; building in-house provided advantage. Anomaly detection could start with open-source tools and be refined over time.

This pattern-based approach transformed an overwhelming "use AI for logistics" directive into a structured roadmap with clear priorities, resource requirements, and success criteria for each component.

## 2.8 Success Metrics and KPIs

You cannot manage what you don't measure. Phase I must define specific, measurable success criteria that align with business objectives and enable objective project evaluation (Huyen, 2022).

Define metrics at multiple levels. **Business metrics** measure ultimate business impact: revenue increase, cost reduction, customer satisfaction improvement, fraud loss prevention. **Model performance metrics** assess AI model quality: accuracy, precision, recall, F1-score, AUC-ROC. **Operational metrics** track deployment success: system uptime, prediction latency, user adoption rates.

All three matter, but they measure different things. You can have technically excellent models (high accuracy) that deliver poor business results (negligible revenue impact). You can have promising business metrics that hide operational problems (poor user adoption compensated by other factors).

Set realistic targets. A 5% improvement in customer retention might be modest for a model accuracy metric but transformational for business results. Calibrate expectations against industry benchmarks, pilot studies, and academic research—not vendor marketing materials.

Establish measurement cadence. When will you assess progress? How will you handle early results that miss targets? What constitutes "good enough" versus "needs improvement" versus "abandon project"?

## 2.9 Essential Points to Remember

Phase I Business Understanding makes or breaks AI projects. Rushing through business alignment to get to "the fun stuff" of model building virtually guarantees failure. Take time to formulate specific, measurable business questions. Tie AI capabilities explicitly to strategic business objectives. Build realistic ROI models with comprehensive cost estimates and conservative benefit projections.

Identify and engage all stakeholder groups early. Map their interests, concerns, and influence. Address conflicts before they derail the project. Apply structured go/no-go criteria to decide whether to proceed. Assess technical, operational, economic, and legal feasibility honestly.

Match business problems to specific AI patterns to guide solution design. Define success metrics at business, model performance, and operational levels. Create measurement plans that enable objective project evaluation.

The time invested in Phase I pays enormous dividends. Teams that complete thorough business understanding deliver successful projects at dramatically higher rates than those who rush to coding. Do the hard work upfront.

## 2.10 Practice Questions

### Business Case Evaluation Scenarios

**Question 51:** A retail company proposes using AI to optimize store layouts based on customer foot traffic patterns. They estimate $2M in annual revenue increase. Costs include $500K for initial development, $200K for sensors and infrastructure, and $150K annual maintenance. Is this financially viable?

A) No, costs exceed benefits B) Yes, payback period under 1 year with ongoing positive ROI C) Cannot determine without more information D) No, retail doesn't benefit from AI

**Answer: B) Yes, payback period under 1 year with ongoing positive ROI**

**Explanation:** Initial investment is $700K (dev + infrastructure). First-year benefit is $2M minus $150K maintenance = $1.85M net benefit. Payback occurs in 4-5 months. Subsequent years yield $1.85M annual benefit, making this financially attractive.

---

**Question 52:** A manufacturing company wants to implement predictive maintenance AI. They experience 100 unplanned equipment failures yearly, each costing $50K in downtime and repairs. The AI solution costs $1.5M initially and $300K annually, promising to prevent 60% of failures. What's the break-even prevention rate?

A) 36% B) 45% C) 60% D) 75%

**Answer: C) 36%**

**Explanation: Exact break-even prevention rate (Year 1, incl. capex + opex): 36%**

- Baseline annual failure cost: $100 \times \$50{,}000 = \$5{,}000{,}000$
- Year-1 AI cost: $\$1{,}500{,}000 + \$300{,}000 = \$1{,}800{,}000$
- Break-even rate $r$: $\$1{,}800{,}000/\$5{,}000{,}000 = 0.36 = 36\%$

---

**Question 53:** An insurance company's business case claims AI will reduce claim processing time from 45 minutes to 15 minutes per claim, processing 1,000 claims daily. Adjusters earn $60/hour. Annual cost savings are calculated as:

A) $360,000 , B) $7,500,000 , C) $1,500,000 ,D) Cannot calculate without AI costs

**Answer: B) $7,500,000**

**Explanation:** Time saved = 30 minutes (0.5 hours) per claim.Daily savings = 1,000 claims $\times$ 0.5 hours $\times$ $60/hour = $30,000.Annual savings (using 250 workdays, as implied in the explanation's text) = $30,000 $\times$ 250 = $7,500,000.

---

**Question 54:** A healthcare provider proposes AI for patient scheduling optimization. Benefits include 5% more appointments (revenue) and 10% reduction in no-shows. They generate $50M annual revenue with 15% no-show rate costing $7.5M. Development costs $800K. What's first-year ROI?

A) Negative B) 200% C) 306% D) 800%

**Answer: C) 306%**

**Explanation:**

- ☐ Added revenue = 5% of $50M = **$2.5M**.
- ☐ No-show reduction = 10% of $7.5M = **$0.75M**.
- ☐ Total Year-1 benefit = **$3.25M**; investment = **$0.8M**.
- ☐ ROI = ($3.25M – $0.8M)/$0.8M = **306%**.

---

**Question 55:** A business case assumes AI will improve customer satisfaction scores by 15%, leading to 10% customer retention increase. Executive asks: "Are these assumptions linked or independent?" How should you respond?

A) They're independent; combine them for 25% total benefit B) They're linked; satisfaction improvement leads to retention C) Neither metric matters for business case D) Retention doesn't affect satisfaction

**Answer: B) They're linked; satisfaction improvement leads to retention**

**Explanation:** Causality matters in business cases. Satisfaction likely drives retention, so these aren't independent benefits to sum. Improved satisfaction causes retention improvement. Avoid double-counting benefits when one causes the other. Document causal relationships and quantify the ultimate business impact (retention).

---

**Question 56:** During business case review, a finance executive notes: "Your revenue projections assume AI performs as well in production as in testing. What if production accuracy is 20% lower?" This is an example of:

A) Rejecting the business case B) Appropriate sensitivity analysis C) Unnecessary pessimism D) Micromanagement

**Answer: B) Appropriate sensitivity analysis**

**Explanation:** Good business cases include sensitivity analysis showing outcomes under various scenarios (optimistic, expected, pessimistic). Testing performance often exceeds production performance. Analyzing business impact at different accuracy levels demonstrates thorough planning and risk awareness.

---

**Question 57:** A marketing team's business case focuses entirely on increased revenue from AI-driven personalization. What critical cost categories might they be overlooking?

A) Nothing; revenue is the only consideration B) Data infrastructure, ML platform, talent, maintenance, compute costs C) Only compute costs matter D) Revenue projections are always accurate

**Answer: B) Data infrastructure, ML platform, talent, maintenance, compute costs**

**Explanation:** Comprehensive business cases include all costs: initial development, data infrastructure, ML platforms, specialized talent, ongoing maintenance, compute/cloud costs, retraining, monitoring, and opportunity costs. Focusing only on benefits while ignoring substantial costs creates misleading projections.

---

**Question 58:** A business case projects 3-year payback for an AI initiative. The CEO responds: "Our strategic plan may change in 18 months. How does this affect the business case?" What's the key consideration?

A) Three years is always acceptable B) Risk of strategic misalignment if business direction changes C) CEO shouldn't question business cases D) Payback period doesn't matter

**Answer: B) Risk of strategic misalignment if business direction changes**

**Explanation:** AI investments require long-term commitment. If strategy might change within the payback period, the investment risks becoming irrelevant. Consider strategic risk, flexibility to pivot, and whether shorter-payback alternatives exist. This isn't rejecting AI—it's ensuring alignment with business strategy timelines.

---

**Question 59:** Two AI projects are proposed: Project A has $5M NPV with 70% confidence; Project B has $3M NPV with 95% confidence. Using risk-adjusted value, which is better?

A) Project A (higher NPV) B) Project B (higher confidence) C) Project A: $5M × 0.70 = $3.5M > Project B: $3M × 0.95 = $2.85M D) Cannot compare different projects

**Answer: C) Project A with risk-adjusted NPV of $3.5M**

**Explanation:** Risk-adjusted value = NPV × probability of success. Project A: $3.5M expected value. Project B: $2.85M expected value. While Project B has higher confidence, Project A's higher upside compensates for lower confidence, yielding better risk-adjusted return.

---

**Question 60:** A business case assumes deployment in 6 months. Development takes 9 months. How does this affect the business case?

A) Doesn't affect it B) Delays benefits by 3 months, reducing NPV and extending payback C) Improves the business case D) Only costs are affected

**Answer: B) Delays benefits by 3 months, reducing NPV and extending payback**

**Explanation:** Delayed deployment pushes benefits further into the future, reducing net present value (future benefits are worth less). It extends payback period and increases cumulative costs before benefits begin. Timelines materially impact business case economics and must be realistic.

---

## Stakeholder Conflict Resolution

**Question 61:** The data science team wants to use deep learning for a classification problem. The IT operations team refuses, saying they can't support GPU infrastructure. How should the project manager resolve this?

A) Force IT to support GPUs B) Cancel the project C) Explore alternatives: simpler models, cloud GPUs, or demonstrate business value justifying infrastructure investment D) Tell data science to use only simple models

**Answer: C) Explore alternatives**

**Explanation:** Effective conflict resolution finds middle ground. Options include: (1) Test whether simpler models achieve acceptable performance; (2) Use cloud GPUs avoiding on-premises infrastructure; (3) Build business case showing GPU investment ROI justifies costs. Forcing or canceling are last resorts.

---

**Question 62:** Marketing wants the AI recommendation system deployed in 2 weeks for a major campaign. Engineering says 6 weeks minimum is needed for proper testing. The CMO escalates to the CEO. As project lead, what do you recommend?

A) Deploy in 2 weeks as Marketing demands B) Deploy a limited MVP in 2 weeks with fallback to rules-based system, full deployment in 6 weeks C) Refuse deployment entirely D) Let the CEO decide without input

**Answer: B) Deploy limited MVP with fallback system**

**Explanation:** This balances business urgency with technical prudence. A limited MVP with rule-based fallback lets Marketing proceed while mitigating risks. Full deployment follows proper testing. Explain tradeoffs clearly: early deployment has limitations and risks; stakeholders can make informed decisions.

---

**Question 63:** The compliance team requires human review of all AI decisions. The operations team says this eliminates efficiency gains, defeating the AI's purpose. How do you reconcile this?

A) Ignore compliance B) Cancel AI deployment C) Implement human review for high-risk/uncertain predictions only, auto-approve high-confidence predictions D) Automate everything despite compliance

**Answer: C) Implement risk-based human review**

**Explanation:** Hybrid approaches balance compliance and efficiency. High-confidence predictions auto-approve; borderline or high-stakes predictions trigger human review. This maintains compliance (humans review risky decisions) while preserving efficiency gains (automation handles clear-cut cases).

---

**Question 64:** Business stakeholders want AI deployed immediately. Data scientists say data quality is insufficient and needs 3 months of cleaning. Business responds: "We've operated on this data for years; it's fine." How should the project manager respond?

A) Deploy immediately as business wants B) Explain that ML models are more sensitive to data quality than human processes; demonstrate quality issues' impact on model performance C) Refuse to deploy ever D) Blame the data scientists

**Answer: B) Explain ML's sensitivity to data quality with demonstrations**

**Explanation:** Education bridges understanding gaps. Demonstrate how data quality issues affect model performance—show examples of errors the model would make. Quantify the risk and impact. Data quality that suffices for human use often fails for ML. Use proof-of-concept results to show the difference between good and poor data quality.

---

**Question 65:** Sales wants AI to predict deal closure probability. Finance wants AI to forecast revenue. Engineering estimates 12 weeks to build one model, 18 weeks for both. How should priorities be set?

A) Build both simultaneously B) Executive decision based on business priority C) Let teams fight it out D) Build neither

**Answer: B) Executive decision based on business priority**

**Explanation:** Resource conflicts require executive prioritization based on strategic importance, urgency, and value. Project managers facilitate the decision by providing: (1) Relative business value of each use case; (2) Resource requirements; (3) Timeline implications; (4) Possibility of staged delivery. Leadership decides based on organizational priorities.

---

**Question 66:** Data scientists want access to customer PII for model training. Privacy/legal team says GDPR prohibits this. Data scientists argue the model won't work without real data. What's the solution?

A) Give data scientists full PII access B) Cancel the project C) Use anonymization, synthetic data, or differential privacy techniques to protect privacy while enabling model development D) Ignore GDPR

**Answer: C) Use privacy-preserving techniques**

**Explanation:** Technical solutions exist: (1) Anonymization/pseudonymization removes direct identifiers; (2) Synthetic data maintains statistical properties without real PII; (3) Differential privacy adds noise protecting individuals; (4) Aggregation/generalization removes granular PII. Privacy and ML can coexist with proper techniques.

---

**Question 67:** The business sponsor wants a project completed in Q1. The data engineering team needs Q1 entirely for infrastructure setup, allowing modeling only in Q2. What's the best approach?

A) Compress infrastructure work to 1 month B) Start modeling without proper infrastructure C) Reset expectations: explain dependencies and propose realistic timeline D) Cancel the project

**Answer: C) Reset expectations with realistic timeline**

**Explanation:** Unrealistic deadlines cause failure. Explain dependencies clearly: proper infrastructure enables successful modeling; rushing infrastructure creates technical debt hampering later phases. Propose phased delivery if possible. Stakeholders need honest assessments, not false commitments that will fail.

---

**Question 68:** During a project, the original business sponsor leaves the company. The new sponsor questions the project's value and threatens cancellation. What should the project manager do?

A) Continue without sponsor support B) Re-engage the new sponsor: present business case, demonstrate progress, address concerns C) Immediately cancel the project D) Escalate to CEO

**Answer: B) Re-engage new sponsor**

**Explanation:** New stakeholders need relationship building. Present the business case, share progress and early results, understand their concerns and priorities, and align the project with their objectives. The previous sponsor's departure doesn't invalidate the project—it requires renewed stakeholder engagement.

---

**Question 69:** IT operations says AI model inference latency of 500ms is unacceptable; they need sub-100ms. Data science responds this requires a simpler model with 5% lower accuracy. Who's right?

A) IT operations; latency always matters most B) Data science; accuracy always matters most C) Both; this is a performance-business value tradeoff requiring stakeholder input D) Neither

**Answer: C) This is a tradeoff requiring stakeholder discussion**

**Explanation:** Technical tradeoffs require business decisions. Present options clearly: (1) Current model: 500ms, higher accuracy; (2) Optimized model: <100ms, slightly lower accuracy. Let business stakeholders weigh user experience (latency) versus decision quality (accuracy) based on use case importance.

---

**Question 70:** A domain expert insists the AI should never override their judgment, but business leaders want full automation. How do you reconcile this?

A) Full automation; ignore the expert B) No automation; keep manual process C) Hybrid: AI provides recommendations, expert makes final decisions with ability to override D) Let them argue forever

**Answer: C) Hybrid AI recommendation with expert override**

**Explanation:** Human-in-the-loop systems balance automation benefits with expert judgment. AI handles routine cases and provides decision support; experts focus on complex cases and can override AI when their judgment differs. This respects expertise while gaining efficiency and often performs better than either AI or humans alone.

---

## ROI Calculation and Justification

**Question 71:** A company invests $2M in an AI project. Year 1 benefits: $600K. Year 2: $800K. Year 3: $1M. Year 4: $1.2M. Discount rate: 10%. What's the NPV?

A) $1.6M B) $777,543K C) $200K D) Negative

**Answer: B) $777,543K**

1. **Explanation:** Discount each year's benefit at 10%:
    - Year 1: $600,000 / 1.10 = **$545,454.55**
    - Year 2: $800,000 / 1.21 = **$661,157.02**
    - Year 3: $1,000,000 / 1.331 = **$751,314.80**
    - Year 4: $1,200,000 / 1.4641 = **$819,616.15**
2. Sum present values and subtract the investment:
    - PV of benefits = **$2,777,542.52**
    - NPV = $2,777,542.52 – $2,000,000 = **$777,542.52**

**Rounded result: $777,543** (or **$0.78M**).

---

**Question 72:** An AI project costs $500K and generates $150K annual savings indefinitely. What's the simple payback period?

A) 1.5 years B) 2.3 years C) 3.3 years D) 5 years

**Answer: C) 3.3 years**

**Explanation:** Payback period = Initial investment / Annual savings = \$500K / \$150K = 3.33 years. This simple calculation ignores time value of money but provides quick assessment of how long until investment is recovered.

---

**Question 73:** A business case projects \$1M annual savings from AI. The CFO asks: "Are these cost savings or cost avoidance?" Why does this distinction matter?

A) It doesn't matter B) Cost savings are actual budget reductions; cost avoidance is prevented future costs C) Cost avoidance is always better D) They're identical

**Answer: B) Cost savings reduce budgets; cost avoidance prevents future costs**

**Explanation:** Cost savings directly reduce operating budgets (headcount, infrastructure). Cost avoidance prevents future costs that would otherwise occur (prevented failures, avoided fines). Finance treats these differently—savings appear in P&L immediately; avoidance affects future budget planning. Business cases should clarify which benefits are which.

---

**Question 74:** You're calculating ROI for an AI initiative. Which costs should be included?

A) Only initial development costs B) Only ongoing operational costs C) Development, infrastructure, talent, maintenance, compute, opportunity costs, and training D) Nothing; ROI only considers benefits

**Answer: C) All costs including development, infrastructure, talent, maintenance, compute, opportunity costs**

**Explanation:** Comprehensive ROI includes: initial development, data infrastructure, ML platforms, data acquisition/cleaning, specialized talent (salary premiums), ongoing maintenance, cloud/compute costs, model retraining, monitoring systems, change management/training, and opportunity cost of team time. Incomplete cost accounting overstates ROI.

---

**Question 75:** An AI project's business case shows 200% ROI over 3 years. During Year 2, a competitor launches a superior product rendering your AI less valuable. This illustrates:

A) Business cases are useless B) Competitive/market risk affecting realized ROI C) Three-year horizons are too long D) AI never works

**Answer: B) Competitive/market risk affecting ROI**

**Explanation:** Business cases make assumptions about competitive environment, market conditions, and technology landscape. External changes (competitor actions, market shifts, regulatory changes) can materially affect realized ROI. Good business cases acknowledge these risks and include contingency planning.

---

**Question 76:** Marketing calculates that AI personalization will increase conversion rates by 2 percentage points (from 3% to 5%). With 10M annual visitors and $100 average order value, what's the incremental annual revenue?

A) $20M B) $50M C) $2M D) $200M

**Answer: A) $20M**

**Explanation:** Incremental conversions: 10M visitors × 2% increase = 200,000 additional orders. Incremental revenue: 200,000 × $100 = $20M. This shows how small percentage improvements at scale generate substantial value, justifying AI investments.

---

**Question 77:** A business case projects $3M benefits but $2M costs, showing $1M net benefit. The finance director asks: "What's your confidence level on these numbers?" How should you respond?

A) "100% confident" B) Provide range estimates: conservative, expected, optimistic scenarios C) Refuse to quantify uncertainty D) Guess randomly

**Answer: B) Provide range estimates with confidence levels**

**Explanation:** Responsible business cases acknowledge uncertainty through scenarios: Conservative (70% of projected benefits, 120% of projected costs), Expected (base case),

Optimistic (130% benefits, 90% costs). This shows how outcomes vary under different assumptions, enabling risk-informed decisions.

---

**Question 78:** You're comparing two AI approaches: Option A costs $500K with 80% success probability; Option B costs $1.2M with 95% success probability. Both deliver $3M value if successful. Which is better on risk-adjusted basis?

A) Option A: $500K cost, $2.4M expected value ($3M × 80%) B) Option B: $1.2M cost, $2.85M expected value ($3M × 95%) C) Option A: $1.9M net expected value ($2.4M - $500K) D) Option B: $1.65M net expected value ($2.85M - $1.2M)

**Answer: C) Option A with $1.9M net expected value**

**Explanation:** Risk-adjusted analysis: Option A expected net value: ($3M × 0.80) - $500K = $1.9M. Option B: ($3M × 0.95) - $1.2M = $1.65M. Despite lower confidence, Option A's cost advantage yields higher risk-adjusted returns. Risk tolerance affects this decision—risk-averse organizations might prefer Option B's higher confidence.

---

**Question 79:** An AI project replaces a manual process taking 100 FTE (full-time employees). HR says, "We can't actually reduce headcount due to hiring freezes and attrition. We'll redeploy people to other work." How does this affect the business case?

A) Benefits unchanged; labor savings still count B) Direct cost savings disappear; benefits become productivity increases in other areas C) Project should be cancelled D) Doesn't matter

**Answer: B) Direct cost savings disappear; benefits become productivity increases**

**Explanation:** Redeployment changes benefit type from cost reduction to productivity improvement. If displaced workers perform other valuable work, there's still benefit—but it's harder to quantify than direct salary savings. Business cases must honestly reflect benefit realization mechanisms.

---

**Question 80:** A 5-year AI project has benefits starting small (Year 1: $500K) and growing (Year 5: $3M) as adoption increases. Should you use simple average annual benefit or year-by-year analysis?

A) Simple average is fine B) Year-by-year analysis with benefit ramp-up C) Assume maximum benefits from day one D) Doesn't matter

**Answer: B) Year-by-year analysis showing benefit ramp-up**

**Explanation:** AI projects rarely deliver full benefits immediately. Adoption curves, learning periods, and iterative improvements mean benefits grow over time. Year-by-year analysis accurately captures this reality, providing realistic NPV and payback calculations. Averaging or assuming immediate full benefits overstates early returns.

---

## Pattern Selection for Business Problems

**Question 81:** A bank wants to reduce customer churn by identifying at-risk customers 3 months before they leave, enabling proactive retention offers. Which AI pattern applies?

A) Recognition B) Predictive analytics C) Anomaly detection D) Conversational AI

**Answer: B) Predictive analytics**

**Explanation:** This is forecasting future outcomes (churn) based on current/historical data. Predictive analytics models learn patterns associated with churn and assign churn probability scores to current customers, enabling proactive intervention.

---

**Question 82:** A hospital wants a system that automatically prioritizes incoming patient X-rays by urgency, flagging critical cases (fractures, tumors) for immediate radiologist review. Which pattern?

A) Conversational AI B) Recognition (image classification) C) Goal-driven systems D) Hyper-personalization

**Answer: B) Recognition (image classification)**

**Explanation:** Identifying conditions in medical images is a recognition task—the system classifies images into categories (urgent, routine) or detects specific findings (fractures, masses). Computer vision/CNNs handle this pattern.

---

**Question 83:** A logistics company wants to optimize delivery routes considering traffic, weather, delivery windows, vehicle capacity, driver schedules, and fuel costs. Which pattern fits best?

A) Recognition B) Predictive analytics C) Goal-driven (optimization) systems D) Anomaly detection

**Answer: C) Goal-driven (optimization) systems**

**Explanation:** This is sequential decision-making to achieve objectives (minimize cost, maximize deliveries) while satisfying constraints. Goal-driven systems/optimization algorithms find near-optimal solutions in complex constraint spaces.

---

**Question 84:** An e-commerce site wants to tailor every aspect of the shopping experience—homepage layout, product displays, search rankings, promotions, and email timing—individually for each of their 10 million customers. Which pattern?

A) Predictive analytics B) Hyper-personalization C) Recognition D) Anomaly detection

**Answer: B) Hyper-personalization**

**Explanation:** Delivering individualized experiences based on detailed user profiles and behaviors is hyper-personalization. While this uses predictive analytics components, the defining feature is end-to-end personalization of the entire customer experience.

---

**Question 85:** A credit card company wants to detect fraudulent transactions in real-time as they occur, flagging unusual patterns that don't match the cardholder's typical behavior. Which pattern applies?

A) Predictive analytics B) Recognition C) Anomaly detection D) Goal-driven systems

**Answer: C) Anomaly detection**

**Explanation:** Identifying unusual patterns that deviate from normal behavior is anomaly detection. The system learns each cardholder's typical transaction patterns and flags significant deviations (unusual locations, amounts, merchants) as potential fraud.

---

**Question 86:** A customer service center wants a virtual assistant that can understand customer questions, search knowledge bases, and provide helpful responses through natural conversation. Which pattern is primary?

A) Conversational AI B) Recognition only C) Predictive analytics D) Anomaly detection

**Answer: A) Conversational AI**

**Explanation:** Natural language understanding, dialogue management, and response generation are conversational AI capabilities. While the system uses recognition (understanding text) and may use other patterns internally, the primary pattern enabling the interaction is conversational AI.

---

**Question 87:** A social media platform wants to recommend content to users based on their interests, connections' activity, trending topics, and engagement patterns. Which pattern combination is most appropriate?

A) Recognition only B) Hyper-personalization + Predictive analytics C) Anomaly detection only D) Goal-driven systems

**Answer: B) Hyper-personalization + Predictive analytics**

**Explanation:** Content recommendation combines predictive analytics (predicting engagement) with hyper-personalization (tailoring to individuals). Many real AI systems combine multiple patterns—this question illustrates that complex problems often need pattern combinations.

---

**Question 88:** A manufacturing plant wants to predict equipment failures 24-48 hours in advance based on sensor data, enabling scheduled maintenance before breakdowns occur. Which pattern?

A) Anomaly detection B) Predictive analytics C) Recognition D) Conversational AI

**Answer: B) Predictive analytics**

**Explanation:** Forecasting future events (failures) based on current sensor readings is predictive analytics. While anomaly detection identifies unusual current patterns, predictive analytics specifically forecasts future outcomes, enabling proactive intervention.

---

**Question 89:** A security system needs to identify individuals from video feeds for access control, comparing captured images against a database of authorized personnel. Which pattern?

A) Conversational AI B) Recognition (facial recognition) C) Predictive analytics D) Goal-driven systems

**Answer: B) Recognition (facial recognition)**

**Explanation:** Identifying individuals from images is a recognition task—specifically facial recognition, a computer vision application. The system matches detected faces against stored templates to identify or verify individuals.

---

**Question 90:** A ride-sharing company wants to dynamically adjust pricing based on real-time supply (available drivers), demand (ride requests), traffic conditions, events, and weather to balance marketplace. Which pattern?

A) Recognition B) Goal-driven (dynamic optimization) C) Anomaly detection D) Conversational AI

**Answer: B) Goal-driven (dynamic optimization)**

**Explanation:** Real-time decision-making to achieve market equilibrium objectives (match supply with demand, maximize utilization, ensure service availability) is a goal-driven system. Dynamic pricing algorithms make sequential decisions optimizing marketplace outcomes.

---

## Project Scoping Situations

**Question 91:** A business unit wants "AI to improve customer experience." As project manager, what's your first action?

A) Start building models immediately B) Define specific, measurable objectives: which aspects of experience, how to measure, what success looks like C) Hire more data scientists D) Buy more GPUs

**Answer: B) Define specific, measurable objectives**

**Explanation:** Vague goals like "improve customer experience" doom projects. Phase I Business Understanding requires translating broad aspirations into specific, measurable objectives: "Reduce average customer service response time from 4 hours to 1 hour" or "Increase NPS from 7.2 to 8.0." Specificity enables success measurement and guides solution design.

---

**Question 92:** During scoping, you discover the business problem requires integrating data from 15 different systems, none of which talk to each other. How does this affect the projcct?

A) Ignore integration complexity B) Massively increases scope, timeline, and resources needed for data integration C) Data integration is trivial D) Cancel immediately

**Answer: B) Massively increases scope, timeline, and resources**

**Explanation:** Data integration is often 60-80% of AI project effort. Fifteen disconnected systems require identifying data schemas, resolving inconsistencies, handling duplicates, creating pipelines, and ensuring data quality. Scope must account for this integration work before modeling begins.

---

**Question 93:** A stakeholder insists on including 20 different use cases in one AI project. What's the risk and recommended approach?

A) No risk; scope is always flexible B) Scope creep increases failure risk; prioritize use cases, build MVP for highest-value case, expand iteratively C) Build all 20 simultaneously D) Refuse to consider any use cases

**Answer: B) Scope creep increases failure risk; prioritize and build iteratively**

**Explanation:** Trying to solve 20 problems simultaneously creates impossible complexity. Apply MVP thinking: identify the highest-value, most feasible use case; deliver that successfully; demonstrate value; expand iteratively based on learnings. Focused delivery beats sprawling scope.

---

**Question 94:** During business understanding, you realize the proposed AI solution addresses a symptom rather than the root cause of a business problem. What should you do?

A) Build what's requested B) Discuss root cause analysis; potentially reframe the problem C) Say nothing D) Cancel the project

**Answer: B) Discuss root cause and potentially reframe the problem**

**Explanation:** Good project management includes challenging requirements when appropriate. If the real problem is poor product quality (root cause) rather than complaints (symptom), treating symptoms with AI-powered complaint handling doesn't solve the underlying issue. Help stakeholders address root causes.

---

**Question 95:** A project sponsor wants the AI system to achieve "human-level performance." How should you respond?

A) Promise human-level performance B) Define specific metrics: what aspects of performance, measured how, compared to which humans? C) Say AI will surpass humans D) Refuse the project

**Answer: B) Define specific metrics and performance targets**

**Explanation:** "Human-level performance" is vague and often misleading. Humans vary widely in performance. Define specific metrics (accuracy, precision, recall, F1), measurement methods, evaluation data, and acceptable performance levels. Replace fuzzy goals with measurable targets.

---

**Question 96:** You're scoping a fraud detection project. Business wants to catch 95% of fraud. Investigation reveals current human detection catches 45% of fraud. What does this tell you?

A) 95% is easy to achieve B) 95% is ambitious but realistic; scope must include sufficient data, features, and iteration C) Humans are terrible; AI will easily achieve 95% D) Give up

**Answer: B) 95% is ambitious; scope appropriately**

**Explanation:** Doubling current detection rates (45% to 95%) is ambitious. This requires excellent data, relevant features, sophisticated models, and iteration. Scope must include time for experimentation, sufficient data collection, and realistic milestones (60% by month 3, 75% by month 6, 90% by launch).

---

**Question 97:** During scoping, legal counsel says the proposed AI system may violate privacy regulations. What should happen?

A) Ignore legal concerns B) Immediately stop all work until legal approval C) Work with legal to understand requirements and adjust scope to ensure compliance D) Cancel forever

**Answer: C) Work with legal to understand requirements and adjust scope for compliance**

**Explanation:** Legal/regulatory compliance isn't optional. Engage legal early to understand requirements, assess compliance risks, and adjust scope as needed (anonymization, consent management, audit trails, explainability). Better to address compliance during scoping than face enforcement after deployment.

---

**Question 98:** A business unit wants AI deployed across all 50 regional offices simultaneously. IT recommends starting with 3 pilot sites. Who's right?

A) Business; go big or go home B) IT; pilots reduce risk and enable learning before full deployment C) Neither D) Deploy to zero sites

**Answer: B) IT; pilots reduce risk and enable learning**

**Explanation:** Phased rollout through pilots is best practice. Pilots surface integration issues, usability problems, edge cases, and training needs with limited blast radius. Learnings inform full deployment, dramatically increasing success probability versus big-bang launches.

---

**Question 99:** During scoping, you discover the success metric (customer satisfaction) takes 6 months to measure after AI deployment. How does this affect the project?

A) Doesn't affect scoping B) Requires interim proxy metrics providing faster feedback C) Makes the project impossible D) Ignore measurement

**Answer: B) Requires interim proxy metrics for faster feedback**

**Explanation:** Six-month measurement delay prevents rapid iteration. Identify proxy metrics providing faster feedback: system usage, engagement metrics, process efficiency, or leading indicators correlating with ultimate success metric. Balance lagging indicators (ultimate goal) with leading indicators (rapid feedback).

---

**Question 100:** A project sponsor says: "We need this AI system to be perfect—zero errors." How should you respond?

A) Promise perfection B) Explain AI systems are probabilistic, not deterministic; define acceptable error rates and risk mitigation C) Say AI is always perfect D) Cancel the project

**Answer: B) Explain probabilistic nature and define acceptable error rates**

**Explanation:** AI systems make probabilistic predictions; perfection is impossible. Educate stakeholders about inherent uncertainty, error types (false positives vs. false negatives), and tradeoffs. Define acceptable error rates balancing business impact with technical feasibility. Design human oversight for uncertain predictions.

---

## Advanced Scenarios

**Question 101:** Your team proposes an AI project requiring 8 months. Business says competitors will launch similar capabilities in 6 months. How should scope be adjusted?

A) Work faster; compress 8 months into 6 B) Launch MVP in 6 months with core capabilities; iterate afterward C) Give up D) Ignore competitive pressure

**Answer: B) Launch MVP in 6 months with core capabilities**

**Explanation:** Time-to-market pressure requires prioritization. Identify minimum viable features delivering competitive parity; launch those in 6 months; add enhancements post-launch. Better to have good-enough capabilities on time than perfect capabilities late, missing market window.

---

**Question 102:** During stakeholder interviews, you discover three departments have different, conflicting definitions of "customer." How does this affect scoping?

A) Ignore definitions B) This is a critical data governance issue requiring resolution before proceeding C) Use all three definitions D) Pick one randomly

**Answer: B) Critical data governance issue requiring resolution**

**Explanation:** Inconsistent fundamental definitions prevent coherent AI projects. "Customer" might mean account holder, household, individual user, or business entity depending on department. Resolve definitional conflicts during business understanding; establish master data management; ensure shared understanding.

---

**Question 103:** A business case assumes AI will process 1,000 transactions daily. After scoping, you learn actual volume is 50,000 daily. What's the impact?

A) No impact B) Massive impact on infrastructure, costs, architecture, and timeline C) Makes the project easier D) Slightly increases budget

**Answer: B) Massive impact on infrastructure, costs, architecture**

**Explanation:** 50× volume difference transforms project scope. Infrastructure must handle scale (distributed systems, load balancing). Costs multiply (compute, storage). Architecture changes (batch to real-time, simple to scalable). Timeline extends for proper load testing. Accurate requirements are essential.

---

**Question 104:** You're scoping a recommendation system. Marketing wants recommendations updated every 5 seconds as user behavior changes. What's your response?

A) Promise 5-second updates B) Question business value of 5-second vs. hourly or daily updates; balance freshness requirements with technical complexity and cost C) Say it's impossible D) Build it without questioning

**Answer: B) Question business value and balance requirements with complexity**

**Explanation:** Real-time requirements dramatically increase complexity and cost. Challenge assumptions: Would hourly updates deliver similar business value? Daily updates? Understand whether 5 seconds is essential or assumed. Sometimes educating stakeholders about cost-benefit tradeoffs leads to more pragmatic requirements.

---

**Question 105:** During business understanding, you realize the proposed AI project would automate jobs held by 50 employees. Union representatives express concerns. What should happen?

A) Ignore union concerns B) Engage HR and union leadership in change management planning; address workforce impacts proactively C) Cancel immediately D) Hide automation impacts

**Answer: B) Engage HR and union leadership in change management**

**Explanation:** Workforce impacts are legitimate stakeholder concerns requiring transparent management. Engage early: discuss retraining, redeployment, timeline, transition support. Change management addresses human dimensions proactively. Ignoring concerns creates resistance; transparency builds trust.

---

**Question 106:** A sponsor wants the AI project to "transform our business model." As project manager, what's your concern?

A) No concern; transformation is good B) Scope is overly ambitious and vague; need specific, achievable objectives C) AI can't transform businesses D) Transformation is too easy

**Answer: B) Overly ambitious, vague scope requiring specific objectives**

**Explanation:** "Transform business model" is too broad for single project scoping. Break into specific initiatives: automate process X, personalize experience Y, optimize operation Z. Transformation comes from accumulating successful specific projects, not attempting everything simultaneously.

---

**Question 107:** During scoping, IT says AI must integrate with a legacy system that's 20 years old with poor documentation. How does this affect the project?

A) Ignore legacy integration B) Increases technical risk and timeline; may require legacy system experts or modernization C) Makes project impossible D) Legacy systems are easy

**Answer: B) Increases technical risk and timeline**

**Explanation:** Legacy integration is often complex: limited/outdated documentation, obsolete technologies, stability concerns, lack of APIs, and scarce expertise. Scope must include discovery time, integration testing, risk mitigation (fallback systems), and possibly legacy system expertise or modernization work.

---

**Question 108:** A business unit wants AI capabilities but refuses to share their data with central IT due to "competitive concerns." How do you scope the project?

A) Force data sharing B) Explore federated learning, edge deployment, or data governance agreements protecting proprietary information C) Cancel the project D) Ignore data needs

**Answer: B) Explore federated learning, edge deployment, or governance agreements**

**Explanation:** Internal data hoarding is common. Solutions include: federated learning (models train locally, share only weights), edge deployment (systems deployed within business unit), or data governance frameworks establishing access controls and use restrictions protecting sensitive information.

---

**Question 109:** You discover that "success" for the AI project means different things to different stakeholders. What must happen during Phase I?

A) Ignore differences B) Establish shared success criteria and metrics with stakeholder alignment C) Let each stakeholder define success differently D) Cancel the project

**Answer: B) Establish shared success criteria with stakeholder alignment**

**Explanation:** Conflicting success definitions guarantee perceived failure—someone will be disappointed. Phase I must align stakeholders on shared success criteria, or explicitly acknowledge different objectives with clear prioritization. Document agreements; revisit if conflicts emerge.

---

**Question 110:** A project requires capabilities (NLP, computer vision, optimization) that no single vendor provides. How does this affect scoping?

A) Use only one vendor regardless B) Multi-vendor integration increases complexity; scope must include integration effort C) Don't use AI D) Build everything in-house

**Answer: B) Multi-vendor integration increases complexity**

**Explanation:** Best-of-breed approaches using multiple specialized vendors deliver better capabilities but increase integration complexity. Scope must include: evaluating/selecting vendors, integration work, ensuring interoperability, managing multiple contracts, and handling support coordination. The capability benefit must justify integration costs.

---

**Stakeholder Management Scenarios**

**Question 111:** Data scientists complete excellent technical work but business stakeholders don't understand the results. Communication breakdown leads to project cancellation despite success. What went wrong?

A) Data scientists failed technically B) Lack of translation between technical results and business value C) Business stakeholders are unreasonable D) AI doesn't work

**Answer: B) Lack of translation between technical and business language**

**Explanation:** Technical excellence is necessary but insufficient. Project managers must translate technical accomplishments (95% AUC-ROC) into business terms (catch 95% of fraud, saving $2M annually). Bridge communication gaps; ensure data scientists present results in business-relevant terms stakeholders understand.

---

**Question 112:** A C-level executive becomes an AI project champion, demanding weekly updates and pushing for acceleration. Other executives feel excluded. What should the project manager do?

A) Focus only on the champion B) Proactively engage all C-level stakeholders; establish consistent communication rhythm C) Ignore other executives D) Tell the champion to stop

**Answer: B) Proactively engage all C-level stakeholders**

**Explanation:** Executive champion is valuable but insufficient. Maintain stakeholder balance by establishing consistent communication with all relevant executives. Share updates broadly; solicit input from multiple perspectives; prevent perception of favoritism; build broad executive support rather than dependence on single champion.

---

**Question 113:** Business stakeholders keep adding requirements mid-project. How should the project manager respond?

A) Accept all new requirements B) Implement change control: assess impact, get approval for scope/timeline/resource changes C) Refuse all changes D) Ignore stakeholders

**Answer: B) Implement change control process**

**Explanation:** Scope creep kills projects. Change control doesn't reject changes—it makes tradeoffs explicit. For each new requirement, assess impact on scope, timeline, resources, and risk. Present options: add time/budget, remove other features, or defer to future release. Let stakeholders make informed decisions.

---

**Question 114:** The data science team and business unit have stopped communicating after a conflict. As project manager, what's your role?

A) Take sides B) Facilitate communication; mediate conflict; rebuild collaboration C) Let them work it out D) Cancel the project

**Answer: B) Facilitate communication and mediate conflict**

**Explanation:** Project managers are communication facilitators. When conflict breaks down collaboration: mediate discussions, clarify misunderstandings, find common ground, establish communication norms, and rebuild trust. Effective conflict resolution is essential project management.

---

**Question 115:** An influential but uninformed stakeholder spreads misconceptions about AI capabilities, creating unrealistic expectations. How do you address this?

A) Ignore it B) Educate through one-on-one sessions, demonstrations, and realistic capability discussions C) Publicly embarrass them D) Give up

**Answer: B) Educate through one-on-one sessions and demonstrations**

**Explanation:** Misinformation is best addressed through patient education. Schedule one-on-one time; demonstrate actual capabilities and limitations; show proof-of-concept results; provide realistic examples. Avoid public confrontation; private education is more effective.

---

**Question 116:** A critical stakeholder refuses to participate in interviews, saying they're too busy. Their input is essential. What should the project manager do?

A) Proceed without their input B) Escalate to their manager; offer flexible participation options (async, written, short sessions) C) Cancel the project D) Force participation

**Answer: B) Escalate to their manager and offer flexible participation**

**Explanation:** Stakeholder engagement is essential. Try flexible approaches: written questionnaires, asynchronous input, brief 15-minute sessions, or delegation to their team. If unsuccessful, escalate to their manager emphasizing criticality. Make participation as easy as possible while securing essential input.

---

**Question 117:** Business stakeholders have unrealistic expectations about timeline (want results in 1 month; realistic timeline is 6 months). How should the project manager handle this?

A) Promise 1 month B) Present realistic timeline with clear dependencies and activities; explain why shortcuts fail C) Say nothing and miss deadlines D) Quit

**Answer: B) Present realistic timeline with clear explanations**

**Explanation:** Honesty is essential. Break down timeline: data collection (6 weeks), cleaning (4 weeks), modeling (8 weeks), evaluation (2 weeks), deployment prep (4 weeks). Show dependencies. Explain risks of rushing (poor data quality, inadequate testing). Sometimes stakeholders accept realistic timelines when they understand the work involved.

---

**Question 118:** Two business units want the same AI capability but with different requirements (one wants speed, one wants accuracy). How do you resolve this?

A) Build one system satisfying neither B) Explore solutions: different models for each, configurable system, or one system with the unit that values it more C) Build nothing D) Flip a coin

**Answer: B) Explore solutions addressing both needs**

**Explanation:** Options include: (1) Two specialized models optimized for each use case; (2) One configurable system with speed/accuracy tradeoff settings; (3) Prioritize the use case with higher business value. Facilitate discussion; present options; help stakeholders reach agreement.

---

**Question 119:** Your executive sponsor suddenly leaves the organization mid-project. What immediate actions should the project manager take?

A) Continue unchanged B) Identify new sponsor; brief them on project status, business case, and critical decisions needed C) Panic D) Cancel project

**Answer: B) Identify and brief new sponsor immediately**

**Explanation:** Sponsor transitions are critical moments. Quickly identify the new sponsor (their manager, another executive); provide comprehensive briefing (business case, progress, risks, pending decisions); rebuild relationship; secure continued commitment. Don't assume continuity—new sponsors need engagement.

---

**Question 120:** Stakeholders disagree on whether to prioritize model accuracy or model explainability. How should this be resolved?

A) Always choose accuracy B) Always choose explainability C) Let business context decide: high-stakes decisions need explainability; low-stakes might prioritize accuracy D) Choose randomly

**Answer: C) Let business context and stakeholder priorities determine the balance**

**Explanation:** Accuracy-explainability tradeoffs depend on use case. Loan decisions require explainability (regulatory, fairness). Movie recommendations tolerate black boxes (low stakes). Healthcare balances both. Facilitate stakeholder discussion of business context, regulatory requirements, and risk tolerance to reach appropriate balance.

---

# Chapter 3.0: Phase II Data Understanding

Let me tell you the uncomfortable truth about AI projects: they live or die based on data quality, and most organizations dramatically overestimate how good their data actually is. You can have the world's best data scientists, unlimited computing power, and brilliant algorithms, but if your data is incomplete, inaccurate, or unrepresentative, you're building on quicksand (Provost & Fawcett, 2013).

Phase II Data Understanding is where you confront that reality. This phase assesses whether adequate data exists, whether you can access it, and whether it's suitable for your AI objectives. Skip this assessment or gloss over data problems, and you'll discover them months later when your expensive models fail in production.

## 3.1 Data Source Identification

Before you can assess data quality, you need to identify what data exists and where it lives. This sounds straightforward—"we have a customer database and transaction system"—until you start asking detailed questions.

Data often resides in multiple systems that don't talk to each other. Your customer relationship management (CRM) system has one view of customers. Your e-commerce platform has another. Your customer service ticketing system has a third. Financial systems, marketing platforms, supply chain databases, and operational systems each contain relevant information, but integrating them requires significant effort.

The CPMAI framework directs teams to create a comprehensive data inventory: identify all potentially relevant data sources, document what data each system contains, assess data granularity and temporal coverage, determine access mechanisms and restrictions, and identify data owners and governance policies (PMI, 2025).

**Case Example: The Financial Services Discovery**

Jamal, a data strategist at a regional bank, was leading a project to predict small business loan default risk. The business case assumed they had "plenty of historical loan data."

Jamal started mapping data sources and uncovered a mess. Loan origination data lived in one system, payment history in another, and customer interactions in a third. Loans from acquired banks used different identifiers and data structures. Business financial statements existed as scanned PDFs in a document management system, not structured data. Credit bureau data was pulled at origination but not updated throughout the loan lifecycle.

Most critically, the bank had approved 10,000 small business loans over five years, but only 200 had reached full maturity. The rest were still active. This limited the training data for default prediction dramatically—you can't train a model to predict defaults when you don't know which current loans will eventually default.

Jamal's data inventory revealed that the "plenty of data" assumption was false. The project needed either to expand data sources significantly (external benchmarking data, alternative data sources) or to narrow scope to a prediction problem addressable with available data (early warning signs of potential default rather than ultimate default prediction).

## 3.2 Data Quality Assessment

Once you know what data exists, you need to assess its quality across multiple dimensions. The CPMAI framework emphasizes evaluating data completeness, accuracy, consistency, timeliness, and relevance (Cognilytics, 2024).

**Completeness** measures whether required data fields are populated. If your customer records are missing email addresses for 40% of customers, any model depending on email engagement data faces immediate problems. Check completeness at field level (what percentage of records have this field populated?) and record level (what percentage of records have all required fields?).

**Accuracy** assesses whether data correctly represents reality. This is harder to measure than completeness. You might discover that address data is often outdated, income estimates are self-reported and inflated, or product categorizations are inconsistent. Accuracy problems often surface only when you start using data and results don't make sense.

**Consistency** evaluates whether data agrees across systems and over time. Does a customer's purchase history total match their financial transaction records? When data about the same entity exists in multiple systems, do those records align? Inconsistency suggests integration problems or process issues creating unreliable data.

**Timeliness** considers whether data is current enough for your use case. Historical data describing customer behavior from five years ago may not predict current behavior if market conditions, products, or customer demographics have changed substantially.

**Relevance** determines whether available data actually relates to your prediction target. Just because you have data doesn't mean it's useful. A retailer building a demand forecasting model discovered they had extensive customer demographic data but minimal information about actual product preferences and purchase drivers.

## 3.3 Training Data Requirements

AI models need specific types of data in sufficient quantities. Phase II must determine whether you can meet these requirements given available data (Géron, 2019).

For **supervised learning**, you need labeled examples showing correct answers. A fraud detection model needs transactions labeled "fraudulent" or "legitimate." A medical diagnosis model needs patient records with confirmed diagnoses. If labels don't exist, someone must create them—often requiring subject matter expert time for thousands or millions of examples.

**Sample size** requirements vary by problem complexity and modeling approach. Simple linear models might work with hundreds of examples. Deep neural networks often need millions. Rule of thumb: you need at least 10 times as many examples as you have features, and more is better. But this is just a starting point; some problems need far more data for acceptable performance.

**Class balance** matters for classification problems. If you're predicting rare events (1% of transactions are fraudulent), naive models can achieve 99% accuracy by always predicting "not fraudulent." You need either enough examples of rare classes to enable learning, techniques for handling imbalanced data, or redefinition of the problem.

**Feature availability** determines what your models can learn. To predict customer churn, you need features that correlate with churn risk: engagement metrics, purchase frequency, customer service interactions, price sensitivity indicators. If these features don't exist or can't be calculated from available data, model performance will suffer.

### Case Example: The Manufacturing Limitation

Priya was developing predictive maintenance models for a manufacturing facility. The business case projected major savings from predicting equipment failures before they occurred.

Phase II data assessment revealed challenging constraints. The factory had sensors on critical equipment collecting temperature, vibration, and pressure readings, but only for the past 18 months. Before that, data collection was spotty.

More problematic: maintenance records were inconsistent. Technicians sometimes logged repairs in the work order system, sometimes recorded them on paper, and sometimes didn't document minor fixes at all. Linking equipment failures to sensor readings required matching timestamps, but many maintenance records lacked precise timing.

Most critically, the equipment was highly reliable. Only 23 failures had occurred in 18 months—not nearly enough to train robust predictive models. Priya recognized they couldn't build what the business case envisioned. Instead, she proposed starting with anomaly detection to identify unusual equipment behavior that might indicate developing problems. This approach needed fewer labeled failure examples and could demonstrate value while data collection improved.

## 3.4 Data Availability and Accessibility

Having data and being able to access it are different things. Phase II must confirm you can actually get the data needed for model development and ongoing operations (PMI, 2025).

**Technical access** covers whether you can connect to data sources, extract data in usable formats, and move it to environments where data scientists work. Legacy systems sometimes lack APIs or export capabilities. Extracting data might require custom integration work, considerable time, and IT resources.

**Legal access** involves permissions and compliance requirements. Privacy regulations like GDPR restrict what data you can collect, how you can use it, and how long you can retain it. Health data falls under HIPAA in the United States, imposing strict controls. Financial data has regulatory requirements. You might technically have data but legally be unable to use it for AI purposes without consent or anonymization.

**Organizational access** navigates internal politics and policies. Different departments own different data. They may resist sharing due to concerns about data quality, competitive advantages, or loss of control. Data governance policies might require approval processes that delay access for months.

**Temporal access** considers when data becomes available. Real-time predictions require real-time data. If you need to predict customer behavior but data is only refreshed nightly, you can't provide immediate predictions. Batch predictions might be acceptable, or you might need infrastructure changes to enable real-time data flow.

## 3.5 The DIKUW Pyramid

The DIKUW Pyramid—Data, Information, Knowledge, Understanding, Wisdom—provides a framework for thinking about how raw data transforms into actionable intelligence (Ackoff, 1989).

**Data** represents raw facts and observations: sensor readings, transaction records, customer demographics, user clicks. Data alone lacks context and meaning.

**Information** emerges when you organize and contextualize data: daily sales totals, average customer purchase values, patterns in user behavior. Information answers "what" and "when" questions.

**Knowledge** develops when you understand relationships and patterns: recognizing that sales spike before holidays, understanding that certain demographics prefer certain products, identifying that user behavior changes on mobile versus desktop. Knowledge answers "how" questions.

**Understanding** synthesizes knowledge into broader insights: comprehending why certain patterns exist, predicting likely outcomes in different scenarios, recognizing cause-and-effect relationships. Understanding answers "why" questions.

**Wisdom** applies understanding to make good decisions: choosing optimal strategies, balancing competing priorities, anticipating consequences. Wisdom answers "what should we do" questions.

AI projects move through this pyramid. You start with data, extract information through analysis, build knowledge through pattern recognition, develop understanding through modeling, and enable wisdom through decision support. But you can't skip levels. Without solid data, information is unreliable. Without good information, knowledge is suspect. Poor knowledge leads to flawed understanding and bad decisions.

## 3.6 Data Gap Identification

Phase II often reveals gaps between available data and required data. Identifying these gaps early allows you to address them before model development (Huyen, 2022).

**Missing features** occur when data elements that would improve predictions don't exist. A retail recommendation system might lack data about products customers viewed but didn't purchase—important information for understanding preferences. An HR attrition model might need information about manager quality, but employee surveys about managers aren't conducted.

**Historical gaps** arise when you need historical data that wasn't collected. Predicting equipment failures requires failure history. If maintenance records are incomplete or equipment is new, historical data doesn't exist. You might need to collect data for months or years before modeling is feasible.

**Quality gaps** emerge when data exists but quality is insufficient. Customer addresses might be 60% accurate—not good enough for geographic analysis. Product categorizations might be inconsistent across categories. Data entry errors might be common enough to corrupt analysis.

**Integration gaps** appear when relevant data exists in disconnected systems that can't easily be linked. Customer service data and sales data might describe the same customers but use different identifiers, making integration difficult without master data management infrastructure.

Addressing gaps requires prioritization. Some gaps are critical—without closing them, the project can't succeed. Others reduce model performance but don't block progress. Still others would be nice to address but aren't worth the effort. Phase II should categorize gaps by importance and feasibility of resolution.

## 3.7 Exploratory Data Analysis

Exploratory Data Analysis (EDA) provides initial insights into data characteristics, distributions, relationships, and potential problems. This preliminary analysis informs modeling strategies and identifies data preparation needs (Tukey, 1977).

Start with **descriptive statistics**: means, medians, standard deviations, minimums, maximums. Do distributions make sense? Are there impossible values (negative ages, purchase amounts larger than any possible transaction)? Are there unexpected patterns (sudden changes in data collection, seasonal effects, systematic biases)?

Examine **missing data patterns**. Is data missing completely at random, or are there patterns? If high-income customers are more likely to decline providing data, missing data isn't random—it's systematically biased, which affects model reliability.

Investigate **relationships between variables**. Do features correlate with your prediction target? Do features correlate with each other (multicollinearity)? Correlation analysis helps identify potentially useful predictors and redundant features.

Look for **outliers and anomalies**. Are there data points that don't fit patterns? Sometimes these are errors that need correction. Sometimes they're rare but legitimate cases that need special handling. Sometimes they're the exact examples you want to study—fraud cases, equipment failures, high-value customers.

**Case Example: The E-Commerce Insight**

Diego was analyzing customer data for a churn prediction model. His initial EDA revealed surprising patterns. Purchase frequency followed a bimodal distribution—two distinct peaks rather than one. This suggested two different customer segments with different shopping behaviors.

Further investigation revealed the pattern reflected occasional shoppers (purchasing 1-2 times per year) and frequent buyers (purchasing monthly or more). These segments had completely different churn drivers. Occasional shoppers churned when they couldn't find specific items. Frequent buyers churned when competition offered better prices on their regular purchases.

This insight, discovered during Phase II data exploration, led Diego to recommend building separate models for each segment rather than one model for all customers. The segmented approach delivered substantially better predictions because it matched models to customer behavior patterns.

## 3.8 Data Volume and Variety Assessment

Modern AI projects often involve big data—datasets too large to fit in memory on a single machine. Phase II must assess whether you're dealing with big data challenges and whether you have infrastructure to handle them (Marz & Warren, 2015).

**Volume** measures data size. Are you working with thousands of rows, millions, billions? Can traditional tools handle this volume, or do you need distributed computing frameworks like Spark? How much will data grow over time?

**Velocity** considers how quickly data arrives. Is this a one-time analysis of historical data, or do you need to process continuous data streams? Can you do batch processing overnight, or do predictions need to be real-time?

**Variety** examines data types and structures. Are you working with structured data in databases, semi-structured data like JSON or XML, unstructured data like text or images, or all of the above? Different data types require different processing approaches.

Many organizations discover during Phase II that their data infrastructure is inadequate. They're trying to build AI on legacy databases designed for operational transactions, not analytical queries. They lack data lakes for storing diverse data types. They don't have compute infrastructure for large-scale model training.

These infrastructure gaps don't necessarily block projects, but they need to be addressed. You might need to invest in cloud computing resources, implement data lakes or warehouses, or start with smaller-scale pilots while building infrastructure.

## 3.9 What You've Learned

Phase II Data Understanding transforms assumptions about data into concrete assessments of what actually exists and whether it's adequate for AI objectives. This phase prevents teams from investing months in model development only to discover fundamental data problems.

Comprehensive data source identification reveals where relevant data lives and how difficult it will be to access and integrate. Data quality assessment across completeness, accuracy, consistency, timeliness, and relevance dimensions exposes problems early. Training data requirements analysis determines whether you have sufficient labeled examples, appropriate sample sizes, balanced classes, and necessary features.

Data availability and accessibility assessment confirms whether you can legally and technically access required data when needed. The DIKUW Pyramid provides a framework for understanding how data transforms into actionable intelligence. Data gap identification reveals missing features, historical gaps, quality issues, and integration challenges that need resolution.

Exploratory data analysis provides initial insights into data distributions, relationships, missing patterns, and outliers that inform modeling strategies. Volume and variety assessment determines whether you're dealing with big data challenges requiring specialized infrastructure.

Do this work thoroughly in Phase II. The time invested in understanding data prevents much larger problems in later phases.

## 3.10 Practice Questions

### Data Quality Problem Scenarios

**Question 121:** You discover that 30% of customer records have missing email addresses. Investigation reveals emails are only collected at checkout, and 30% of customers browse without purchasing. Is this a data quality problem?

A) Yes, this is always a quality problem B) No, this is systematic missingness reflecting business process; document it C) Delete all records without emails D) Fabricate email addresses

**Answer: B) Systematic missingness reflecting business process**

**Explanation:** Not all missing data indicates quality problems. This missingness has a clear business reason (customers who don't purchase don't provide emails). Document this pattern; understand implications for model training (emails available only for purchasers); consider if this creates bias. Systematic missingness differs from random quality issues.

---

**Question 122:** A dataset has customer ages ranging from 0 to 150 years. What's your first step?

A) Accept all values as valid B) Investigate extreme values: are they errors or legitimate outliers? C) Delete all records with unusual ages D) Replace all ages with the median

**Answer: B) Investigate extreme values**

**Explanation:** Extreme values warrant investigation, not automatic deletion. Ages of 0 might be missing data coded as zero; 150 could be data entry errors (transposed digits) or truly old customers. Profile the data: how many records have these extremes? Look for patterns. Distinguish errors (fix or remove) from legitimate outliers (keep).

---

**Question 123:** You're analyzing transaction data and discover that 99% of transactions are under $100, but 1% are over $1,000,000. How should you treat these high-value transactions?

A) Delete them as outliers B) Investigate: are they legitimate high-value transactions, errors, or fraud? C) Always keep all data unchanged D) Replace them with median values

**Answer: B) Investigate legitimacy**

**Explanation:** Context determines treatment. For B2B sales, $1M transactions might be normal. For consumer retail, they're likely errors or fraud. Investigation reveals truth: verify against source systems, check for data entry errors (extra zeros), review business logic. Understanding context prevents wrongly deleting legitimate data or keeping erroneous data.

---

**Question 124:** Two data sources contain customer information. Source A shows 50,000 customers; Source B shows 48,000. Which source is "correct"?

A) Source A has more records, so it's correct B) Neither—investigate the discrepancy: different time periods, definitions, or filtering? C) Source B is always correct D) Randomly pick one

**Answer: B) Investigate the discrepancy**

**Explanation:** Count differences require understanding, not assumption. Possible causes: different extraction times (2,000 customers added between extracts), different customer definitions (active vs. all), different filters (one excludes test accounts), or data quality issues. Understanding why sources differ informs which to use and how to reconcile them.

---

**Question 125:** A dataset has a "customer_type" field with values: "Gold", "gold", "GOLD", "Platinum", "platnum" (misspelling). What data quality issue is this?

A) Missing data B) Inconsistent encoding and typos C) Outliers D) Insufficient data

**Answer: B) Inconsistent encoding and typos**

**Explanation:** Inconsistent capitalization and misspellings create artificial categories. "Gold", "gold", and "GOLD" should be one category. "platnum" is a typo for "Platinum". Standardize through case normalization and spell correction. This prevents models from treating identical categories as different due to formatting inconsistencies.

---

**Question 126:** You discover that 15% of transaction timestamps are in the future. What's the most likely cause and solution?

A) Time travel; this is normal B) Time zone inconsistencies or system clock errors; investigate and correct C) The data is perfect D) Delete all future timestamps

**Answer: B) Time zone inconsistencies or system clock errors**

**Explanation:** Future timestamps indicate technical issues: mixed time zones (some UTC, some local), incorrect system clocks, or data entry errors. Investigate patterns (all from certain systems/locations?). Correct by standardizing time zones, fixing clock drift, or removing irreconcilable records. Document time handling for future data collection.

---

**Question 127:** A field that should contain only numerical product IDs (like "12345") contains entries like "N/A", "Unknown", "TBD", and "---". How should you handle this?

A) Convert all to 0 B) Recognize these as missing value indicators; standardize to NULL C) Keep as-is D) Delete the entire field

**Answer: B) Recognize as missing value indicators and standardize**

**Explanation:** Different systems encode missing data differently ("N/A", "Unknown", empty strings, special characters). Standardize all missing value indicators to NULL for consistency. This enables proper missing data handling during modeling. Documenting which values represent missingness is essential data understanding.

---

**Question 128:** You're analyzing e-commerce data and notice that "revenue" field has negative values for 5% of transactions. Is this necessarily an error?

A) Yes, always an error B) Possibly legitimate (returns/refunds) or errors; investigate context C) Ignore negative values D) Take absolute value of everything

**Answer: B) Possibly legitimate or errors depending on context**

**Explanation:** Context determines legitimacy. Negative revenue might represent returns, refunds, or adjustments—legitimate business events. Or they might be data errors. Investigate: Do negative values align with known returns? Are amounts reasonable? Understanding business processes reveals whether to keep (legitimate) or fix (errors).

---

**Question 129:** A dataset's "last_purchase_date" field shows many customers with purchases dated 1900-01-01. What does this likely indicate?

A) Customers purchased in 1900 B) Default/placeholder value for NULL; investigate why nulls exist C) This is perfect data D) System error in random dates

**Answer: B) Default/placeholder value for missing data**

**Explanation:** Dates like 1900-01-01, 1970-01-01, or 9999-12-31 commonly indicate missing data or system defaults. Investigate why these customers lack purchase dates: never purchased, data migration issues, or incomplete records? Treat as missing data rather than legitimate 1900 purchases. Understanding placeholder conventions prevents misinterpretation.

---

**Question 130:** You discover duplicate customer records with slightly different information (same person, different addresses/phones). How should you handle this?

A) Keep all duplicates B) Implement deduplication with merge strategy preserving most recent/complete information C) Randomly delete some records D) Ignore duplicates

**Answer: B) Implement deduplication with merge strategy**

**Explanation:** Duplicates corrupt analysis and training. Deduplication requires: (1) Identifying duplicates (fuzzy matching on name, DOB, etc.); (2) Merge strategy (keep most recent address, most complete record, or create composite); (3) Preserving audit trail. Balance precision (not merging distinct people) with recall (catching true duplicates).

---

## Source Selection Dilemmas

**Question 131:** You need customer demographic data. Source A (CRM) has 95% completeness but is 6 months old. Source B (real-time customer portal) has 60% completeness but current data. Which should you choose?

A) Always choose Source A (more complete) B) Always choose Source B (more current) C) Depends on use case: if recency matters more than completeness, use B; if completeness critical, use A D) Don't use either

**Answer: C) Depends on use case and priorities**

**Explanation:** No universal "better" source. For time-sensitive applications (fraud detection, real-time personalization), recency trumps completeness—use Source B. For stable attributes (demographics for segmentation), completeness might matter more—use Source A. Consider combining sources: real-time for time-sensitive features, complete for stable features.

---

**Question 132:** Three sources provide product pricing data: (1) Internal ERP system, (2) E-commerce website scrapes, (3) Vendor-provided feeds. Which is most authoritative?

A) Always website scrapes B) Internal ERP is authoritative for your pricing; websites for competitor pricing C) Always vendor feeds D) All equally authoritative

**Answer: B) Internal ERP for your pricing; external sources for competitors**

**Explanation:** Authority depends on what you're measuring. Internal ERP is authoritative for your company's prices (master data source). Website scrapes show what customers see (might differ due to promotions). Vendor feeds show supplier prices. Match source to question: your prices (ERP), customer-facing prices (website), supplier costs (vendor feeds).

---

**Question 133:** You need transaction data going back 5 years. The production database has 2 years; data warehouse has 5 years but different schema. What's the best approach?

A) Use only production database (2 years) B) Use only data warehouse (5 years) C) Combine both: map data warehouse historical data to current schema, append recent production data D) Give up

**Answer: C) Combine both sources with schema mapping**

**Explanation:** Combining sources provides full history. Map data warehouse schema to current production schema (field name changes, new fields with defaults, deprecated fields). Validate the mapping with overlapping time periods (do 2-year overlap match?). This effort yields 5 years of data, improving model performance versus 2-year limited history.

---

**Question 134:** For training a fraud detection model, you have: (1) Last 3 months of data (1M transactions), or (2) Last 3 years (5M transactions) with different fraud patterns due to system changes. Which is better?

A) Always more data (3 years) B) Recent data (3 months) if old patterns are obsolete C) Can't decide without understanding pattern stability D) Neither will work

**Answer: C) Depends on pattern stability**

**Explanation:** More data isn't always better if patterns changed. If system changes made old fraud patterns irrelevant, recent data is better. If core fraud patterns remain stable despite system changes, historical data adds value. Analyze: Do fraud patterns in years 1-2 still occur in year 3? Test models trained on different time windows to empirically determine optimal history.

---

**Question 135:** A vendor offers to provide enriched customer data (demographics, psychographics) for 80% of your customers. Cost: $50K. Your current data has basic demographics for 95% of customers. Should you purchase?

A) Always buy more data B) Evaluate: will enrichment improve model performance enough to justify cost? C) Never buy external data D) Buy it regardless of value

**Answer: B) Evaluate value versus cost**

**Explanation:** Data purchases require ROI analysis. Test with a sample: does enriched data improve model performance? By how much? What's the business value of that improvement? Compare benefit to $50K cost. Consider alternatives: can you derive similar features from existing data? External data makes sense when value exceeds cost and alternatives don't exist.

---

**Question 136:** Two databases contain employee data. HR database is authoritative but updates weekly. Payroll database updates daily but sometimes has errors. For a real-time application, which should you use?

A) HR database (authoritative) B) Payroll database (more frequent) C) Hybrid: use payroll for currency with HR database reconciliation/validation D) Neither

**Answer: C) Hybrid approach with validation**

**Explanation:** Balance authority with frequency. Use payroll database for real-time needs but validate against HR database during processing. Flag discrepancies for investigation. Schedule weekly reconciliation to correct payroll errors. This provides needed frequency while maintaining data quality through validation against authoritative source.

---

**Question 137:** You're building a model requiring weather data. Options: (1) Free government data (coarse spatial resolution), (2) Commercial data ($10K/year, fine-grained). How do you decide?

A) Always choose free data B) Start with free data; evaluate if resolution limitation impacts model performance; upgrade if needed C) Always buy commercial data D) Don't use weather data

**Answer: B) Start with free data; evaluate impact before purchasing**

**Explanation:** Start cost-effectively. Build models with free data; evaluate performance. If weather granularity limits performance and better resolution would materially improve results (test with sample commercial data), the purchase justifies cost. Many projects find free data sufficient; others need precision worth paying for. Evidence-based decisions beat assumptions.

---

**Question 138:** A third-party data provider offers social media sentiment data about your brand. They won't disclose methodology or data sources. Should you use this data?

A) Yes, any data is good data B) No, without methodology transparency, you can't assess quality or bias C) Use it without questions D) Buy it immediately

**Answer: B) No, lack of transparency prevents quality assessment**

**Explanation:** Data sources require transparency for responsible use. Without methodology details, you can't assess: collection methods (sampling bias?), sentiment scoring (validated algorithms?), data freshness, or quality controls. Opaque data sources introduce unknown risks. Insist on methodology disclosure or seek alternative providers with transparency.

---

**Question 139:** Your company has data in multiple languages (English, Spanish, Mandarin). You need training data for a text classification model. Should you:

A) Use only English data B) Use all languages with appropriate NLP models for each C) Machine-translate everything to English D) Abandon multilingual markets

**Answer: B) Use all languages with appropriate NLP models**

**Explanation:** Multilingual data reflects customer reality. Use language-specific or multilingual models (mBERT, XLM-R) that handle multiple languages. Machine translation introduces errors and loses nuance. If resources constrain you, prioritize languages by business value, but don't ignore major customer segments. Proper multilingual handling delivers better global performance.

---

**Question 140:** For customer lifetime value prediction, you have: (1) Transaction history (5 years), (2) Website behavior (2 years), (3) Customer service interactions (1 year). Should you use only overlapping time periods (1 year) or all available data?

A) Use only 1-year overlap B) Use all available data with proper handling of missingness before each feature's availability C) Use only transaction data (longest history) D) Use only customer service data (most recent)

**Answer: B) Use all available data with proper missing value handling**

**Explanation:** Use all valuable information. For customers existing 5 years, use full transaction history. For website features, mark as missing before 2 years ago. Models handle missing data (imputation, indicators). Using only overlapping periods discards valuable information. Properly handling temporal availability of different data sources maximizes model performance.

---

## Data Sufficiency Evaluation

**Question 141:** You're building a fraud detection model. You have 100,000 transactions with 100 fraud cases (0.1% fraud rate). Is this sufficient data?

A) Yes, 100,000 is always enough B) The absolute number of fraud cases (100) is more critical than total transactions C) No, never enough data D) Can't determine

**Answer: B) Absolute number of fraud cases (100) is critical**

**Explanation:** For imbalanced problems, minority class size matters most. 100 fraud examples might be insufficient for learning complex fraud patterns, despite 100,000 total transactions. Consider: oversampling, synthetic data generation (SMOTE), anomaly detection approaches, or collecting more fraud examples (historical data, other sources). Model complexity must match minority class size.

---

**Question 142:** A stakeholder asks: "How much data do we need?" What's the best response?

A) "1 million records" B) "It depends on problem complexity, feature count, model type, and data quality" C) "More is always better" D) "As little as possible"

**Answer: B) "It depends on multiple factors"**

**Explanation:** No universal answer. Factors include: problem complexity (simple patterns need less data), feature count (more features need more data), model type (deep learning needs more than linear models), class imbalance (rare classes need sufficient examples), and data quality (noisy data needs more examples). Start with available data; let performance guide need for more.

---

**Question 143:** You have 10,000 labeled images for training a computer vision model. A colleague suggests using transfer learning from ImageNet instead of training from scratch. Why?

A) Transfer learning is always slower B) Transfer learning leverages patterns learned from millions of images, compensating for limited training data C) 10,000 images is insufficient for any approach D) Training from scratch is always better

**Answer: B) Transfer learning compensates for limited training data**

**Explanation:** Transfer learning uses knowledge from large datasets (ImageNet's millions of images) and adapts it to your specific task. This dramatically reduces data requirements—models that need millions of examples to train from scratch often perform well with thousands when using transfer learning. For limited data scenarios, transfer learning is often the only viable approach.

---

**Question 144:** Your dataset has 50 features and 1,000 examples. A data scientist warns about the "curse of dimensionality." What's the concern?

A) No concern; this is fine B) Too many features relative to examples increases overfitting risk C) Too few examples; collect exactly 50 examples D) Dimensionality doesn't matter

**Answer: B) High feature-to-example ratio increases overfitting**

**Explanation:** Rules of thumb suggest examples should significantly exceed features (often 10× or more). 1,000 examples for 50 features is marginal. High-dimensional spaces become sparse—data points are too far apart for meaningful pattern learning. Solutions: feature selection (reduce to most important features), regularization, or collect more data. The ratio of features to examples critically impacts model performance.

---

**Question 145:** You're building a seasonal demand forecasting model. You have 2 years of data (24 months). Is this sufficient to model annual seasonality?

A) Yes, perfect amount B) Marginal—2 cycles of annual seasonality is minimal; 3-5 years preferred C) Too much data D) Seasonality doesn't need historical data

**Answer: B) Marginal—prefer 3-5 years for robust seasonality learning**

**Explanation:** Time series models need multiple cycles to learn seasonal patterns reliably. 2 years provides only 2 observations of each season (2 Decembers, 2 Januaries). Unusual events in either year (weather, promotions) might be mistaken for normal patterns. 3-5 years provides better seasonality estimates. With 2 years, be cautious about confidence in seasonal patterns; document this limitation.

---

**Question 146:** A rare disease affects 1 in 100,000 people. You have health records for 50,000 people. How many positive cases do you expect, and is this sufficient for supervised learning?

A) 500 cases; plenty for modeling B) 0-1 cases expected; insufficient for supervised learning—consider alternative approaches C) 50,000 cases; perfect D) Can't calculate

**Answer: B) 0-1 expected cases; insufficient for supervised learning**

**Explanation:** Expected positive cases: 50,000 × (1/100,000) = 0.5 cases. You might have zero or one positive case—completely insufficient for supervised learning. Alternatives: unsupervised anomaly detection, one-class classification, or acquiring much larger datasets (millions of records). Some problems require massive data due to extreme rarity; recognize when data is fundamentally insufficient.

---

**Question 147:** You're validating data sufficiency by splitting into train (70%), validation (15%), and test (15%) sets. With 1,000 total examples, how many test examples do you have, and is this concerning?

A) 150 test examples; adequate for reliable performance estimates B) 150 test examples; concerning if test set doesn't represent population diversity C) 150 test examples; always perfect D) Can't determine test set size

**Answer: B) 150 test examples; potentially insufficient for reliable estimates**

**Explanation:** 150 test examples might not adequately represent population diversity, especially with imbalanced classes or multiple subgroups. Test set confidence intervals widen with smaller sample sizes. If you have 10 fraud cases in 1,000 examples, test set might have only 1-2 fraud cases—unreliable for estimating fraud detection performance. Consider k-fold cross-validation for more robust estimates with limited data.

---

**Question 148:** A colleague suggests collecting more data will always improve model performance. Is this true?

A) Yes, always B) No, beyond a point, more data yields diminishing returns; other factors (features, algorithms) may matter more C) Never collect more data D) Data volume doesn't affect performance

**Answer: B) Diminishing returns; other factors may matter more**

**Explanation:** Learning curves typically show rapid improvement with initial data, then plateau. Beyond the plateau, marginal performance gains per additional example shrink. At some point, better features, algorithm selection, or hyperparameter tuning improve performance more than more data. Understand your position on the learning curve; if performance plateaued despite more data, invest elsewhere.

---

**Question 149:** You have sufficient data volume but it's all from January-March. You're building a model for year-round use. What's the concern?

A) No concern; volume is sufficient B) Temporal coverage gap; seasonal patterns outside January-March are absent C) Too much data D) Three months is always enough

**Answer: B) Temporal coverage gap creates seasonality blind spots**

**Explanation:** Data sufficiency includes temporal representativeness, not just volume. Models trained only on Q1 data lack patterns from other seasons (holiday shopping, summer trends). Deploy year-round models on year-round data, or explicitly limit scope: "This model is calibrated for Q1; recalibrate for other quarters." Temporal gaps create performance issues when deployed outside training period coverage.

---

**Question 150:** How can you determine if you have sufficient data without actually building the model?

A) You can't; always build first B) Analyze learning curves from pilot models; if performance plateaus, data is sufficient C) Count to 1 million records D) Guess randomly

**Answer: B) Pilot models and learning curves indicate sufficiency**

**Explanation:** Build simple pilot models on increasing data subsets. Plot performance versus data size (learning curve). If curve plateaus, additional data yields minimal improvement—you have sufficient data. If curve still climbs steeply, more data would help. This empirical approach beats theoretical estimates. Learning curves guide data collection investment decisions.

---

## Data Acquisition Strategies

**Question 151:** Your project needs customer email engagement data, but the email platform is a separate SaaS tool with no direct database access. How do you acquire this data?

A) Give up on email data B) Explore API access, scheduled exports, or third-party integration platforms C) Manually copy data D) Hack into the system

**Answer: B) Explore API access, scheduled exports, or integration platforms**

**Explanation:** Modern SaaS platforms typically provide data access through: (1) REST APIs for programmatic access, (2) Scheduled export features (CSV/JSON downloads), (3) Integration platforms (Zapier, Segment) connecting systems, or (4) Data warehouse connectors (Fivetran, Stitch). Evaluate options based on data freshness needs, volume, and technical capabilities. APIs enable real-time access; exports work for batch processing.

---

**Question 152:** You need competitor pricing data. Web scraping their public website is technically feasible but legal counsel expresses concerns. What's the appropriate approach?

A) Scrape anyway B) Consult legal on terms of service, copyright, and data protection laws; explore alternatives (third-party data providers) C) Never collect competitor data D) Ignore legal concerns

**Answer: B) Consult legal and explore compliant alternatives**

**Explanation:** Web scraping legality depends on: website terms of service, copyright law, data protection regulations (GDPR), and Computer Fraud and Abuse Act implications. Legal review is essential. Alternatives include: third-party competitive intelligence services (legally obtained data), public datasets, or purchasing competitor products to collect pricing legitimately. Compliance protects the organization from liability.

---

**Question 153:** A data source requires manual data entry from paper forms. You need 50,000 records. What's the most efficient acquisition strategy?

A) Have data scientists manually enter everything B) Outsource to data entry services or use OCR (optical character recognition) with manual verification C) Forget the data D) Reduce sample size to 100 records

**Answer: B) Outsource to data entry services or use OCR**

**Explanation:** Manual entry doesn't scale. Options: (1) OCR scans paper forms, extracts text automatically, requires manual verification for accuracy (hybrid approach); (2) Outsource to specialized data entry services (cost-effective, quality controls); (3) Crowdsource via platforms like Amazon Mechanical Turk for simple forms. Balance cost, speed, and accuracy. Automation (OCR) plus human verification often optimal.

---

**Question 154:** You discover valuable data locked in PDFs (invoices, contracts, reports). How do you extract structured data from unstructured PDF documents?

A) Manually read and enter data B) Use PDF parsing libraries (PyPDF2, Tabula) for tables; NLP for text extraction; document AI services for complex layouts C) Ignore PDF data D) Convert PDFs to images and give up

**Answer: B) Use PDF parsing tools and NLP**

**Explanation:** PDF extraction strategies: (1) Simple PDFs: use libraries like PyPDF2 (text), Tabula (tables); (2) Complex layouts: Document AI services (Google Document AI, AWS Textract) use ML to understand document structure; (3) Scanned PDFs: OCR first, then parse; (4) Entity extraction: NLP extracts specific fields (amounts, dates, parties). Approach depends on PDF complexity and volume.

---

**Question 155:** Your data acquisition strategy depends on third-party API with rate limits: 100 requests per minute. You need 1 million records. How do you handle this?

A) Make requests as fast as possible and hope for the best B) Implement rate-limiting logic with backoff; batch requests; estimate 7 days for full acquisition C) Give up D) Ignore rate limits

**Answer: B) Implement rate-limiting logic respecting constraints**

**Explanation:** Respect rate limits to avoid being blocked. Calculate timing: 1M records ÷ 100 per minute = 10,000 minutes ≈ 7 days continuous operation. Implement: exponential backoff for rate limit errors, request batching (multiple records per request if supported), parallel workers up to limit, persistent queue to resume after interruptions. Plan acquisition timeline accordingly.

---

**Question 156:** You need real-time customer behavior data for model training, but your analytics platform has a 24-hour delay. How do you address this?

A) Wait 24 hours for all data B) Implement streaming data pipeline capturing events in real-time; use message queue (Kafka) for model training C) Don't train models D) Guess customer behavior

**Answer: B) Implement streaming data pipeline**

**Explanation:** Real-time requirements need real-time infrastructure. Solutions: (1) Event streaming platform (Kafka, Kinesis) captures events as they occur; (2) Stream processing (Spark Streaming, Flink) aggregates features; (3) Feature store maintains real-time features for training and serving. While batch delayed data works for some use cases, truly real-time applications need streaming architecture.

---

**Question 157:** A potential data source contains sensitive PII that your project doesn't need (full SSNs, when only last 4 digits needed). How should you acquire this data?

A) Acquire full data and figure it out later B) Acquire only necessary fields; minimize PII collection following data minimization principles C) Acquire everything possible D) Don't acquire any data

**Answer: B) Acquire only necessary data following minimization principles**

**Explanation:** Privacy by design and GDPR's data minimization principle: collect only data necessary for defined purposes. Acquiring full SSNs when you only need last 4 digits violates this principle, increases breach risk, and may violate regulations. Design acquisition to collect minimal necessary data. This reduces privacy risk, simplifies compliance, and limits breach exposure.

---

**Question 158:** Your company operates globally. Should you centralize all customer data in one region for model training, or keep it distributed by region?

A) Always centralize B) Consider regulatory requirements (GDPR restrictions on data transfers); federated learning if centralization prohibited C) Always distribute D) Ignore regulations

**Answer: B) Consider regulatory requirements; use federated learning if needed**

**Explanation:** GDPR and similar regulations restrict cross-border data transfers. Options: (1) If compliant, centralize for simpler training; (2) If restricted, use federated learning (models train locally in each region, only model weights centralize); (3) Train region-specific models. Legal requirements constrain technical architecture. Early legal consultation prevents expensive rework.

---

**Question 159:** You're acquiring historical data from a legacy system being decommissioned in 3 months. What's the appropriate urgency and approach?

A) No rush; acquire slowly B) High priority—extract all valuable data before decommissioning; validate extraction quality C) Wait until after decommissioning D) Don't acquire legacy data

**Answer: B) High priority extraction before decommissioning**

**Explanation:** Decommissioning deadlines create urgency. Strategy: (1) Prioritize data extraction (move it to the front of project timeline); (2) Extract more than immediately needed (future uses); (3) Validate extraction quality while source system available for comparison; (4) Document data dictionary, business rules; (5) Create archive for future access. Once decommissioned, data may be irrecoverable.

---

**Question 160:** A data vendor offers a free trial of their dataset before purchase. How should you use this trial period?

A) Don't bother with trials B) Test with your use case: build pilot model, measure performance improvement, validate data quality and coverage C) Look at data once and decide D) Buy without trial

**Answer: B) Test with your use case during trial**

**Explanation:** Free trials enable evidence-based purchasing. During trial: (1) Build pilot model with trial data; measure performance versus baseline; (2) Assess data quality (accuracy, completeness); (3) Validate coverage (does it cover your customer base?); (4) Test integration ease; (5) Calculate ROI based on measured improvement. Make data purchasing decisions based on demonstrated value, not vendor claims.

---

## Exploratory Data Analysis Situations

**Question 161:** During EDA, you plot a histogram of customer ages and see two distinct peaks (one around age 25, another around age 55). What does this bimodal distribution suggest?

A) Data error B) Potentially distinct customer segments with different characteristics C) Normal distribution D) Ignore it

**Answer: B) Potentially distinct customer segments**

**Explanation:** Bimodal distributions suggest mixture of subpopulations. Two age peaks might indicate: (1) Young professionals (25) and established households (55) with different behaviors; (2) Different product lines attracting different demographics. Investigate: do these segments differ in other features? Might warrant separate models or segmentation. Distributions reveal population structure valuable for modeling.

---

**Question 162:** You create a correlation matrix and discover two features with 0.95 correlation. What's the implication?

A) Perfect—keep both features B) Multicollinearity; consider removing one feature or combining them C) Always remove all correlated features D) Correlations don't matter

**Answer: B) Multicollinearity; consider removing one or combining**

**Explanation:** High correlation between features (multicollinearity) causes problems: (1) Redundant information; (2) Unstable coefficient estimates in linear models; (3) Difficulty interpreting feature importance. Options: (1) Remove one correlated feature; (2) Combine via PCA; (3) Use regularization; (4) If different concepts, keep both for tree-based methods (less sensitive to multicollinearity). The right choice depends on model type and interpretability needs.

---

**Question 163:** During EDA, you discover that 80% of your target variable (customer churn) is "No" and only 20% is "Yes". What's the implication for modeling?

A) No implication B) Class imbalance; will need special handling (resampling, class weights, appropriate metrics) C) Perfect balance D) Too balanced

**Answer: B) Class imbalance requiring special handling**

**Explanation:** 80/20 imbalance (while not extreme) requires consideration: (1) Standard training might ignore minority class; (2) Accuracy is misleading metric (predicting "No" always gives 80% accuracy); (3) Solutions include oversampling minority, undersampling majority, class weighting, or balanced algorithms. Focus on precision, recall, F1-score, and AUC-ROC. EDA revealing imbalance guides modeling strategy.

---

**Question 164:** You plot feature distributions across train and test sets and notice significant differences. What does this indicate?

A) Perfect data B) Train-test distribution mismatch; test performance may not reflect production C) Normal variation D) Ignore differences

**Answer: B) Train-test mismatch; test estimates may be misleading**

**Explanation:** Train and test distributions should be similar—both representing production population. Significant differences indicate: (1) Non-random splitting; (2) Temporal drift (test is from later period); (3) Sampling bias. Consequences: test performance estimates won't generalize to production. Fix: ensure proper random stratified splitting or temporal splitting respecting time series structure. Validate splitting preserves distribution similarity.

---

**Question 165:** During EDA, you discover that a "customer_satisfaction" feature has values of 1-5, but 60% of values are 5. What might this indicate?

A) Customers are extremely satisfied B) Response bias (satisfied customers more likely to respond) or survey design issues C) Perfect distribution D) Data error only

**Answer: B) Response bias or survey design issues**

**Explanation:** Highly skewed satisfaction ratings suggest: (1) Response bias (satisfied customers respond; dissatisfied don't); (2) Survey timing (immediately after purchase vs. later); (3) Social desirability bias; (4) Rating scale design. Investigate: response rates by satisfaction level, survey methodology. Understand that observed satisfaction might not represent true population. Consider: is this feature reliable for modeling?

---

**Question 166:** You calculate summary statistics and notice the mean income is $75K but the median is $45K. What does this tell you?

A) Calculation error B) Right-skewed distribution with high-income outliers pulling mean up C) Left-skewed distribution D) Perfect normal distribution

**Answer: B) Right-skewed distribution with high-income outliers**

**Explanation:** Mean > median indicates right-skewed distribution (long tail of high values). High-income outliers inflate the mean while median (50th percentile) is robust to outliers. Income distributions typically right-skewed (most people earn moderate amounts; few earn very high amounts). For skewed data, median often better represents "typical" value than mean. Consider log transformation for modeling.

---

**Question 167:** You create scatter plots of features against the target variable. One feature shows no relationship (random cloud). Should you:

A) Keep it; more features are always better B) Consider removing it; irrelevant features add noise and increase overfitting risk C) Force a relationship D) Ignore all features

**Answer: B) Consider removing; irrelevant features harm models**

**Explanation:** Features with no relationship to target add noise without signal. Consequences: (1) Increase overfitting risk; (2) Slow training; (3) Complicate interpretation; (4) Reduce model performance. Feature selection removes irrelevant features. Caveat: univariate analysis might miss interaction effects. Balance removal of clearly irrelevant features with preserving features that might have value in combinations.

---

**Question 168:** During temporal EDA, you plot monthly sales and notice regular spikes every December. What's the implication for modeling?

A) Ignore seasonality B) Must incorporate seasonality into features or model architecture C) Remove December data D) Seasonality doesn't matter

**Answer: B) Incorporate seasonality into modeling**

**Explanation:** Regular seasonal patterns (December spikes, likely holidays) must be captured for accurate forecasting. Approaches: (1) Feature engineering: month indicators, is_holiday flags; (2) Time series models explicitly handling seasonality (SARIMA); (3) Yearly lags as features. Ignoring strong seasonality guarantees poor forecasts. EDA revealing seasonality guides feature engineering and model selection.

---

**Question 169:** You're analyzing text data and create word frequency distributions. You find "the", "a", "is" dominate. What should you do?

A) Keep all words equally B) Remove stop words (common words with little meaning); focus on informative terms C) Remove all common words D) Use only stop words

**Answer: B) Remove stop words; focus on informative terms**

**Explanation:** Stop words (the, a, is, of) appear frequently but carry little meaning for most text classification tasks. Removing them: (1) Reduces feature space; (2) Emphasizes informative words; (3) Improves signal-to-noise ratio. Standard NLP preprocessing includes stop word removal. However, context matters—for some tasks (authorship attribution, grammar checking), stop words contain signal. Typical text classification benefits from removal.

---

**Question 170:** During EDA, you use dimensionality reduction (PCA, t-SNE) to visualize high-dimensional data in 2D. You see clear cluster separation. What does this suggest?

A) Data is one homogeneous group B) Natural clusters/segments exist; unsupervised learning or segmentation might be valuable C) Visualization error D) Clusters don't mean anything

**Answer: B) Natural clusters suggest segmentation opportunities**

**Explanation:** Clear cluster separation in dimensionality reduction indicates: (1) Distinct subpopulations with different characteristics; (2) Potential for unsupervised clustering (K-means, hierarchical); (3) Different segments might need different models. Investigate: what distinguishes clusters? Do they align with known categories or reveal new segments? Cluster visualization guides modeling strategy (global model vs. segment-specific models).

---

**Question 171:** You create box plots for a numerical feature by target class (churn vs. non-churn). The boxes completely overlap. What does this indicate?

A) Feature perfectly predicts target B) Feature has little univariate discriminatory power for target C) Data error D) Feature is the best predictor

**Answer: B) Feature has little univariate predictive power**

**Explanation:** Overlapping distributions across target classes mean the feature alone doesn't differentiate outcomes. Feature values are similar for both churners and non-churners. This doesn't mean the feature is useless (might have value in interactions with other features) but suggests limited standalone predictive power. Prioritize features showing clear separation for feature selection.

---

**Question 172:** During EDA, you discover that 90% of observations have feature values within a narrow range, but 10% span a much wider range. How should you handle this for modeling?

A) Delete the 10% outliers B) Consider standardization/normalization; outliers might need special treatment depending on context C) Keep everything unchanged D) Delete the 90% majority

**Answer: B) Consider scaling; context-dependent outlier handling**

**Explanation:** Scaling (standardization, min-max normalization) prevents features with large ranges from dominating distance-based algorithms. For the 10% outliers: investigate legitimacy (errors vs. genuine extreme values). Options: (1) Keep if legitimate; (2) Winsorize (cap at percentiles); (3) Log transform if right-skewed; (4) Model robustness (tree-based methods less sensitive). Context determines treatment.

---

**Question 173:** You analyze missing data patterns and discover that when Feature A is missing, Features B, C, and D are also always missing. What's the implication?

A) Coincidence B) Suggests systematic missingness (might all come from same optional form/source) C) Random missingness D) No pattern exists

**Answer: B) Systematic missingness indicating common source**

**Explanation:** Correlated missingness suggests features come from the same data source or process (optional survey section, secondary database, specific customer segment). Understanding missingness patterns informs handling: (1) Missing indicator features capture information; (2) Might represent distinct customer segment (those who don't complete optional fields); (3) Imputation strategies should consider pattern. Missingness itself can be informative signal.

---

**Question 174:** During EDA, you plot model features over time and notice gradual shifts in distributions. What's the concern?

A) No concern B) Indicates potential concept drift; models may degrade over time and need retraining C) Perfect stability D) Ignore temporal patterns

**Answer: B) Concept drift; models will need maintenance**

**Explanation:** Feature distribution shifts over time (concept drift) mean patterns learned during training diverge from current patterns. Consequences: model performance degrades without retraining. EDA revealing drift informs: (1) Retraining frequency needs; (2) Monitoring requirements; (3) Drift detection systems. Anticipating drift during EDA enables proactive maintenance planning rather than reactive responses to production failures.

---

**Question 175:** You're performing EDA on image data and sample 100 random images for manual inspection. You notice 20% are blurry or corrupted. What should you do?

A) Ignore quality issues B) Implement systematic quality filtering; estimate cleaning effort; assess impact on training data volume C) Delete all images D) Use corrupted images anyway

**Answer: B) Implement systematic quality filtering**

**Explanation:** Image quality issues require systematic handling: (1) Implement automated quality checks (blur detection, corruption detection); (2) Estimate percentage requiring removal; (3) Assess impact on data volume (if removing 20% leaves sufficient data); (4) Consider quality-based weighting. Poor quality images harm model training. EDA sampling revealing quality issues guides systematic cleaning strategy.

---

**Question 176:** During EDA, you calculate inter-rater reliability for labeled data and discover annotators agree only 60% of the time. What's the implication?

A) This is excellent agreement B) Low agreement indicates label quality issues; need clearer guidelines or expert labeling C) Perfect labels D) Ignore agreement

**Answer: B) Low agreement indicates label quality problems**

**Explanation:** Low inter-rater agreement (60% is concerning, depending on task complexity) suggests: (1) Ambiguous labeling guidelines; (2) Task difficulty; (3) Insufficient annotator training; (4) Subjective/ambiguous cases. Consequences: unreliable labels lead to poor model performance. Solutions: clarify guidelines, provide examples, re-label by experts, measure/track agreement, remove ambiguous cases. Label quality determines model performance ceiling.

---

**Question 177:** You perform cross-tabulation (crosstab) analysis and discover that certain feature combinations never appear in your data (e.g., no male customers aged 18-25 in certain locations). What's the concern?

A) No concern B) Sparse regions in feature space; model will not learn patterns for these combinations C) Perfect data D) Impossible combinations

**Answer: B) Sparse feature space regions; model won't learn absent patterns**

**Explanation:** Absent feature combinations (sparse feature space) mean models can't learn patterns for these combinations—they've never seen them. If production data includes these combinations, model predictions are unreliable extrapolations. Options: (1) Collect data covering sparse regions; (2) Document limitations (model doesn't apply to male 18-25 customers in location X); (3) Use domain knowledge to fill gaps. EDA revealing sparsity identifies model blind spots.

---

**Question 178:** During EDA on time series data, you discover irregular sampling intervals (some hourly, some daily, some weekly). How should you handle this?

A) Keep irregular sampling B) Resample to consistent intervals (interpolation for upsampling, aggregation for downsampling) C) Delete irregular data D) Ignore timing

**Answer: B) Resample to consistent intervals**

**Explanation:** Most time series algorithms assume regular intervals. Irregular sampling requires preprocessing: (1) Upsample (interpolate missing values) to finest interval; (2) Downsample (aggregate) to coarser interval; or (3) Use algorithms handling irregular sampling. Choice depends on application: high-frequency applications need fine intervals; long-term forecasting can use coarse intervals. Consistent intervals simplify modeling.

---

**Question 179:** You create a confusion matrix on training data showing perfect classification (no errors). What's the likely issue?

A) Perfect model; deploy immediately B) Severe overfitting or data leakage C) Model is production-ready D) This is ideal

**Answer: B) Severe overfitting or data leakage**

**Explanation:** Perfect training performance typically indicates problems: (1) Overfitting (model memorized training data); (2) Data leakage (target information leaked into features); (3) Target itself used as feature. Real-world data has noise; perfect training accuracy is suspicious. Check: validation performance (will be much worse if overfitting), feature engineering (any leakage?), model complexity (too complex for data size?). Investigate before celebrating perfect performance.

---

**Question 180:** During EDA, you discover that your target variable (conversion) has drastically different rates across different data sources (5% from Source A, 45% from Source B). What should you investigate?

A) Nothing; this is normal B) Source differences: different populations, label definitions, or data quality issues C) Perfect consistency D) Ignore source differences

**Answer: B) Investigate source differences in population or labeling**

**Explanation:** Dramatic target rate differences across sources suggest: (1) Different populations (Source A is cold leads, Source B is warm leads); (2) Different definitions ("conversion" means different things); (3) Data quality issues in one source. Investigate: what defines each source? Are populations comparable? Are labels defined consistently? Understanding source differences informs: combining versus separate analysis, weighting, or model segmentation.

---

# Chapter 4.0: Phase III Data Preparation

Here's the uncomfortable reality: data scientists spend 60-80% of their time preparing data, not building sophisticated models (Lohr, 2014). Phase III Data Preparation is where the real work happens—cleaning messy data, handling missing values, engineering features, creating pipelines, and transforming raw information into something machine learning algorithms can actually use. Skip corners here, and even the most advanced models will fail.

## 4.1 Data Cleaning Fundamentals

**Data cleaning** removes errors, inconsistencies, and inaccuracies that corrupt model training. Real-world data is messy. Systems capture incorrect values, users make data entry errors, integrations fail silently, and processes change over time leaving artifacts in your data (Wickham & Grolemund, 2017).

Start by identifying **missing values**. Are fields blank, marked as NULL, recorded as placeholder values like -999 or "N/A"? Missing data patterns matter. If data is missing completely at random—random technical glitches, occasional user skips—that's one scenario. If data is systematically missing—high-income customers decline to provide certain information—that creates bias requiring different handling.

Common approaches for missing data include deletion (removing records with missing values), mean/median/mode imputation (replacing missing values with statistical measures), forward/backward fill (using previous or next values for time series), and model-based imputation (predicting missing values using other features). No single approach works universally; the right choice depends on your data patterns and model requirements (Little & Rubin, 2019).

**Outlier detection** identifies values that deviate significantly from typical patterns. Sometimes outliers are errors—a customer age of 250 years, a purchase amount of $10 million for a consumer product. Sometimes they're legitimate but rare cases—an actual high-value customer, a genuine equipment failure. Blindly removing outliers can eliminate exactly the cases you want to study, while keeping errors corrupts training.

Statistical methods like z-scores (values more than 3 standard deviations from mean) or IQR (interquartile range) help identify outliers. Domain expertise determines whether they're errors or legitimate data requiring special handling.

**Duplicate detection** finds and resolves repeated records. Sometimes duplicates are obvious—identical records entered twice. Sometimes they're subtle—same customer with slight name variations, same transaction recorded in multiple systems with different timestamps. Fuzzy matching algorithms help identify near-duplicates that exact matching misses.

**Case Example: The Healthcare Data Reality**

Dr. Keisha Washington led data preparation for a hospital readmission prediction model. The raw data was a disaster. Patient weight fields contained values in pounds, kilograms, and occasionally tons (data entry errors). Height had similar issues—inches, centimeters, and impossible values like 12 feet. Date fields mixed formats—MM/DD/YYYY, DD/MM/YYYY, YYYY-MM-DD—causing systemic errors when parsed incorrectly.

Diagnosis codes used two different classification systems (ICD-9 and ICD-10) depending on when data was recorded. Medication names appeared as brand names, generic names, and abbreviations with no consistent mapping. Patient identifiers had multiple formats after system migrations, making it difficult to link records for the same patient across time.

Dr. Washington's team spent three months building comprehensive data cleaning pipelines. They created validation rules to identify impossible values, standardized units of measurement, harmonized date formats, mapped diagnosis codes to consistent classifications, and built patient matching algorithms to link records across different identifier systems.

The cleaned dataset was 30% smaller than the raw data after removing duplicates and irreconcilable records, but model performance improved dramatically. Garbage in, garbage out—no amount of sophisticated modeling overcomes bad data.

## 4.2 Data Labeling Strategies

**Data labeling** assigns ground truth labels to training examples for supervised learning. You're teaching the model what correct answers look like. Label quality directly determines model performance—if your labels are wrong or inconsistent, your model learns incorrect patterns (Roh et al., 2019).

**Manual labeling** by human annotators provides high-quality labels but scales poorly. A medical imaging model might need radiologists to label thousands of scans as showing specific conditions. Each label requires expert time and attention. For complex tasks requiring deep expertise, manual labeling may be the only viable approach, but it's expensive and slow.

**Crowdsourced labeling** through platforms like Amazon Mechanical Turk or specialized services scales better but requires quality control. Multiple annotators label the same examples; consensus or majority voting determines final labels. Clear instructions, training examples, and quality checks ensure consistency. This works well for tasks where non-experts can provide reliable labels—identifying objects in images, categorizing text sentiment, transcribing audio.

**Semi-supervised learning** combines small amounts of labeled data with large amounts of unlabeled data. The model learns from labeled examples, makes predictions on unlabeled data, and incorporates high-confidence predictions as additional training data. This approach works when labeling is expensive but unlabeled data is abundant.

**Active learning** strategically selects which examples to label. Instead of randomly labeling data, the model identifies examples where it's most uncertain or where labels would provide maximum information gain. Human annotators focus on these high-value examples. This can reduce labeling requirements by 50-90% while maintaining model performance (Settles, 2012).

**Label consistency** challenges emerge when multiple annotators disagree. For subjective tasks—sentiment analysis, content moderation, quality assessment—even experts often disagree. Measuring inter-annotator agreement using metrics like Cohen's kappa or Fleiss' kappa quantifies consistency. Low agreement suggests unclear guidelines, ambiguous cases, or inherently subjective judgments requiring different modeling approaches.

## 4.3 Feature Engineering Decisions

**Feature engineering** creates new input variables from raw data to help models learn patterns more effectively. Good features make learning easier; poor features make it harder or impossible (Zheng & Casari, 2018).

**Numerical features** might need transformation. Skewed distributions benefit from log transforms to normalize. Large-magnitude differences between features (age ranges 0-100, income ranges $0-$1,000,000) require scaling or normalization so no single feature dominates due to magnitude alone. Common scaling approaches include min-max scaling (mapping values to 0-1 range), z-score standardization (mean 0, standard deviation 1), and robust scaling (using median and interquartile range to handle outliers).

**Categorical features** need encoding for most algorithms. **One-hot encoding** creates binary columns for each category—a "color" feature with values red/blue/green becomes three columns (is_red, is_blue, is_green) with 1 indicating presence and 0 indicating absence. This works well for low-cardinality features but creates too many columns for high-cardinality features like zip codes or product IDs. **Target encoding** replaces categories with statistics related to the target variable—mean target value for each category—but risks overfitting.

**Temporal features** extract information from timestamps. A raw timestamp contains year, month, day, hour, day of week, and derived features like "is weekend," "is holiday," "days since last purchase," "time of day category." These engineered features often predict better than raw timestamps.

**Interaction features** capture relationships between variables. A model predicting house prices might multiply square footage by number of bedrooms—larger houses with more bedrooms command premium prices. E-commerce models might combine category and price range—luxury electronics behave differently than budget electronics. Domain knowledge guides which interactions to create.

**Case Example: The Retail Revenue Breakthrough**

Omar, a data scientist at a specialty retailer, was building customer lifetime value prediction models. Initial models using basic features (purchase history, demographics) performed poorly, explaining only 35% of variance in customer value.

Omar engineered sophisticated features from raw data. From purchase history, he created recency (days since last purchase), frequency (purchases per month), monetary value (average purchase amount), trend features (is spending increasing or decreasing), and product diversity (number of different categories purchased). From website behavior, he engineered browsing intensity (pages viewed per session), search patterns (branded vs. generic searches), and engagement signals (video watches, review reads, wish list additions).

He created interaction features combining customer characteristics with seasonal patterns—holiday spending varies dramatically by customer segment. He built look-back windows capturing behavior over different timeframes—7-day patterns differ from 90-day patterns.

The enhanced model with engineered features explained 78% of variance—more than double the initial performance. The improvement came not from sophisticated algorithms but from better representing business reality through thoughtful feature engineering.

## 4.4 Data Anonymization and Privacy

**Data anonymization** protects personally identifiable information (PII) while preserving data utility for machine learning. Privacy regulations like GDPR and CCPA impose strict requirements for handling personal data (Voigt & Von dem Bussche, 2017).

**Direct identifiers** like names, social security numbers, email addresses, and phone numbers must be removed or encrypted. Simply deleting these fields is straightforward but insufficient—indirect identification through combinations of other attributes remains possible.

**Pseudonymization** replaces direct identifiers with pseudonyms or tokens. A customer ID becomes a randomly generated identifier with no connection to the original person. The mapping between real identities and pseudonyms is stored separately under strict access controls. This allows data analysis while protecting identity.

**K-anonymity** ensures each individual's data is indistinguishable from at least k-1 other individuals. If you know someone's zip code, birthdate, and gender, k-anonymity guarantees at least k records share these attributes. This prevents singling out specific individuals. Achieving k-anonymity requires generalization (broadening categories—specific ages become age ranges) or suppression (removing records that can't be anonymized without losing too much information).

**Differential privacy** adds calibrated noise to data or query results so no individual's presence or absence significantly affects outcomes. This mathematical guarantee prevents inference about specific individuals even with unlimited computational resources and auxiliary information. Privacy budget parameters control the privacy-utility tradeoff—more privacy means more noise and less accuracy (Dwork & Roth, 2014).

**Synthetic data generation** creates artificial datasets that maintain statistical properties of real data without containing actual individual records. Models learn patterns from real data and generate synthetic examples following the same distributions. This provides privacy protection while enabling analysis and sharing.

## 4.5 Data Pipeline Creation

**Data pipelines** automate the flow of data from sources through transformation steps to model training and inference. Pipelines ensure reproducibility, enable monitoring, and prevent training-serving skew where data processed differently in development versus production causes model failures (Paleyes et al., 2022).

Pipelines typically include **data ingestion** (extracting from sources), **validation** (checking data quality and schema compliance), **cleaning** (handling missing values and outliers), **transformation** (feature engineering and encoding), **splitting** (train/validation/test sets), and **versioning** (tracking data versions used for each model).

**Pipeline orchestration** tools like Apache Airflow, Prefect, or cloud-native services schedule and monitor pipeline execution. They handle dependencies between steps, retry failures, alert on errors, and log execution details for debugging.

**Data validation** at pipeline entry prevents garbage data from corrupting models. Define schemas specifying expected data types, valid value ranges, required fields, and relationship constraints. Validate incoming data against schemas before processing. Data that fails validation goes to error handling—human review, automated correction, or rejection.

**Reproducibility** requires tracking exactly which data versions and transformation steps produced each model. Version control for data (using tools like DVC or MLflow) enables reproducing any model by rerunning pipelines with specific data versions and parameters.

## 4.6 Handling Imbalanced Datasets

**Imbalanced datasets** occur when target classes have drastically different frequencies—fraud detection where 0.1% of transactions are fraudulent, medical diagnosis where rare diseases affect 1% of patients, equipment failure prediction where failures are infrequent. Standard training approaches fail because models can achieve high accuracy by always predicting the majority class (Chawla et al., 2002).

**Resampling techniques** modify the training dataset to balance classes. **Oversampling** creates additional minority class examples by duplicating existing ones or generating synthetic examples using techniques like SMOTE (Synthetic Minority Over-sampling Technique) that creates new examples by interpolating between existing minority class instances. **Undersampling** removes majority class examples to match minority class frequency, though this discards potentially useful information.

**Class weighting** adjusts the loss function during training to penalize minority class misclassifications more heavily than majority class errors. Most machine learning frameworks support class weights that tell the model to care more about rare classes. This changes model behavior without modifying the dataset.

**Ensemble methods** specifically designed for imbalanced data combine multiple models trained on different balanced subsets of data. Techniques like Balanced Random Forest or EasyEnsemble create multiple training sets by undersampling the majority class, train models on each subset, and combine predictions.

**Evaluation metric selection** matters critically for imbalanced data. Accuracy is misleading—a model predicting "not fraud" for all transactions achieves 99.9% accuracy on a 0.1% fraud rate dataset while being completely useless. Precision, recall, F1-score, and area under the ROC curve (AUC-ROC) provide better assessment of model performance on minority classes.

### Case Example: The Fraud Detection Fix

Chen, a data scientist at a payment processor, built fraud detection models on a dataset with 0.08% fraud rate—80 fraudulent transactions per 100,000. Initial models achieved 99.92% accuracy by predicting all transactions as legitimate, catching zero fraud.

Chen tried oversampling with SMOTE to create synthetic fraud examples, balancing the training set to 50/50 fraud/legitimate. Model performance improved but suffered from overfitting—excellent validation metrics but poor production performance. The synthetic examples didn't capture the full diversity of real fraud patterns.

She switched to a combination approach: moderate oversampling (bringing fraud rate to 5%), class weighting (10x penalty for missing fraud), and ensemble methods (training multiple models on different data samples). She focused on precision-recall curves rather than accuracy, optimizing for catching 80% of fraud while keeping false positives manageable.

The final model caught 83% of fraud with a false positive rate of 0.3%—36 legitimate transactions flagged per 10,000. This balance worked for the business, which could handle manual review of flagged transactions while preventing most fraud losses.

## 4.7 Train Test Validation Splits

**Data splitting** divides data into separate sets for training, validation, and testing. This prevents overfitting and enables honest assessment of model performance on unseen data (Hastie et al., 2009).

**Training data** (typically 60-80% of data) trains the model—adjusting parameters to minimize prediction errors on these examples. The model sees these examples during learning.

**Validation data** (10-20% of data) evaluates models during development. You use validation performance to compare different algorithms, tune hyperparameters, and make architectural decisions. Models don't train directly on validation data, but you use validation performance to guide development choices, which creates indirect exposure.

**Test data** (10-20% of data) provides final, unbiased performance assessment. You don't touch test data until model development is complete. Test performance estimates how the model will perform on truly unseen real-world data.

**Random splitting** works when data points are independent and identically distributed. Randomly assign each example to train, validation, or test sets. This works for many problems but fails when data has temporal dependencies or natural groupings.

**Temporal splitting** preserves time order for time series data. Training data comes from early time periods, validation from middle periods, test from recent periods. This simulates real deployment where you train on historical data to predict future outcomes. Random splitting would leak future information into training, creating unrealistically optimistic performance estimates.

**Stratified splitting** maintains class proportions across splits. If your overall data is 10% positive class, stratified splitting ensures each split is approximately 10% positive class. This is especially important for imbalanced datasets where random splitting might create test sets with too few minority class examples for reliable evaluation.

**Group splitting** keeps related examples together. For medical data, all records for the same patient should be in the same split—otherwise the model learns to recognize specific patients rather than general patterns. For retail, all purchases by the same customer stay together. This prevents data leakage where information about test examples influences training.

## 4.8 Data Format Standardization

**Data standardization** creates consistent formats and representations across all data sources. Inconsistent formats cause integration failures, model errors, and maintenance headaches (Rahm & Do, 2000).

**Numerical precision** needs standardization. Are amounts stored as integers, decimals to 2 places, or full floating point? Currency calculations particularly require careful handling to avoid rounding errors that compound over time.

**Categorical encoding** must be consistent. If "color" is encoded as integers during training (red=0, blue=1, green=2) but strings during inference ("red", "blue", "green"), the model receives nonsensical inputs. Document and enforce encoding schemes.

**Date/time standardization** prevents common errors. Use ISO 8601 format (YYYY-MM-DD HH:MM:SS) throughout. Store dates in UTC and convert to local time zones only for display. Mixing time zones or date formats creates subtle bugs that are difficult to debug.

**Text preprocessing** includes lowercasing, removing punctuation, handling special characters, and tokenization. These decisions need consistency across training and inference. If you lowercase text during training, you must lowercase during inference.

**Null handling** requires explicit decisions. Are missing values represented as NULL, empty string, zero, or special codes like -999? Different systems may use different conventions; standardize during data preparation.

## 4.9 Generative AI for Data

**Generative AI** is transforming data preparation workflows. Large language models can clean text data, label examples, generate synthetic training data, and augment existing datasets (Radford et al., 2019).

**Automated data cleaning** uses LLMs to standardize formats, correct errors, and fill missing values. An LLM can recognize that "New York," "NY," "NYC," and "New York City" refer to the same location and standardize to one format. It can identify and correct typos, parse messy addresses, and extract structured information from unstructured text.

**Label generation** leverages LLMs to create training labels at scale. For sentiment analysis, an LLM can label thousands of text examples with sentiment scores. For text classification, it can categorize documents. While not perfect, LLM-generated labels can bootstrap training when human labels are unavailable or expensive. Human review of a sample validates label quality.

**Synthetic data generation** creates training examples that don't exist in real data. For privacy-sensitive domains, synthetic data provides alternatives to real data. For rare cases, synthetic examples augment limited real examples. For testing edge cases, synthetic data ensures coverage of unusual scenarios.

**Data augmentation** generates variations of existing examples. For text, LLMs can paraphrase sentences, translate to other languages and back, or generate similar examples. This expands training data and improves model robustness.

Caution is warranted. Generative AI can introduce biases, create unrealistic patterns, or generate incorrect labels. Always validate LLM outputs, use confidence thresholds, and maintain human oversight for critical applications.

## 4.10 Chapter Highlights

Phase III Data Preparation transforms raw data into training-ready inputs through comprehensive cleaning, labeling, feature engineering, and pipeline construction. Data cleaning handles missing values, outliers, duplicates, and inconsistencies that corrupt training. Labeling strategies balance quality and scale using manual annotation, crowdsourcing, semi-supervised learning, and active learning.

Feature engineering creates new variables that help models learn patterns effectively through transformations, encodings, temporal features, and interactions. Data anonymization protects privacy through pseudonymization, k-anonymity, differential privacy, and synthetic data generation while maintaining utility.

Data pipelines automate workflows from ingestion through transformation to ensure reproducibility and prevent training-serving skew. Handling imbalanced datasets requires resampling, class weighting, ensemble methods, and appropriate evaluation metrics. Train/test/validation splits enable unbiased performance assessment using random, temporal, stratified, or group splitting strategies based on data characteristics.

Data format standardization creates consistency across sources for numerical precision, categorical encoding, date/time handling, text preprocessing, and null representation. Generative AI assists with automated cleaning, label generation, synthetic data creation, and augmentation while requiring validation and human oversight.

The 60-80% of project time spent on data preparation directly determines model success. Invest this time wisely.

# 4.11 Practice Questions

## Data Preprocessing Challenges

**Question 181:** You have a dataset where 40% of records have missing values in the "income" field. Investigation reveals missing values are not random—high-income customers declined to provide income. How should you handle this?

A) Delete all records with missing income B) Impute with median income C) Create a missing indicator feature; impute with median; consider missingness itself as informative D) Replace all with zero

**Answer: C) Missing indicator feature + imputation; missingness is informative**

**Explanation:** When missingness is not random (Missing Not At Random - MNAR), it carries information. High-income individuals declining to share income is systematic. Solutions: (1) Create binary "income_missing" feature capturing this pattern; (2) Impute missing values for algorithms requiring complete data; (3) Use algorithms handling missingness natively (tree-based). The missing indicator preserves signal that absence itself is meaningful.

---

**Question 182:** You're preprocessing text data for sentiment analysis. Should you remove punctuation and convert to lowercase?

A) Always remove punctuation and lowercase B) Depends on task: for sentiment, keeping "!" and "?" might help; for topic modeling, removal often helps C) Never preprocess text D) Remove all text

**Answer: B) Task-dependent preprocessing**

**Explanation:** Text preprocessing isn't one-size-fits-all. Sentiment analysis benefits from preserving excitement ("!") or questioning ("?") that indicate sentiment intensity. However, capitalization is usually normalized. Topic modeling often removes punctuation focusing on words. Consider: Does punctuation carry meaning for your task? Test both approaches; let performance guide decisions.

---

**Question 183:** Your dataset has "date" fields in multiple formats: "MM/DD/YYYY", "YYYY-MM-DD", "DD-MMM-YYYY". What's the preprocessing strategy?

A) Keep all formats B) Standardize all to single format (ISO 8601: YYYY-MM-DD); parse each format explicitly C) Delete date fields D) Ignore format differences

**Answer: B) Standardize to single format with explicit parsing**

**Explanation:** Inconsistent date formats cause errors. Standardization process: (1) Identify all formats present (EDA); (2) Parse each format explicitly using format strings; (3) Convert to standard format (ISO 8601 recommended for consistency); (4) Validate parsed dates (check for impossible dates); (5) Handle ambiguous dates (is "01/02/2020" Jan 2 or Feb 1?). Standardization prevents misinterpretation and enables temporal feature engineering.

---

**Question 184:** You have a "name" field with values like "John Smith", "MARY JONES", "bob brown", and " Alice Cooper ". What preprocessing is needed?

A) Keep as-is B) Standardize: trim whitespace, title case or lowercase consistently, handle special characters C) Delete all names D) Uppercase everything randomly

**Answer: B) Standardize capitalization and clean whitespace**

**Explanation:** Name inconsistencies create artificial variation. Preprocessing: (1) Strip leading/trailing whitespace; (2) Normalize internal whitespace (multiple spaces → single); (3) Consistent capitalization (title case for display, lowercase for matching); (4) Handle special characters/punctuation consistently; (5) Consider name parsing (first/last) if needed. Standardization enables reliable matching and analysis.

---

**Question 185:** Your numeric features have different scales: age (0-100), income ($0-$1M), transaction count (0-10,000). Should you scale them?

A) Never scale features B) Scale for distance-based algorithms (KNN, neural networks, SVM); often unnecessary for tree-based models C) Always use original scales D) Delete features with large scales

**Answer: B) Scale for distance-based algorithms; context-dependent for others**

**Explanation:** Scaling necessity depends on algorithm: (1) Distance-based (KNN, SVM, neural networks): require scaling—large-scale features dominate; (2) Tree-based (random forest, gradient boosting): scale-invariant—splits handle different scales naturally; (3) Linear models: scaling doesn't affect predictions but helps coefficient interpretation. Common methods: standardization (mean=0, std=1), min-max (0-1), robust scaling (using median/IQR).

---

**Question 186:** You're preprocessing categorical features with high cardinality (1,000 unique zip codes). One-hot encoding creates 1,000 new columns. What alternatives exist?

A) One-hot encode everything B) Target encoding, frequency encoding, embedding, or grouping rare categories C) Delete all categorical features D) Keep as strings

**Answer: B) Use high-cardinality encoding techniques**

**Explanation:** High-cardinality challenges need special handling: (1) Target encoding: replace categories with target mean (risk: overfitting—use cross-fold encoding); (2) Frequency encoding: replace with occurrence frequency; (3) Embedding: learn dense representations (neural networks); (4) Grouping: combine rare categories into "Other"; (5) Hash encoding: map to fixed dimension. Each has tradeoffs; test multiple approaches for your use case.

---

**Question 187:** Your dataset has duplicate records that aren't exact matches: "John Smith, 123 Main St" vs "J. Smith, 123 Main Street". How do you identify and handle these fuzzy duplicates?

A) Only remove exact duplicates B) Use fuzzy matching (Levenshtein distance, Jaro-Winkler) to identify similar records; define similarity threshold; merge carefully C) Keep all records regardless D) Randomly delete records

**Answer: B) Use fuzzy matching with careful merging**

**Explanation:** Fuzzy deduplication process: (1) Blocking: group candidates by shared attributes (same street number); (2) Similarity scoring: calculate string similarity

(Levenshtein, Jaro-Winkler, Soundex for names); (3) Thresholding: define similarity cutoff balancing precision/recall; (4) Manual review: inspect matches near threshold; (5) Merging strategy: keep most complete/recent record. Balance removing true duplicates versus incorrectly merging distinct entities.

---

**Question 188:** You have transaction amounts in multiple currencies (USD, EUR, GBP). For a global analysis, how should you preprocess this?

A) Keep multiple currencies B) Convert all to common currency using appropriate exchange rates; document conversion date/rates C) Delete non-USD transactions D) Ignore currency differences

**Answer: B) Convert to common currency with documented rates**

**Explanation:** Currency standardization: (1) Choose base currency (USD, EUR, or reporting currency); (2) Obtain exchange rates (historical rates for past data, current for recent); (3) Convert all amounts to base currency; (4) Document conversion (rates used, conversion date); (5) Consider storing both original and converted for audit trail. For time series, use period-appropriate historical rates, not current rates.

---

**Question 189:** Your dataset has a "phone_number" field with inconsistent formatting: "(555) 123-4567", "555-123-4567", "5551234567", "+1-555-123-4567". What preprocessing is needed?

A) Keep all formats B) Extract digits only; optionally standardize to common format; validate length C) Delete phone numbers D) Use random format

**Answer: B) Extract digits and standardize format**

**Explanation:** Phone number standardization: (1) Strip non-digits (parentheses, spaces, hyphens); (2) Handle country codes (strip or standardize); (3) Validate length (US: 10 digits); (4) Format consistently for storage (storage: digits only; display: formatted); (5) Flag invalid numbers. Standardization enables matching, validation, and analysis. Consider international number variations if applicable.

---

**Question 190:** You discover that a continuous variable (age) was mistakenly stored as text. Values include "25", "thirty-five", "N/A", "refused". How do you preprocess this?

A) Keep as text B) Parse numeric strings to numbers; map text numbers to digits; handle special values as missing; validate results C) Delete the entire field D) Convert all to zero

**Answer: B) Parse with text-to-number mapping and missing value handling**

**Explanation:** Mixed format parsing: (1) Convert numeric strings ("25") to integers; (2) Map written numbers ("thirty-five" → 35) using dictionary or NLP; (3) Handle missing indicators ("N/A", "refused") as NULL; (4) Validate converted values (reasonable age range?); (5) Create quality flag for confidence. This rescues data from poor storage practices. Consider data quality issues: why did this happen? Can source be fixed?

---

**Question 191:** Your preprocessing pipeline needs to scale features using training set statistics (mean, std). When applying to test/production data, should you recalculate statistics?

A) Yes, recalculate statistics for each dataset B) No, use training set statistics on all data to prevent data leakage C) Use test set statistics only D) Don't scale test data

**Answer: B) Use training set statistics to prevent leakage**

**Explanation:** Critical principle: test/production data can't influence preprocessing. Scaling must use training set statistics only: (1) Calculate mean/std on training data; (2) Apply same transformation to validation/test/production data; (3) This prevents information leakage from test to training. Recalculating on test data leaks information and creates optimistic performance estimates. Save training statistics for production use.

---

**Question 192:** You're preprocessing a dataset where "0" in certain fields means "missing" rather than zero. How should you handle this?

A) Treat all zeros as missing B) Investigate which fields use 0 as missing indicator; document; convert to NULL; impute appropriately C) Keep all zeros as literal values D) Delete all zeros

**Answer: B) Investigate, document, convert 0 to NULL where appropriate**

**Explanation:** Domain knowledge reveals encoding conventions. Some systems use 0, -1, 999, or special codes for missing data. Preprocessing: (1) Identify fields where 0 is impossible/meaningless (e.g., customer lifetime value); (2) Document in data dictionary; (3) Convert to explicit NULL; (4) Apply appropriate imputation. Distinguishing true zeros from missing-as-zero is essential for accurate analysis.

---

**Question 193:** Your text data contains URLs, email addresses, and hashtags mixed with regular text. Should you remove or preserve these?

A) Always remove all special elements B) Task-dependent: social media analysis might need hashtags; other tasks remove URLs/emails as noise C) Keep everything unchanged D) Remove all text

**Answer: B) Task-dependent preservation vs. removal**

**Explanation:** Context determines treatment: (1) URL/email: usually remove (noise for most tasks) or replace with tokens (URL_TOKEN); (2) Hashtags: valuable for social media analysis (topic indicators), noise for formal text; (3) Mentions (@username): similar to hashtags. Consider: does element carry meaning for your task? Test with/without; measure impact. Standard NLP preprocessing often removes; specialized tasks preserve.

---

**Question 194:** You're preprocessing datetime features and need to create time-based features. What engineering makes sense?

A) Keep raw datetime only B) Extract: hour, day_of_week, month, quarter, is_weekend, is_holiday, time_since_previous_event C) Delete all temporal information D) Convert all to year only

**Answer: B) Extract multiple temporal features**

**Explanation:** Datetime feature engineering extracts temporal patterns: (1) Cyclical: hour (0-23), day_of_week (0-6), month (1-12); (2) Binary: is_weekend, is_holiday, is_business_hours; (3) Derived: time_since_last_purchase, days_until_deadline; (4) Seasonal: quarter, week_of_year. Consider cyclical encoding (sine/cosine transformations) so hour 23 is close to hour 0. Rich temporal features enable models to learn time-based patterns.

---

**Question 195:** Your dataset has outliers: transaction amounts with a few values 1000× the typical value. Should you always remove outliers?

A) Always remove all outliers B) Investigate: legitimate (VIP customers) vs. errors; remove errors, consider capping/transformation for legitimate outliers C) Keep all outliers unchanged D) Delete the entire dataset

**Answer: B) Investigate legitimacy; context-dependent handling**

**Explanation:** Outlier treatment requires investigation: (1) Legitimate outliers (high-value customers, unusual but real events): keep or cap at percentile (Winsorization); (2) Errors (data entry, system glitches): remove or correct; (3) Fraud/anomalies: might be exactly what you want to detect. Methods: statistical (IQR, z-score), domain rules, visualization. Tree-based models are robust to outliers; linear models sensitive. Context and domain knowledge guide decisions.

---

## Labeling Quality Issues

**Question 196:** You have 10,000 labeled training examples. Audit of 100 random examples reveals 15% have incorrect labels. What should you do?

A) Ignore labeling errors B) Estimate error impact; consider relabeling subset or all data; implement quality controls C) Use all labels as-is D) Delete all labeled data

**Answer: B) Estimate impact and implement quality improvements**

**Explanation:** 15% label error is significant. Actions: (1) Estimate impact: train models on clean (audited 85%) vs. all data—how much does noise hurt?; (2) If substantial impact, prioritize relabeling (at least high-impact examples); (3) Implement quality process: clearer guidelines, multiple annotators, expert review for difficult cases; (4) Focus relabeling on influential examples (high uncertainty, near decision boundaries). Label quality directly limits model performance ceiling.

---

**Question 197:** Three annotators label the same 500 examples. Agreement is only 60% (Cohen's kappa = 0.45). What does this indicate?

A) Perfect labels B) Poor agreement indicates ambiguous guidelines, subjective task, or insufficient training C) Excellent inter-rater reliability D) Ignore agreement

**Answer: B) Poor agreement suggests labeling process issues**

**Explanation:** Kappa < 0.60 indicates moderate or poor agreement. Root causes: (1) Ambiguous labeling guidelines (edge cases undefined); (2) Inherently subjective task (opinion-based); (3) Inadequate annotator training; (4) Task complexity. Solutions: (1) Clarify guidelines with examples; (2) Provide training with discussion; (3) Review disagreements to identify ambiguities; (4) Use expert adjudication for disagreements; (5) Measure/track agreement over time. High agreement is necessary for reliable labels.

---

**Question 198:** You're using crowdsourced labeling (Amazon Mechanical Turk). How do you ensure label quality without ground truth?

A) Trust all crowd labels B) Multiple annotators per example; majority voting; qualification tests; attention checks; statistical quality metrics C) Use single annotator per example D) Ignore quality concerns

**Answer: B) Multiple annotators with quality controls**

**Explanation:** Crowdsourcing quality assurance: (1) Redundancy: 3-5 annotators per example; use majority vote or weighted voting; (2) Qualification: pre-test workers; maintain approved worker pool; (3) Attention checks: honeypot examples with known labels; (4) Agreement tracking: monitor inter-annotator agreement; remove poor performers; (5) Payment incentives: pay fairly to encourage quality. Balance cost (multiple annotators) with reliability needs.

---

**Question 199:** You discover systematic labeling bias: one annotator consistently labels borderline cases differently than others. How should you handle this?

A) Keep all labels as-is B) Review annotator's labels; provide retraining; consider relabeling by reliable annotators; track individual performance C) Delete all their labels immediately D) Ignore the bias

**Answer: B) Review, retrain, selectively relabel with tracking**

**Explanation:** Systematic bias needs correction: (1) Identify scope: how many examples affected? (2) Review sample: understand their interpretation; (3) Retrain: clarify guidelines; provide examples; (4) Decision: relabel if bias significant, or weight their labels lower; (5) Monitor: track future agreement to verify improvement. Preserve institutional knowledge (understand why they differed) while ensuring quality. Balance correction cost vs. bias impact.

---

**Question 200:** Labels were created years ago for a different purpose. Reusing for your new task, you discover 30% don't align with your current task definition. What do you do?

A) Use all old labels B) Relabel relevant subset for new task; document definition changes; potentially map old to new if possible C) Abandon the project D) Force old labels to work

**Answer: B) Relabel for new task with documentation**

**Explanation:** Label definitions must match task requirements. Misaligned labels create noise. Options: (1) Full relabeling (expensive, highest quality); (2) Selective relabeling (prioritize important/uncertain examples); (3) Label mapping (if old labels partially align); (4) Hybrid (use old labels as weak supervision, relabel subset). Document definition differences; assess feasibility/cost; proceed based on budget and importance. Using wrong labels guarantees poor performance.

---

**Question 201:** You're labeling medical images for disease detection. Who should do the labeling?

A) Crowdworkers are sufficient B) Medical experts (radiologists) required for accurate diagnosis labels C) Anyone can label medical data D) No labels needed

**Answer: B) Domain experts required for specialized tasks**

**Explanation:** Labeling quality depends on task complexity and stakes. Medical diagnosis requires: (1) Specialized expertise (radiologists for imaging); (2) Licensed professionals for regulatory compliance; (3) Multiple experts with adjudication for difficult cases. Cost is higher but necessary for accuracy and legal liability. Non-expert labeling of specialized domains produces unreliable labels. Match annotator expertise to task requirements.

---

**Question 202:** Your active learning system identifies uncertain predictions for human labeling. However, the human labeler shows "confirmation bias," tending to agree with model predictions. What's the risk?

A) No risk B) Feedback loop: biased labels reinforce model errors; undermines active learning benefits C) Perfect labeling D) Ignore the bias

**Answer: B) Confirmation bias creates harmful feedback loop**

**Explanation:** Confirmation bias in labeling is dangerous: (1) Labeler sees model prediction before labeling; (2) Tends to agree even when wrong; (3) Model retrains on biased labels; (4) Errors reinforce themselves. Solutions: (1) Blind labeling (hide predictions from annotators); (2) Multiple independent annotators; (3) Regular audits comparing labels to ground truth; (4) Explicit training against confirmation bias. Active learning requires unbiased labels to work.

---

**Question 203:** You have both high-quality expert labels (expensive, 1,000 examples) and noisy crowdsourced labels (cheap, 10,000 examples). How should you use both?

A) Use only expert labels B) Use only crowd labels
C) Train on crowd labels; fine-tune on expert labels; or use expert labels for validation/testing D) Ignore both

**Answer: C) Leverage both: bulk training on crowd, refinement on expert, or expert for evaluation**

**Explanation:** Combine complementary data sources: (1) Pre-train on large noisy crowd data (learns general patterns); fine-tune on small expert data (corrects errors); (2) Use expert labels for validation/test sets (reliable performance estimates); crowd for training;

(3) Use expert labels as ground truth to weight crowd labels (confident crowd labels get higher weight). Leverages volume (crowd) and quality (expert) advantages.

---

**Question 204:** During labeling, you realize 20% of examples are ambiguous (could legitimately have multiple correct labels). How should you handle this?

A) Force single label anyway B) Allow multiple labels, use soft labels (probabilities), or remove ambiguous examples C) Delete all ambiguous examples D) Randomly assign labels

**Answer: B) Allow multiple labels or use soft labels**

**Explanation:** Genuine ambiguity requires special handling: (1) Multi-label classification (examples can have multiple labels); (2) Soft labels (probabilities: 60% class A, 40% class B); (3) Remove if fundamentally unclear and rare; (4) Collect additional context/features to disambiguate. Forcing single labels on ambiguous cases adds noise. Models can handle uncertainty better when labels reflect it (soft labels, confidence scores).

---

**Question 205:** You outsource labeling to a vendor who delivers labels with 95% claimed accuracy. How do you verify this claim?

A) Trust the vendor's claim B) Audit random sample; compare to expert labels or multiple independent annotations C) Don't verify D) Accept all labels

**Answer: B) Independent audit with expert verification**

**Explanation:** Verify don't trust. Validation process: (1) Sample randomly (stratified across classes); (2) Have independent experts relabel sample; (3) Calculate actual agreement; (4) If lower than claimed, investigate (systematic issues? sampling bias in their testing?); (5) Contractual remedies if discrepancy significant. Build verification into acceptance criteria. Vendor incentives may not align with your quality needs.

---

**Question 206:** Label definitions evolved during a 6-month labeling project. Early labels used different criteria than late labels. What's the impact?

A) No impact B) Inconsistent labels create noise; model may learn temporal patterns instead of true patterns C) Perfect consistency D) Ignore definition changes

**Answer: B) Inconsistent definitions harm model quality**

**Explanation:** Evolving definitions create systematic inconsistency: (1) Model learns date-dependent patterns instead of stable task patterns; (2) Temporal split creates artificial train-test differences; (3) Performance estimates unreliable. Solutions: (1) Identify definition change point; (2) Relabel early data under new definition; (3) Train only on consistently labeled data; (4) Document definitions with timestamps. Prevention: lock definitions before labeling begins; version control labeling guidelines.

---

**Question 207:** You're using weak supervision (programmatic labeling) to generate training labels automatically from rules. Labels are noisy but cheap. When is this appropriate?

A) Never appropriate B) When you lack labeled data and can tolerate noise; combine multiple weak sources; use strong validation set C) Always use weak supervision D) Only for perfect labels

**Answer: B) Appropriate for label-scarce situations with noise tolerance**

**Explanation:** Weak supervision trades label quality for quantity: (1) Use when manual labeling impractical (millions of examples needed); (2) Combine multiple weak sources (rules, patterns, external knowledge) for denoising; (3) Frameworks (Snorkel) learn from weak sources; (4) Require clean validation set for model evaluation. Works when: signal in weak labels > noise, large-scale training possible, task tolerates some errors. Not suitable for high-stakes applications without validation.

---

**Question 208:** During labeling quality review, you discover that 5% of your "negative" class examples are actually positive cases mislabeled. How significant is this?

A) Not significant B) Very significant—systematic false negatives severely impact minority class learning C) Perfect labels D) Ignore it

**Answer: B) Systematic false negatives critically harm learning**

**Explanation:** 5% mislabeling seems small but is critical: (1) If negative class is 90% of data, you're missing 5% × 90% = 4.5% of total dataset as false negatives; (2) These false negatives are actually positives, harming minority class; (3) Model learns incorrect patterns ("these positive examples should be negative"); (4) Impacts recall more than precision. Systematic label errors in minority class are especially harmful. Prioritize cleaning these mislabels.

---

**Question 209:** You have 100,000 examples to label. Budget allows labeling 10,000. How should you select which examples to label?

A) Random selection B) Strategic selection: diverse examples, uncertain examples, representative samples; iterative active learning C) First 10,000 chronologically D) Label nothing

**Answer: B) Strategic selection maximizing information gain**

**Explanation:** Strategic labeling maximizes limited budget: (1) Diversity sampling (cluster data, sample from each cluster ensures coverage); (2) Uncertainty sampling (label examples where simple model is uncertain); (3) Active learning (iteratively train, identify uncertain, label, repeat); (4) Representative sampling (match population distribution). Random works but strategic approaches learn more from fewer labels. Balance exploration (diverse coverage) with exploitation (informative examples).

---

**Question 210:** After model deployment, you discover production predictions differ from expected because label definitions don't match production use cases. What preventive measure should have been implemented?

A) No prevention needed B) Align label definitions with production use cases during project planning; involve end-users in defining labels C) Labels never need alignment D) Deploy without checking

**Answer: B) Align labels with production use cases from the start**

**Explanation:** Prevention through planning: (1) Define production use case clearly; (2) Involve end-users/stakeholders in label definition; (3) Create labeling examples from realistic scenarios; (4) Validate labels match business logic; (5) Test on production-like data before deployment. Misalignment happens when data scientists define labels in isolation. Collaborative definition prevents costly rework.

---

**Privacy Protection Scenarios**

**Question 211:** Your dataset contains customer SSNs which aren't needed for modeling. What should you do?

A) Keep SSNs for potential future use B) Remove SSNs entirely following data minimization principles C) Keep SSNs but promise not to use them D) Share SSNs with vendors

**Answer: B) Remove SSNs following data minimization**

**Explanation:** Data minimization (GDPR principle): collect only necessary data. SSNs unnecessary for modeling should be removed: (1) Reduces breach risk; (2) Simplifies compliance; (3) Limits liability; (4) Follows privacy best practices. If SSNs needed for identity (joining datasets), use after joining then remove, or use hashed identifiers. Never retain sensitive data "just in case."

---

**Question 212:** You need to share training data with a vendor for model development, but it contains PII. How should you share it?

A) Share raw data with PII B) De-identify: remove direct identifiers, pseudonymize, aggregate sensitive fields; assess re-identification risk C) Don't share any data D) Share publicly

**Answer: B) De-identify before sharing**

**Explanation:** Vendor data sharing requires de-identification: (1) Remove direct identifiers (names, SSNs, emails); (2) Pseudonymize (replace IDs with random tokens); (3) Generalize sensitive fields (exact age → age ranges); (4) Assess re-identification risk (k-anonymity, differential privacy); (5) Contractual protections (data use restrictions, return/destruction requirements). Balance utility (vendors need usable data) with privacy (minimize re-identification risk).

---

**Question 213:** Your dataset includes health information covered by HIPAA. What special considerations apply for data preparation?

A) No special requirements B) HIPAA safe harbor: remove 18 specific identifiers; or expert determination of de-identification; additional security controls C) Health data has no restrictions D) Share freely

**Answer: B) HIPAA requires specific de-identification standards**

**Explanation:** HIPAA Protected Health Information (PHI) requires: (1) Safe Harbor method: remove 18 identifier types (names, dates except year, SSN, medical records numbers, etc.); (2) Expert Determination: statistical expert certifies re-identification risk very small; (3) Access controls: limit who can access PHI; (4) Audit logging: track all PHI access; (5) Business Associate Agreements with vendors. Penalties for violations are severe. Engage privacy/legal counsel.

---

**Question 214:** You're implementing differential privacy for model training. This adds noise to data, reducing model accuracy slightly. Stakeholders question why you're "making the model worse." How do you respond?

A) Remove differential privacy B) Explain it's a necessary privacy-utility tradeoff; quantify privacy guarantees and accuracy impact C) Ignore privacy D) Maximize noise

**Answer: B) Explain privacy-utility tradeoff with quantification**

**Explanation:** Differential privacy education: (1) Explain mathematical privacy guarantees (epsilon/delta parameters); (2) Quantify accuracy impact (typically small with proper tuning); (3) Compare to risk of data breaches/re-identification without privacy; (4) Frame as insurance against privacy risks; (5) Show that small accuracy loss is acceptable tradeoff for strong privacy. Privacy isn't free but cost is usually manageable. Involve stakeholders in setting privacy budget.

---

**Question 215:** You're joining customer data from multiple sources. One source uses email addresses as identifiers; another uses phone numbers. How can you join while minimizing PII exposure?

A) Store all PII permanently B) Hash identifiers (one-way hash of email/phone); use hashed values for joining; delete original PII after join C) Expose all PII to everyone D) Don't join datasets

**Answer: B) Hash identifiers for joining, then delete original PII**

**Explanation:** Privacy-preserving record linkage: (1) Hash identifiers (SHA-256) before joining—deterministic hashing enables matching without exposing raw PII; (2) Join on hashed values; (3) Delete original PII after successful join; (4) Retain only hashed versions if needed for future joins; (5) Salt hashes per use case (prevents cross-context matching). This enables necessary data integration while minimizing PII exposure and breach risk.

---

**Question 216:** A data scientist wants to manually inspect sample records to understand data quality issues, but records contain sensitive customer information. How do you enable this?

A) Provide raw data access B) Create sanitized sample: redact/mask sensitive fields; provide only fields needed for quality assessment C) Deny all access D) Provide everyone full access

**Answer: B) Provide sanitized sample with masked sensitive fields**

**Explanation:** Least privilege for quality review: (1) Identify fields needed for quality assessment; (2) Mask/redact sensitive fields (replace PII with synthetic values, partial masking); (3) Provide minimal sample size sufficient for review; (4) Access controls and audit logging; (5) Time-limited access. Balance enabling necessary work with privacy minimization. Not all quality assessment requires seeing full PII.

---

**Question 217:** You're using synthetic data generation to create privacy-safe training data. What's the key validation?

A) Assume all synthetic data is perfect B) Validate synthetic data preserves statistical properties and model performance while preventing re-identification C) Synthetic data always prevents privacy issues D) Don't validate

**Answer: B) Validate statistical fidelity and privacy protection**

**Explanation:** Synthetic data validation: (1) Statistical similarity: do distributions match real data? (2) Model performance: does model trained on synthetic perform similarly on real test data? (3) Privacy assessment: can synthetic data be linked back to real individuals? (4) Utility testing: does synthetic enable your use case? Poor synthetic data creates privacy without utility; poorly anonymized synthetic risks re-identification despite being "synthetic." Both privacy and utility require validation.

---

**Question 218:** GDPR gives individuals the "right to be forgotten." A customer exercises this right. How do you handle this for ML models already trained on their data?

A) Ignore the request B) Remove customer data; retrain affected models; or use machine unlearning techniques if feasible C) Keep using their data D) Ban all ML

**Answer: B) Remove data and retrain or use unlearning**

**Explanation:** Right to erasure compliance: (1) Identify all data about the individual across systems; (2) Delete personal data; (3) For ML models: full retraining (expensive but complete), machine unlearning (removes individual's influence without full retrain—research area), or document that model doesn't retain individual information (statistical models aggregate patterns). Practical considerations: retraining frequency, model lifecycle. Plan for this during architecture design.

---

**Question 219:** You're analyzing data that includes geolocation information (GPS coordinates). This is highly re-identifiable. How should you handle it?

A) Use exact coordinates B) Geospatial aggregation (zip code level), coordinate truncation, differential privacy on location queries C) Ignore privacy risks D) Publish all coordinates

**Answer: B) Aggregate or add noise to location data**

**Explanation:** Location data is uniquely identifying (home/work locations reveal identity). Privacy techniques: (1) Spatial aggregation: zip code, city level instead of exact coordinates; (2) Coordinate truncation: reduce precision (drop last digits); (3) Differential privacy: add calibrated noise to coordinates; (4) Temporal aggregation: analyze patterns over time rather than point visits. Balance privacy (coarser location) with utility (analysis needs). High-risk data requires stronger protections.

---

**Question 220:** You're implementing k-anonymity (each record indistinguishable from k-1 others). What's a practical implementation challenge?

A) K-anonymity is effortless B) Generalization and suppression reduce data utility; higher k increases privacy but decreases utility C) K-anonymity has no tradeoffs D) Always use k=1

**Answer: B) Privacy-utility tradeoff through generalization/suppression**

**Explanation:** K-anonymity implementation: (1) Identify quasi-identifiers (attributes that combined can identify individuals); (2) Generalize (age 25 → age 20-30); (3) Suppress (remove records that can't be anonymized); (4) Higher k = stronger privacy but more generalization/suppression = less utility. Challenges: determining appropriate k (5, 10, 50?), balancing privacy and utility, computational complexity. K-anonymity isn't perfect (vulnerable to attacks) but provides measurable privacy guarantee.

---

**Question 221:** Your model uses demographic features (age, gender, race) that are protected attributes. Privacy team says you can't use these. How do you respond?

A) Use them anyway B) Discuss fairness implications: removing protected attributes may hide bias; consider fairness-aware approaches C) Delete all demographic data D) Ignore privacy team

**Answer: B) Discuss fairness implications; removing doesn't guarantee fairness**

**Explanation:** Complex tradeoff: (1) Using protected attributes enables bias detection and fairness constraints; (2) Not using them doesn't eliminate bias (correlated features proxy for protected attributes); (3) Fairness-aware ML: use protected attributes during training to enforce fairness, optionally remove from production model; (4) Alternative: train without, but evaluate fairness using protected attributes; (5) Legal context: some jurisdictions restrict use; others require fairness testing. Facilitate informed discussion balancing privacy, fairness, and legal requirements.

---

**Question 222:** You discover that seemingly innocuous features (zip code + birth date) can re-identify 87% of individuals when combined. What's the concept and response?

A) These features are safe individually B) Quasi-identifiers: features that combine for re-identification; require joint protection through generalization or removal C) No re-identification risk D) Ignore the combination

**Answer: B) Quasi-identifiers require joint protection**

**Explanation:** Quasi-identifiers are combinations of attributes that enable re-identification despite no single attribute being directly identifying. Famous study: 87% of U.S. population uniquely identified by {zip code, birth date, gender}. Response: (1) Identify quasi-identifier combinations; (2) Apply k-anonymity or similar to combinations; (3) Generalize (age ranges, coarser geography); (4) Consider removing if not essential. Cannot evaluate privacy one attribute at a time; combinations matter.

---

**Question 223:** A researcher wants to publish a paper using your company's customer data with model results. What privacy considerations apply before publication?

A) Publish raw data and models B) Assess re-identification risk; aggregate results; avoid publishing individual-level predictions; obtain legal review C) No privacy concerns for research D) Publish customer names

**Answer: B) Assess re-identification risk and aggregate appropriately**

**Explanation:** Publication privacy: (1) Aggregate results (summary statistics, group-level findings) rather than individual predictions; (2) Assess if published information enables re-identification (even without names); (3) Differential privacy for published statistics; (4) Legal review (confidentiality agreements, privacy policies); (5) Institutional review board (IRB) approval if human subjects research. Academic freedom balanced with privacy obligations. Case studies require special care (synthetic examples, composites).

---

**Question 224:** You're training models using federated learning where data stays at source (hospitals, devices). What privacy advantages and challenges exist?

A) Perfect privacy with no challenges B) Advantage: raw data never centralized; Challenge: model updates can leak information; use secure aggregation C) No privacy benefits D) Federated learning prevents all attacks

**Answer: B) Reduces but doesn't eliminate privacy risks**

**Explanation:** Federated learning privacy: (1) Advantage: raw data stays local; central server never sees it; (2) Challenge: model updates (gradients) can leak training data information; (3) Protection: secure aggregation (cryptographically combine updates), differential privacy on updates, encrypted communication; (4) Still vulnerable to model inversion attacks (reconstructing training data from models). Better than centralizing raw data but not perfect privacy. Requires additional protections for strong guarantees.

---

**Question 225:** Your privacy policy promises customers their data will only be used for stated purposes. Marketing wants to use customer data (collected for product recommendations) to train a fraud detection model. Is this permissible?

A) Yes, all internal use is fine B) No, likely violates purpose limitation principle; need explicit consent or policy update C) Marketing decides D) Always repurpose data

**Answer: B) Violates purpose limitation without consent/policy update**

**Explanation:** Purpose limitation (GDPR, CCPA): data used only for stated collection purposes. Using recommendation data for fraud detection is different purpose. Compliance options: (1) Obtain explicit consent for new use; (2) Update privacy policy and notify users; (3) Assess if new use is "compatible" (legal analysis); (4) Collect separate consent for fraud detection. Internal use doesn't exempt from purpose limitation. Purpose creep erodes consumer trust and risks regulatory penalties.

---

## Pipeline Design Problems

**Question 226:** Your data preparation pipeline has 15 steps. When failures occur, the entire pipeline reruns from scratch (30 hours). How can you improve this?

A) Accept 30-hour reruns B) Implement checkpointing: save intermediate outputs; restart from last successful checkpoint on failure C) Make pipeline slower D) Give up on pipelines

**Answer: B) Implement checkpointing for fault tolerance**

**Explanation:** Pipeline resilience through checkpointing: (1) Save outputs after expensive steps; (2) On failure, restart from most recent checkpoint; (3) Idempotent steps (same inputs → same outputs) enable safe restarts; (4) Version checkpoints (data versions tied to code versions); (5) Automatic cleanup of old checkpoints. Tools: Airflow, Luigi, Prefect support checkpointing. Reduces costly reruns; enables faster iteration.

---

**Question 227:** Your data pipeline needs to process daily incremental data (yesterday's new records) and also support historical reprocessing (full dataset). How should you design this?

A) Separate pipelines for incremental and full B) Parameterized pipeline: date range parameters; same code handles both incremental and full backfills C) Only support incremental D) Only support full processing

**Answer: B) Parameterized pipeline supporting both modes**

**Explanation:** Flexible pipeline design: (1) Parameters: start_date, end_date defining processing window; (2) Incremental: set window to yesterday; (3) Full reprocessing: set window to all history; (4) Same transformation code handles both; (5) Idempotent logic (reprocessing same data produces same results). Benefits: one codebase to maintain, easy backfills, testing with small windows. Avoids duplicate logic between incremental and batch pipelines.

---

**Question 228:** Your pipeline dependencies form a DAG (directed acyclic graph). One upstream data source is frequently delayed, blocking downstream processing. How do you handle this?

A) Wait indefinitely B) Implement timeouts; fallback to cached/previous data; or skip optional dependencies with graceful degradation C) Delete the delayed source D) Remove all dependencies

**Answer: B) Timeouts, fallbacks, and graceful degradation**

**Explanation:** Handling unreliable dependencies: (1) Timeouts: don't wait forever; (2) Fallback data: use yesterday's data if today's unavailable (document freshness); (3) Optional dependencies: mark non-critical sources as optional; (4) Alerts: notify on delays/failures; (5) SLAs: establish source reliability agreements. Balance: waiting for complete data vs. processing with partial data. Critical dependencies require escalation; optional can degrade gracefully.

---

**Question 229:** Your data preparation pipeline needs to be reproducible: running it with same inputs should produce identical outputs. What practices ensure reproducibility?

A) Randomness everywhere B) Version control code and config; pin dependency versions; set random seeds; document data versions; idempotent operations C) Change code frequently without tracking D) Never document anything

**Answer: B) Version control, pinned dependencies, random seeds, data versioning**

**Explanation:** Reproducibility requirements: (1) Code: version control (Git) with tags for releases; (2) Dependencies: pin package versions (requirements.txt, lock files); (3) Data: version datasets, record which version used; (4) Randomness: set seeds for sampling, shuffling, initialization; (5) Idempotency: reprocessing same data produces same output; (6) Environment: containerization (Docker) captures full environment. Reproducibility enables debugging, auditing, regulatory compliance.

---

**Question 230:** Your pipeline processes 1TB of data daily. Processing takes 10 hours on a single machine, barely meeting deadlines. How do you scale?

A) Buy a bigger single machine B) Parallelize: partition data; process partitions in parallel using distributed computing (Spark, Dask) C) Accept slow processing D) Process less data

**Answer: B) Parallelize with distributed computing**

**Explanation:** Scalability through parallelization: (1) Partition data (by date, customer segments, geographic regions); (2) Process partitions in parallel across multiple workers; (3) Aggregate results; (4) Distributed frameworks: Spark, Dask, Ray handle parallelization details; (5) Horizontal scaling: add more workers for more throughput. 10 hours → 1 hour with 10 workers (minus overhead). Cloud services provide elastic compute for processing spikes.

---

**Question 231:** Your pipeline combines data from real-time streams and batch sources. How do you handle different latencies?

A) Ignore latency differences B) Lambda architecture: separate real-time and batch paths merging results; or Kappa architecture: unified stream processing C) Process everything slowly D) Drop real-time data

**Answer: B) Lambda or Kappa architecture for mixed latencies**

**Explanation:** Mixed latency architectures: (1) Lambda: real-time path (stream processing) and batch path (comprehensive processing) run in parallel, merge for queries; pros: robust; cons: duplicate logic; (2) Kappa: treat batch as special case of streaming; single stream processing handles both; pros: unified; cons: more complex. Choose based on: use case latency requirements, team expertise, infrastructure. Both patterns address fundamental challenge of mixed-latency data sources.

---

**Question 232:** Multiple data scientists need to run experiments with different preprocessing variations. How should you structure the pipeline?

A) Each person writes separate pipeline B) Modular pipeline: configurable preprocessing steps; shared components; version control configurations C) One person controls all processing D) No structure

**Answer: B) Modular, configurable pipeline with shared components**

**Explanation:** Experimentation-friendly pipelines: (1) Modular design: independent preprocessing steps (cleaning, feature engineering, scaling); (2) Configuration files: YAML/JSON defining preprocessing choices; (3) Shared components: reusable functions in library; (4) Version control: code and configs tracked; (5) Branching: experiments in separate branches. Benefits: consistency (shared cleaning logic), experimentation (easy to swap preprocessing), reproducibility (configs capture choices). Avoid monolithic pipelines hard to modify.

---

**Question 233:** Your pipeline generates features for model training and needs to generate the same features for production inference. How do you ensure consistency?

A) Duplicate logic in training and serving B) Shared feature engineering code: train and serve use identical functions; feature store for consistency C) Different logic for each D) Hope they match

**Answer: B) Shared code or feature store ensures consistency**

**Explanation:** Training-serving skew prevention: (1) Shared code: same feature engineering functions for training and production (package as library); (2) Feature store: centralized feature computation and storage; training and serving query same features; (3) Testing: validate training and serving produce identical features for same inputs; (4) Version control: features versioned with models. Separate implementations cause subtle bugs where training and production features differ, causing model failures.

---

**Question 234:** Your pipeline needs to process sensitive data. How do you secure the pipeline?

A) No security needed B) Encryption at rest and in transit; access controls; audit logging; secure secrets management; network segmentation C) Make data public D) Ignore security

**Answer: B) Comprehensive security controls**

**Explanation:** Pipeline security: (1) Encryption: data at rest (disk) and in transit (network); (2) Access control: least privilege—users/services access only needed data; (3) Secrets management: API keys, credentials in secure vaults (not code); (4) Audit logging: track all data access; (5) Network segmentation: isolated networks for sensitive processing; (6) Secure by design: security in architecture, not afterthought. Security incidents are expensive; prevention is investment.

---

**Question 235:** Your pipeline needs to handle schema changes in source data (new fields added, fields renamed). How do you build resilience?

A) Hope schema never changes B) Schema validation; graceful handling of extra fields; clear error messages for missing required fields; schema evolution strategies C) Crash on any schema change D) Ignore schema

**Answer: B) Schema validation with evolution handling**

**Explanation:** Schema resilience: (1) Validation: check incoming data against expected schema; (2) Additive changes: ignore extra fields gracefully (forward compatibility); (3) Required fields: clear errors if missing; (4) Field renames: mapping logic or alias definitions; (5) Schema versioning: track schema changes over time; (6) Backward compatibility: new code handles old schemas. Balance: detecting breaking changes (alert on problems) vs. flexibility (handle expected changes automatically).

---

**Question 236:** You're designing a pipeline for a time-sensitive application (fraud detection). What latency optimizations are important?

A) Batch process overnight B) Stream processing; in-memory computation; pre-computation; database indexing; parallel processing C) Process monthly D) Add intentional delays

**Answer: B) Low-latency architecture: streaming, in-memory, precomputation**

**Explanation:** Latency optimization: (1) Stream processing: process events as they arrive (Kafka, Kinesis); (2) In-memory: cache frequently used data in Redis/Memcached; (3) Pre-computation: calculate expensive features offline, look up at inference; (4) Indexing: optimize database queries; (5) Parallelization: distribute work; (6) Lightweight preprocessing: simple transforms, avoid expensive operations. Tradeoff: complexity and cost for lower latency. Design for latency requirements from start; retrofitting is harder.

---

**Question 237:** Your pipeline needs to coordinate with external APIs that have rate limits (100 requests/minute). How do you design around this?

A) Ignore rate limits B) Rate limiting logic: request throttling, exponential backoff on errors, request batching, caching responses C) Overwhelm the API D) Don't use external APIs

**Answer: B) Implement rate limiting and retry logic**

**Explanation:** API integration resilience: (1) Rate limiting: throttle requests to stay within limits (100/min); (2) Exponential backoff: on rate limit errors, wait increasingly long before retry; (3) Batching: if API supports batch requests, reduce request count; (4) Caching: store responses, reuse when possible; (5) Queuing: buffer requests in queue, process at allowed rate; (6) Monitoring: track API usage, alert on approaching limits. Plan for API constraints in pipeline design; violating limits gets you blocked.

---

**Question 238:** Your data pipeline has grown organically and now has complex dependencies that are hard to understand. How do you improve maintainability?

A) Leave it complex B) Refactor: document dependencies (DAG visualization); modularize; simplify logic; add tests; create documentation C) Make it more complex D) Delete everything

**Answer: B) Refactor for clarity and maintainability**

**Explanation:** Pipeline refactoring: (1) Visualization: draw dependency DAG showing data flow; (2) Modularization: break monoliths into focused components; (3) Documentation: inline comments, README, architecture docs; (4) Testing: unit tests for components, integration tests for pipeline; (5) Simplification: remove unused steps, combine redundant logic; (6) Convention: naming standards, consistent patterns. Technical debt accumulates; periodic refactoring maintains health. Balance: refactoring time vs. ongoing maintenance pain.

---

**Question 239:** You need to migrate your data pipeline from on-premises to cloud. What design considerations change?

A) Everything stays the same B) Leverage managed services; design for elasticity; optimize for cloud costs; security in shared responsibility model; region selection C) Cloud is identical to on-premises D) Don't use cloud

**Answer: B) Cloud-native design with managed services and elasticity**

**Explanation:** Cloud migration considerations: (1) Managed services: use cloud databases, storage, processing (less operational burden); (2) Elasticity: auto-scale compute for processing spikes; (3) Cost optimization: shut down resources when idle, use spot/preemptible instances; (4) Security: understand shared responsibility model, configure properly; (5) Regions: data residency, latency considerations; (6) Cloud-native patterns: serverless, containers. Cloud enables patterns infeasible on-premises but requires different thinking.

---

**Question 240:** Your pipeline generates artifacts (processed datasets, feature sets) that need to be versioned and tracked. How do you implement data versioning?

A) Overwrite files constantly B) Version control: timestamp artifacts, use data versioning tools (DVC, lakeFS), tie data versions to model versions C) Never track versions D) Delete all artifacts

**Answer: B) Data versioning with tools and metadata tracking**

**Explanation:** Data versioning practices: (1) Timestamp artifacts: include date/time or version number in filenames; (2) Data versioning tools: DVC (version control for data), lakeFS (Git for data lakes); (3) Metadata: track data lineage (source, transformations, code version used); (4) Immutability: don't modify artifacts, create new versions; (5) Model-data linking: record which data version trained which model. Enables reproducibility, debugging (compare model trained on data v1 vs. v2), rollback (previous data version available).

---

## Data Transformation Decisions

**Question 241:** You have a highly skewed numerical feature (99% values between 0-100, 1% between 1000-10000). What transformation is appropriate?

A) No transformation B) Log transformation to reduce skewness; or cap outliers; test both approaches C) Delete the feature D) Square the values

**Answer: B) Log transformation or outlier capping**

**Explanation:** Skewed data transformations: (1) Log transform: log(x+1) compresses large values, spreads small values; makes distribution more normal; (2) Box-Cox: family of power transformations; (3) Winsorization: cap at percentiles (e.g., cap above 99th percentile); (4) Test impact: train models with raw vs. transformed; measure performance. Linear models benefit more from normalization; tree-based models are robust to skewness. Context and empirical testing guide choice.

---

**Question 242:** You need to encode a cyclical feature (hour of day: 0-23). Simple numeric encoding treats hour 23 and hour 0 as far apart, when they're actually adjacent. How do you handle this?

A) Use hours as-is (0-23) B) Cyclical encoding: sin/cos transformation captures cyclical nature C) Delete time features D) Random encoding

**Answer: B) Sin/cos encoding for cyclical features**

**Explanation:** Cyclical encoding: (1) Hour: hour_sin = sin(2π × hour/24), hour_cos = cos(2π × hour/24); (2) This maps to unit circle: hour 23 and hour 0 are close; (3) Preserves cyclical relationships; (4) Applicable to: day of week, month, angle measurements. Alternative: treat as categorical with one-hot encoding (loses ordinality). Sin/cos is elegant solution preserving both cyclical nature and ordinality. Test vs. simple numeric encoding.

---

**Question 243:** You have text with emoticons and emoji (😊, 💐, ❤️). Should you remove them during preprocessing?

A) Always remove B) Depends on task: for sentiment analysis, emoji carry strong signal; consider emoji-to-text conversion or emoji embeddings C) Keep but ignore D) Replace all with generic token

**Answer: B) Task-dependent; emoji are informative for sentiment**

**Explanation:** Emoji in NLP: (1) Sentiment analysis: emoji carry strong sentiment signal (😊 = positive, 😞 = negative); (2) Approaches: keep as-is (modern tokenizers handle emoji), convert to text ("😊" → "happy face"), use emoji embeddings; (3) Remove for: formal text, where emoji are noise; (4) Research shows emoji improve sentiment classification. Don't automatically remove—consider whether they provide signal for your task.

**Question 244:** You need to create features from transactional data (multiple transactions per customer). How do you aggregate transactions into customer-level features?

A) Use only most recent transaction B) Aggregate functions: count, sum, mean, std, min, max, recency, frequency; time-based windows C) Randomly select transactions D) Delete transaction data

**Answer: B) Multiple aggregation functions and windows**

**Explanation:** Transaction aggregation features: (1) Count: number of transactions (last 30/90/365 days); (2) Sum: total amount spent; (3) Statistical: mean, median, std of transaction amounts; (4) Temporal: days since last transaction, average frequency; (5) Categorical: most common merchant category, diversity of categories; (6) Windows: 7-day, 30-day, 90-day, lifetime. Rich features capture customer behavior patterns. RFM (Recency, Frequency, Monetary) is classic framework.

---

**Question 245:** You have a feature that's 99.9% zeros (rare events). Should you keep this feature?

A) Always delete sparse features B) Keep if informative; consider binary transformation (occurred/didn't occur) or special handling for sparse features C) Delete all zero values D) Ignore sparsity

**Answer: B) Keep if informative; consider transformations**

**Explanation:** Sparse feature handling: (1) Assess informativeness: do non-zero values predict target?; (2) Binary: convert to indicator (1 if ever occurred, 0 otherwise); (3) Count-based: frequency of events; (4) Some algorithms handle sparsity well (tree-based); others need care (distance-based); (5) Storage: sparse matrices efficient for sparse data. Don't delete solely due to sparsity—sparse features can be highly predictive (fraud flags, rare events). Test predictive value before removing.

---

**Question 246:** You need to create interaction features (products of feature pairs). With 100 features, there are 4,950 possible pairs. Should you create all interactions?

A) Create all 4,950 interactions B) Selective interactions: domain knowledge guides which pairs; automated feature selection; tree-based feature importance C) No interactions ever D) Random interaction selection

**Answer: B) Selective interactions guided by domain knowledge and importance**

**Explanation:** Interaction feature engineering: (1) Domain-guided: business logic suggests meaningful interactions (price × quality, income × debt); (2) Automated discovery: train tree-based model, examine splits revealing interactions; (3) Polynomial features: scikit-learn creates interactions; (4) Dimensionality: all pairs create curse of dimensionality; (5) Selection: use Lasso, feature importance to prune. Creating all interactions often overfits. Start with domain knowledge; expand based on evidence.

---

**Question 247:** You're creating features from time series data. What temporal features should you engineer?

A) Only current value B) Lags, rolling statistics (mean, std), rate of change, trend, seasonality decomposition, time since events C) Delete all time series D) Only use timestamps

**Answer: B) Comprehensive temporal feature engineering**

**Explanation:** Time series features: (1) Lags: previous values (t-1, t-7, t-30); (2) Rolling windows: moving average, rolling std (7-day, 30-day); (3) Rate of change: percent change from previous period; (4) Trend: linear trend over window; (5) Seasonality: day of week effects, monthly patterns; (6) Time since: days since last purchase, event; (7) Aggregates: max/min over window. Rich temporal features enable learning from sequence patterns.

---

**Question 248:** You have hierarchical categorical data (Country > State > City). How should you encode this?

A) Flatten to single feature B) Separate features for each level; consider hierarchical encoding or aggregated features at each level C) Use only most granular level (City) D) Delete geographic data

**Answer: B) Separate features per level with aggregations**

**Explanation:** Hierarchical categorical encoding: (1) Separate features: country, state, city with appropriate encoding (one-hot, target encoding); (2) Aggregated stats: average target by country, by state, by city; (3) Hierarchical embeddings: learn representations respecting hierarchy; (4) Benefits: models can learn at different granularities (national trends, state patterns, city specifics); (5) Missing levels: handle incomplete hierarchies (city without state). Hierarchy contains valuable structure; preserve it.

---

**Question 249:** You need to create features from user clickstream data (sequences of pages visited). How do you engineer features from sequences?

A) Use only last click B) Sequence length, common patterns (n-grams), session duration, bounce rate, paths to conversion, sequential features C) Delete clickstream data D) Random sequence sampling

**Answer: B) Multiple sequence-based features**

**Explanation:** Clickstream features: (1) Sequence metrics: length, duration, unique pages; (2) Patterns: common n-grams (frequent 2-3 page sequences); (3) Behavior: bounce rate, repeat visits, depth; (4) Conversion: steps to conversion, abandonment points; (5) Temporal: time between clicks, session timing; (6) Content: page categories, product views; (7) Advanced: sequence embeddings (RNN-based). Sequences contain rich behavioral signals requiring thoughtful feature engineering.

---

**Question 250:** You're creating features for a recommendation system. What user-item interaction features are valuable?

A) Only user features B) User history (items interacted with), item popularity, user-item affinity scores, temporal patterns, contextual features C) Only item features D) No features needed

**Answer: B) User history, item stats, interactions, temporal/contextual factors**

**Explanation:** Recommendation features: (1) User: interaction history, preferences, demographics; (2) Item: popularity, ratings, categories, attributes; (3) User-item: past interactions, affinity scores, similarity to user's previous items; (4) Temporal: time of day, seasonality, trends; (5) Context: device, location, session; (6) Collaborative: similar users' preferences; (7) Content: item attributes matching user preferences. Rich feature engineering beyond just collaborative filtering often improves recommendations.

# Chapter 5.0: Phase IV Model Development

Phase IV is where data science gets exciting—you're finally building models. But here's what separates successful projects from failures: understanding that model development is systematic experimentation, not magic. You test hypotheses, compare approaches, iterate on architectures, and make principled decisions based on performance data. Teams that approach this haphazardly—trying random algorithms hoping something works—waste time and rarely achieve production-quality results.

## 5.1 Algorithm Selection Fundamentals

Choosing the right algorithm starts with understanding your problem type. **Classification** assigns examples to discrete categories—is this email spam or legitimate, will this customer churn or stay, does this image show a cat or dog. **Regression** predicts continuous numerical values—what price will this house sell for, how many units will we sell next month, what temperature will the equipment reach. **Clustering** finds natural groupings in unlabeled data—customer segments, product categories, anomaly detection (Géron, 2019).

For **binary classification**, common starting points include logistic regression (simple, interpretable, fast), decision trees (handle non-linear relationships, easy to explain), random forests (ensemble of trees, robust, good default choice), gradient boosting machines (XGBoost, LightGBM—often best performance), and support vector machines (effective for moderate-sized datasets with clear margins between classes). Neural networks work but often overkill for simpler problems.

**Multi-class classification** extends binary classification to multiple categories. One-vs-rest creates binary classifiers for each class. One-vs-one creates classifiers for each pair of classes. Many algorithms naturally handle multi-class problems—decision trees, random forests, neural networks.

For **regression**, linear regression provides baselines despite simplicity. Regularized variants (Lasso, Ridge, Elastic Net) prevent overfitting when you have many features. Tree-based methods (random forests, gradient boosting) handle non-linear relationships without manual feature engineering. Neural networks model complex non-linear patterns but need more data and careful tuning (James et al., 2013).

**Case Example: The Insurance Algorithm Journey**

Fatima, leading model development for insurance risk assessment, needed to predict claim amounts for policies. Her team evaluated multiple approaches systematically.

Linear regression served as baseline—simple, fast, interpretable. It achieved moderate performance ($R^2 = 0.58$) but struggled with non-linear patterns in claim severity. Adding polynomial features improved slightly ($R^2 = 0.62$) but risked overfitting.

Random forests handled non-linearity naturally, achieving $R^2 = 0.71$ without manual feature engineering. The ensemble approach proved robust to outliers and missing values. However, models were harder to explain to underwriters who needed to understand risk factors driving predictions.

Gradient boosting (XGBoost) delivered best performance ($R^2 = 0.78$) through iterative refinement. The model captured complex interactions between risk factors. The team used SHAP values to explain predictions, satisfying interpretability requirements while maintaining performance.

Neural networks achieved similar performance ($R^2 = 0.77$) but required more tuning, longer training, and provided less interpretability. For this problem, XGBoost offered the best performance-interpretability-maintainability balance.

## 5.2 Supervised versus Unsupervised Learning

**Supervised learning** trains on labeled examples where correct answers are known. You show the model inputs and desired outputs; it learns the mapping between them. This works when you have labeled training data and want to predict specific outcomes on new examples. Most business applications—credit scoring, demand forecasting, fraud detection, customer churn—use supervised learning (Hastie et al., 2009).

**Unsupervised learning** finds patterns in unlabeled data without predefined correct answers. The algorithm discovers structure, groupings, or relationships that weren't explicitly provided. This works when labels don't exist, labeling is impractical, or you want to explore data structure before defining prediction tasks.

**Clustering** groups similar examples together. K-means algorithm partitions data into k clusters by iteratively assigning points to nearest cluster centers and updating centers. Hierarchical clustering builds nested cluster trees showing relationships at different granularity levels. DBSCAN finds dense regions separated by sparse areas, identifying arbitrarily shaped clusters and outliers.

**Dimensionality reduction** compresses high-dimensional data into lower dimensions while preserving important structure. Principal Component Analysis (PCA) finds orthogonal directions of maximum variance. t-SNE creates two or three-dimensional visualizations of high-dimensional data preserving local neighborhood structure. These techniques help visualize data, reduce noise, and improve computational efficiency (Cunningham & Ghahramani, 2015).

**Anomaly detection** identifies unusual examples that differ significantly from normal patterns. This can be framed as unsupervised (finding statistical outliers) or supervised (training on labeled normal/abnormal examples). Unsupervised approaches include isolation forests, one-class SVM, and autoencoders that learn to reconstruct normal data and flag examples with high reconstruction error.

**Semi-supervised learning** combines small labeled datasets with large unlabeled datasets. The model learns from labeled examples, makes predictions on unlabeled data, and incorporates high-confidence predictions as additional training data. This works when labeling is expensive but unlabeled data is abundant.

## 5.3 Deep Learning Architecture Decisions

**Deep learning** uses neural networks with many layers enabling the learning of hierarchical representations. Early layers learn simple patterns, deeper layers combine those into complex concepts (Goodfellow et al., 2016).

**Feedforward neural networks** (also called multilayer perceptrons) connect neurons in layers without cycles. Information flows forward from input through hidden layers to output. These work for tabular data but don't leverage structure in images, text, or sequences.

**Convolutional neural networks (CNNs)** excel at image processing. Convolutional layers apply learned filters that detect features like edges, textures, or patterns. Pooling layers reduce spatial dimensions. CNNs automatically learn hierarchical visual features—early layers detect edges, middle layers recognize parts, deep layers identify objects. Architectures like ResNet, VGG, and EfficientNet provide proven building blocks (He et al., 2016).

**Recurrent neural networks (RNNs)** process sequential data like text or time series. Hidden states maintain memory of previous inputs, enabling the network to learn temporal dependencies. Long Short-Term Memory (LSTM) and Gated Recurrent Units (GRU) variants handle long-term dependencies better than vanilla RNNs. These work for language modeling, machine translation, and time series forecasting.

**Transformer architectures** revolutionized natural language processing through attention mechanisms that weigh the importance of different input elements. BERT, GPT, and their variants power modern LLMs. Transformers also show promise for computer vision (Vision Transformers) and other domains (Vaswani et al., 2017).

Architecture decisions include **network depth** (how many layers), **width** (how many neurons per layer), **activation functions** (ReLU, sigmoid, tanh), **regularization** (dropout, batch normalization, weight decay), and **optimization** (Adam, SGD with momentum). These choices interact complexly; systematic experimentation guides decisions.

## 5.4 Transfer Learning and Pretrained Models

**Transfer learning** leverages knowledge learned on one task to improve learning on a related task. Instead of training from scratch, you start with a model pretrained on large datasets and adapt it to your specific problem (Pan & Yang, 2010).

For **computer vision**, models pretrained on ImageNet (millions of labeled images) learn general visual features applicable to many tasks. You take a pretrained model like ResNet or EfficientNet, remove the final classification layer, and add new layers for your specific task. Fine-tune the entire network or just train the new layers depending on dataset size and similarity to the pretraining task.

For **natural language processing**, pretrained language models like BERT, RoBERTa, or GPT capture language patterns from vast text corpora. You adapt these to specific tasks—sentiment analysis, named entity recognition, question answering—through fine-tuning on task-specific data. This requires far less data than training from scratch and delivers better performance.

**When to use transfer learning**: Small dataset (thousands rather than millions of examples), similar problem to available pretrained models, limited computing resources, and faster time-to-production desired. Transfer learning often delivers better results with 10-100x less data than training from scratch.

**When to train from scratch**: Large dataset available, problem very different from available pretrained models, specialized architecture requirements, or sufficient computing resources for full training. Starting from scratch gives maximum flexibility but requires more data and compute.

**Case Example: The Medical Imaging Success**

Dr. Ravi Kumar needed to detect rare conditions in chest X-rays. His hospital had only 800 labeled images of the specific condition—insufficient for training deep CNNs from scratch, which typically need hundreds of thousands of images.

Ravi used transfer learning with a ResNet model pretrained on ImageNet. Even though ImageNet contains everyday objects rather than medical images, the early layers learned general visual features (edges, textures, patterns) applicable to X-rays.

He froze the early layers, replaced the final classification layer, and fine-tuned the last few layers on his medical dataset. This approach achieved 89% sensitivity and 94% specificity—performance comparable to radiologists—with only 800 training images. Training from scratch with this limited data would have failed completely, achieving barely better than random guessing.

Transfer learning enabled practical AI deployment in a specialized domain where large labeled datasets are difficult to obtain.

## 5.5 Hyperparameter Tuning Strategies

**Hyperparameters** are configuration settings chosen before training starts—learning rate, number of layers, regularization strength, tree depth, number of estimators. Unlike model parameters learned during training, hyperparameters must be set explicitly. Optimal values dramatically affect performance (Feurer & Hutter, 2019).

**Manual tuning** relies on intuition and experience. You try different values, observe performance, and iteratively adjust. This works for simple models or when you have deep expertise, but doesn't scale to high-dimensional hyperparameter spaces.

**Grid search** defines ranges for each hyperparameter and evaluates all combinations. If you test 3 learning rates, 4 tree depths, and 5 regularization values, grid search trains 60 models (3×4×5). This is exhaustive but computationally expensive and doesn't scale beyond a few hyperparameters.

**Random search** samples random hyperparameter combinations from defined distributions. Surprisingly, random search often finds good solutions faster than grid search because it explores more unique values per hyperparameter. Bergstra and Bengio (2012) showed random search more efficient than grid search for most problems.

**Bayesian optimization** builds a probabilistic model of the objective function and uses it to select promising hyperparameter combinations to evaluate. Each evaluation updates the model, guiding search toward optimal regions. This typically finds good hyperparameters with fewer evaluations than random or grid search. Tools like Optuna or Hyperopt implement Bayesian optimization.

**Learning rate** particularly affects neural network training. Too high causes training instability or divergence. Too low means slow convergence requiring excessive training time. Learning rate schedules that decay rate during training or cyclical learning rates that vary rate within ranges often improve results.

**Regularization strength** controls the tradeoff between fitting training data and maintaining model simplicity. Too little regularization causes overfitting. Too much causes underfitting by oversimplifying the model.

## 5.6 Overfitting and Underfitting Prevention

**Overfitting** occurs when models learn training data too well, including noise and irrelevant patterns, harming performance on new data. The model memorizes rather than generalizes. **Underfitting** happens when models are too simple to capture underlying patterns, performing poorly on both training and test data (Goodfellow et al., 2016).

**Regularization techniques** prevent overfitting by penalizing model complexity. L1 regularization (Lasso) adds penalty proportional to absolute weight values, encouraging sparse models with many zero weights. L2 regularization (Ridge) penalizes squared weight values, encouraging small weights overall. Elastic Net combines both.

**Dropout** randomly deactivates neurons during training, forcing the network to learn robust features that don't depend on specific neurons. This prevents co-adaptation where neurons become overly reliant on each other. Dropout rates of 0.2-0.5 are typical.

**Early stopping** monitors validation performance during training and stops when performance stops improving, preventing the model from overfitting as training continues. Keep the model checkpoint with best validation performance rather than the final training state.

**Data augmentation** artificially expands training data through transformations that preserve labels. For images: rotation, flipping, cropping, color adjustments. For text: synonym replacement, back-translation, paraphrasing. More training data reduces overfitting.

**Cross-validation** provides more reliable performance estimates by training and evaluating on multiple data splits. K-fold cross-validation divides data into k subsets, trains k models using each subset once for validation and others for training, and averages performance. This uses all data for both training and validation, providing robust estimates especially with limited data.

**Model complexity management** matches model capacity to data size and problem complexity. Simpler models (fewer parameters, shallower networks) suit smaller datasets. Complex models need more data to avoid overfitting. Start simple and increase complexity only when necessary.

## 5.7 Ensemble Methods

**Ensemble methods** combine multiple models to achieve better performance than individual models. The wisdom of crowds principle applies—diverse models make different errors; averaging predictions often improves accuracy (Dietterich, 2000).

**Bagging** (Bootstrap Aggregating) trains multiple models on different random subsets of training data and averages predictions. Random forests use bagging with decision trees, training hundreds of trees on bootstrapped samples and averaging their predictions. This reduces variance and improves stability.

**Boosting** trains models sequentially, each focusing on examples previous models handled poorly. AdaBoost adjusts example weights, increasing weight for misclassified examples so subsequent models focus on them. Gradient boosting (XGBoost, LightGBM, CatBoost) fits models to residual errors, iteratively improving predictions. Boosting reduces bias and often delivers best performance on tabular data.

**Stacking** trains diverse models and combines them using a meta-model that learns optimal combination weights. Base models might include logistic regression, random forest, neural network, and gradient boosting. A meta-model (often simple logistic regression or linear model) learns to weight base model predictions optimally.

**Voting** combines model predictions through simple rules—majority vote for classification, average for regression. This works when base models have similar performance but make different mistakes.

**Case Example: The Credit Risk Ensemble**

Li, a risk analyst, was developing credit default prediction models. Individual algorithms showed varying strengths: logistic regression provided interpretability and calibrated probability estimates, gradient boosting achieved highest discrimination, neural networks captured non-linear patterns, and decision trees handled missing values naturally.

Rather than choosing one algorithm, Li built an ensemble. She trained each algorithm on the full dataset, validated performance, and combined predictions using stacking. A meta-model (logistic regression) learned optimal weights for each base model—boosting received 45% weight, neural network 30%, logistic regression 15%, decision tree 10%.

The ensemble outperformed any individual model, achieving AUC-ROC of 0.84 versus 0.81 for the best individual model. The improvement came from combining diverse strengths while mitigating individual weaknesses. The ensemble proved more robust to data distribution changes than single models.

## 5.8 Model Training Approaches

**Batch training** processes all training data in each iteration. The model sees every example before updating parameters. This provides stable gradient estimates but requires loading full datasets into memory and can be slow for large datasets.

**Mini-batch training** divides data into small batches (32, 64, 128, 256 examples) and updates parameters after each batch. This balances gradient estimate quality with computational efficiency. Mini-batch training enables training on datasets larger than memory by streaming batches.

**Online learning** updates models continuously as new data arrives. Each example or small batch updates parameters immediately. This works for streaming data or when distributions change over time. Models adapt to new patterns without complete retraining.

**Learning rate schedules** adjust learning rate during training. Step decay reduces rate at predetermined epochs. Exponential decay continuously decreases rate. Cosine annealing varies rate in cycles. Adaptive methods (Adam, RMSprop) automatically adjust per-parameter rates based on gradient history. Proper scheduling often improves final performance and training stability.

**Curriculum learning** presents training examples in meaningful order, starting with easier examples and progressing to harder ones. This mimics human learning and can improve convergence and final performance, especially for complex tasks.

## 5.9 Model Artifacts Documentation

**Model artifacts** document everything needed to reproduce, deploy, and maintain models. Poor documentation causes failures when models move to production, team members change, or issues arise months later (Sculley et al., 2015).

Document **model architecture**: algorithm choice, hyperparameters, layer configurations, feature engineering steps, preprocessing requirements. Include code, configuration files, and architecture diagrams.

Record **training details**: dataset versions, training duration, compute resources, random seeds for reproducibility, convergence metrics, and any unusual training behaviors or interventions.

Track **performance metrics**: training, validation, and test set performance across relevant metrics. Include confusion matrices, ROC curves, calibration plots, and error analysis showing where models succeed and fail.

Document **dependencies**: software versions (Python, libraries, frameworks), hardware requirements (CPU, GPU, memory), external services or APIs, and any special installation requirements.

Maintain **model cards** summarizing intended use, limitations, performance characteristics, ethical considerations, and update history. This provides stakeholders with essential information for responsible deployment (Mitchell et al., 2019).

**Version control** tracks models, code, data, and configuration through tools like Git (code), DVC (data), and MLflow or Weights & Biases (experiments). This enables reproducing any model version and understanding how models evolved.

## 5.10 Core Concepts Recap

Phase IV Model Development systematically builds and refines machine learning models through principled algorithm selection, architectural decisions, and training approaches. Algorithm selection matches problem types (classification, regression, clustering) with appropriate methods from simple baselines to complex deep learning.

Supervised learning trains on labeled data for prediction tasks. Unsupervised learning discovers patterns in unlabeled data. Deep learning architectures (CNNs, RNNs, transformers) handle specialized data types through hierarchical feature learning. Transfer learning leverages pretrained models to achieve strong performance with limited data.

Hyperparameter tuning optimizes model configuration through grid search, random search, or Bayesian optimization. Overfitting prevention employs regularization, dropout, early stopping, and data augmentation. Ensemble methods combine diverse models to outperform individual approaches.

Training strategies balance efficiency and performance through batch sizes and learning rate schedules. Comprehensive documentation of model artifacts, training details, performance metrics, dependencies, and version control enables reproducibility, deployment, and long-term maintenance.

Successful model development requires systematic experimentation, careful evaluation, and thorough documentation—not hoping random algorithms magically work.

# 5.1 Practice Questions

## Algorithm Selection for Specific Use Cases

**Question 251:** A bank needs to predict loan defaults (binary classification). They require model explainability for regulatory compliance. Which algorithm is most appropriate?

A) Deep neural network with 10 layers B) Logistic regression or gradient boosting with SHAP explanations C) Black box ensemble without explanations D) Random predictions

**Answer: B) Logistic regression or gradient boosting with SHAP**

**Explanation:** Regulatory explainability requirements favor interpretable models. Logistic regression provides inherent interpretability (coefficients show feature impact). If more performance needed, gradient boosting with SHAP values provides strong predictions plus explanations. Deep neural networks, while powerful, are harder to explain for regulatory purposes. Balance performance with interpretability requirements.

---

**Question 252:** You're building an image classification system for manufacturing defect detection. You have 5,000 labeled images. Which approach is best?

A) Train CNN from scratch B) Transfer learning: fine-tune pretrained CNN (ResNet, EfficientNet) C) Logistic regression on raw pixels D) K-nearest neighbors on images

**Answer: B) Transfer learning with pretrained CNN**

**Explanation:** 5,000 images is insufficient for training CNNs from scratch (need millions). Transfer learning leverages ImageNet's knowledge: pretrained models learned general visual features (edges, textures, patterns) from millions of images. Fine-tuning adapts these features to defect detection with limited data. This approach delivers strong performance with moderate data requirements. Training from scratch would severely overfit.

---

**Question 253:** You need to forecast retail sales for 500 products across 50 stores (25,000 time series). Which approach scales best?

A) Individual ARIMA model for each series B) Global model (gradient boosting or deep learning) trained on all series with product/store features C) Manual forecasting for each D) Random predictions

**Answer: B) Global model trained on all series**

**Explanation:** Individual models for 25,000 series are impractical (maintenance burden, limited data per series). Global approaches train one model on all series: (1) Gradient boosting with features (product_id, store_id, historical sales, seasonality); (2) Deep learning (LSTM, Transformer) handling multiple series. Benefits: learns cross-series patterns, works with sparse data per series, easier maintenance. Modern forecasting increasingly uses global models for scale.

---

**Question 254:** You're building a recommendation system for a new streaming service with limited user-item interactions. Which approach handles cold start best?

A) Pure collaborative filtering B) Hybrid: content-based (using item attributes) + collaborative filtering C) Random recommendations D) Only recommend most popular items

**Answer: B) Hybrid approach with content-based component**

**Explanation:** Cold start (new users, new items, limited interactions) challenges pure collaborative filtering. Hybrid solutions combine: (1) Content-based: recommend similar items based on attributes (genre, actors, director); (2) Collaborative: learn from interaction patterns where available; (3) Fallback: use content-based for new users/items, collaborative for established users. This gracefully handles sparsity while leveraging interaction data when available.

---

**Question 255:** You need to detect anomalies in server logs where normal behavior is well-defined but anomalies are rare and varied. Which approach fits?

A) Supervised classification with labeled anomalies B) Unsupervised anomaly detection (Isolation Forest, One-Class SVM, Autoencoders) C) Manual rule-based detection only D) Ignore anomalies

**Answer: B) Unsupervised anomaly detection**

**Explanation:** Rare, varied anomalies make supervised learning impractical (insufficient labeled examples of each anomaly type). Unsupervised approaches learn normal behavior patterns and flag deviations: (1) Isolation Forest: isolates outliers efficiently; (2) One-Class SVM: learns normal data boundary; (3) Autoencoders: learn to reconstruct normal data; high reconstruction error indicates anomalies. These handle unknown anomaly types without requiring labeled examples of each.

---

**Question 256:** You're building a chatbot for customer service. Which AI approach is most appropriate?

A) Simple decision tree B) Large language model (BERT, GPT) fine-tuned on customer service dialogues + retrieval system C) Logistic regression D) Image classification model

**Answer: B) LLM fine-tuned on domain dialogues plus retrieval**

**Explanation:** Modern chatbot architecture combines: (1) Language understanding: fine-tuned LLM (BERT for understanding, GPT for generation) trained on customer service conversations; (2) Retrieval: fetch relevant knowledge base articles; (3) Dialogue management: track conversation context; (4) Intent classification: route to appropriate responses/workflows. Pure rule-based approaches lack flexibility; LLMs provide natural language understanding and generation capabilities essential for conversational AI.

---

**Question 257:** You need to segment customers into groups with similar behavior patterns. You don't have predefined segments. Which algorithm applies?

A) Supervised classification B) Unsupervised clustering (K-means, Hierarchical, DBSCAN) C) Regression D) Time series forecasting

**Answer: B) Unsupervised clustering**

**Explanation:** Discovering natural groupings without predefined categories is clustering: (1) K-means: fast, requires specifying k segments; (2) Hierarchical: creates dendrogram showing nested segments; (3) DBSCAN: finds arbitrary-shaped clusters, handles noise; (4) Gaussian Mixture Models: probabilistic clustering. Evaluate using silhouette scores, domain validation (do segments make business sense?). Clustering discovers structure; supervised learning requires existing labels.

---

**Question 258:** You're predicting house prices (continuous target) with 50 numerical features and 100K examples. Which algorithm is a good starting point?

A) Convolutional neural network B) Gradient boosting (XGBoost, LightGBM) or Random Forest C) Recurrent neural network D) K-means clustering

**Answer: B) Gradient boosting or Random Forest**

**Explanation:** Structured/tabular data with continuous target suits gradient boosting or random forests: (1) Handle non-linear relationships automatically; (2) Robust to outliers and scaling; (3) Feature importance built-in; (4) Strong performance with moderate tuning. XGBoost/LightGBM often win Kaggle competitions on tabular data. Deep learning is overkill for tabular problems with moderate data. Start with gradient boosting; explore neural networks if performance insufficient.

---

**Question 259:** You need to classify medical images but explain predictions to clinicians. Which approach balances performance and explainability?

A) Black box ensemble without explanations B) CNN with attention mechanisms or Grad-CAM visualization showing which image regions influenced decisions C) Pure rule-based system D) Random forest on raw pixels

**Answer: B) CNN with attention/Grad-CAM for explainability**

**Explanation:** Medical imaging requires both accuracy and explainability: (1) CNNs deliver state-of-the-art image classification; (2) Grad-CAM creates heatmaps highlighting influential image regions; (3) Attention mechanisms show what network focused on; (4) SHAP for images provides pixel-level importance. This combination gives clinicians both accurate predictions and visual explanations they can validate against their expertise. Pure rule-based systems can't match CNN performance; pure black boxes lack necessary trust.

---

**Question 260:** You're building a real-time bidding system for online advertising needing sub-100ms predictions. Model accuracy and inference speed both matter. What's the right approach?

A) Complex ensemble of 20 models B) Single lightweight model (logistic regression, small neural network, shallow gradient boosting) optimized for latency C) Deep neural network with 50 layers D) Run offline batch predictions

**Answer: B) Lightweight model optimized for latency**

**Explanation:** Real-time latency requirements constrain model complexity: (1) Simple models: logistic regression (<1ms), shallow neural nets (1-10ms); (2) Model compression: pruning, quantization reduce size/latency; (3) Tradeoff: slight accuracy loss for dramatic speed gains; (4) Infrastructure: GPU acceleration, model serving optimization, caching. Measure both offline accuracy and production latency. Complex ensembles and deep networks can't meet millisecond requirements. Start simple; add complexity only if latency permits.

---

**Question 261:** You need to predict customer churn with imbalanced data (5% churn rate). Which algorithms handle imbalance well?

A) Standard algorithms without adjustments always fail B) Gradient boosting with class weights, or SMOTE + any algorithm, or cost-sensitive learning C) Only use undersampling D) Ignore imbalance

**Answer: B) Class weighting, SMOTE, or cost-sensitive approaches**

**Explanation:** Imbalanced data strategies: (1) Algorithm-level: gradient boosting, random forest with class_weight parameter; (2) Data-level: SMOTE creates synthetic minority examples; (3) Cost-sensitive: penalize minority class errors more heavily; (4) Ensemble: balanced random forest samples balanced subsets; (5) Threshold tuning: optimize decision threshold for precision-recall tradeoff. Algorithms themselves aren't inherently good/bad for imbalance—handling strategy matters. Test multiple approaches.

---

**Question 262:** You're building a sentiment analysis model for social media text (informal language, slang, emojis). Which approach works best?

A) Classical bag-of-words with logistic regression B) Pretrained transformer model (BERT, RoBERTa) fine-tuned on social media data C) Rule-based keyword matching D) Image classification

**Answer: B) Pretrained transformer fine-tuned on social media**

**Explanation:** Social media language challenges require modern NLP: (1) Pretrained transformers (BERT, RoBERTa) understand context, handle slang; (2) Fine-tuning adapts general language knowledge to social media domain; (3) Handles emojis, informal language, context-dependent sentiment; (4) Transfer learning reduces data requirements. Bag-of-words loses context; rule-based can't handle variation. Domain-specific fine-tuning (social media corpus) improves performance over general models.

---

**Question 263:** You need to optimize a complex logistics problem (vehicle routing with constraints) in real-time. Which AI approach applies?

A) Supervised classification B) Reinforcement learning or optimization algorithms (genetic algorithms, simulated annealing) C) Clustering D) Linear regression

**Answer: B) Reinforcement learning or optimization algorithms**

**Explanation:** Sequential decision-making with constraints requires: (1) Reinforcement learning: agent learns optimal routing policies through trial-and-error; (2) Classical optimization: genetic algorithms, simulated annealing, mixed-integer programming; (3) Hybrid: RL for learning heuristics, optimization for final solution. Supervised learning doesn't fit (no input-output pairs for routing). This is goal-driven optimization problem. Modern approaches combine ML (learning patterns) with operations research (constraint handling).

---

**Question 264:** You're building a spam filter needing to adapt quickly to new spam tactics. Which approach enables rapid adaptation?

A) Static model never updated B) Online learning with continuous updates or rapid retraining pipeline C) Manual rules only D) No spam filtering

**Answer: B) Online learning or rapid retraining**

**Explanation:** Spam evolves rapidly requiring adaptive models: (1) Online learning: model updates continuously with new examples (streaming algorithms); (2) Rapid retraining: automated pipeline retrains daily/hourly on recent data; (3) Ensemble: combine stable long-term model with adaptive short-term model; (4) Feedback loop: user reports feed training. Concept drift in spam requires model evolution. Static models become obsolete quickly. Balance stability (avoid false positives on legitimate email) with adaptation (catch new spam).

---

**Question 265:** You need to generate synthetic data (images, text) for data augmentation. Which AI approach generates new data?

A) Classification models B) Generative models (GANs, VAEs, diffusion models for images; GPT for text) C) Regression models D) Clustering

**Answer: B) Generative models (GANs, VAEs, diffusion, GPT)**

**Explanation:** Generative AI creates new data: (1) Images: GANs (generate realistic images), VAEs (probabilistic generation), diffusion models (state-of-the-art image synthesis); (2) Text: GPT, T5 generate synthetic text; (3) Tabular: CTGAN generates synthetic tabular data. Applications: data augmentation (expand limited training data), privacy (synthetic data instead of real), simulation (create test scenarios). Discriminative models (classification, regression) predict existing data; generative models create new data.

---

## Training Strategy Optimization

**Question 266:** Your neural network training loss plateaus after 20 epochs with validation loss still decreasing. Should you stop training?

A) Yes, always stop at 20 epochs B) No, continue training—model can still improve (validation loss decreasing) C) Start over from scratch D) Delete the model

**Answer: B) Continue training while validation improves**

**Explanation:** Training decisions based on validation performance, not training metrics alone: (1) Training loss plateau is normal as model approaches optimum; (2) Validation loss still decreasing means generalization improving; (3) Stop when validation loss plateaus or increases (overfitting); (4) Early stopping monitors validation performance, saves best model. Training loss plateaus don't indicate when to stop—validation performance does. Continue training until validation performance stops improving.

---

**Question 267:** You're training a model with limited data (5,000 examples). How should you split train/validation/test sets?

A) 80/10/10 split B) K-fold cross-validation (5-10 folds) for training/validation; hold out separate test set C) 50/50 split D) Use all data for training

**Answer: B) K-fold cross-validation with held-out test set**

**Explanation:** Limited data requires efficient usage: (1) Hold out test set (15-20%) for final evaluation; (2) K-fold CV on remaining data: each example used for validation once, training K-1 times; (3) Provides robust performance estimates with limited data; (4) Average metrics across folds reduces variance; (5) Final model trains on all non-test data. Simple splits waste data (small validation set has high variance). K-fold maximizes training data usage while maintaining validation.

---

**Question 268:** Your gradient boosting model trains well on your laptop but needs to run on production servers 100× larger datasets. How do you scale training?

A) Use only your laptop B) Distributed gradient boosting (LightGBM distributed mode, XGBoost on Spark, Dask) or cloud-based training C) Reduce data to laptop size D) Give up on large data

**Answer: B) Distributed training on cluster or cloud**

**Explanation:** Scaling gradient boosting: (1) LightGBM: efficient distributed training across multiple machines; (2) XGBoost: Spark/Dask integration for cluster computing; (3) Cloud: managed services (SageMaker, Vertex AI) handle infrastructure; (4) Sampling: train on representative sample if full data infeasible, validate on full data. Distributed frameworks partition data across workers, parallelize tree building. Cloud provides elastic compute for occasional large-scale training.

---

**Question 269:** During training, your model's loss suddenly spikes (increases dramatically). What's the likely cause and solution?

A) This is normal; ignore it B) Learning rate too high causing instability; reduce learning rate or use learning rate scheduling C) Model is perfect D) Increase learning rate

**Answer: B) Learning rate too high; reduce it**

**Explanation:** Loss spikes indicate training instability: (1) Cause: learning rate too high, causing divergence; (2) Solutions: reduce learning rate, use learning rate schedules (decay over time), adaptive optimizers (Adam adjusts per-parameter), gradient clipping (prevent extreme updates); (3) Monitor: track training curves; restart from checkpoint before spike. Stable training shows smooth loss decrease. Large spikes suggest parameter updates too large, overshooting minima.

---

**Question 270:** You're training a deep learning model on GPUs. Training takes 3 days. How can you reduce training time?

A) Be patient; accept 3 days B) Mixed precision training, larger batch sizes, multiple GPUs (data/model parallelism), gradient accumulation, efficient data loading C) Use CPU instead D) Train for only 1 epoch

**Answer: B) Multiple optimization techniques**

**Explanation:** Training acceleration: (1) Mixed precision: float16 instead of float32 (2× speedup on modern GPUs); (2) Batch size: larger batches utilize GPU better (monitor memory); (3) Multiple GPUs: data parallelism (split batches) or model parallelism (split model); (4) Data loading: parallel data loading, preprocessed data; (5) Architecture: efficient models (MobileNet vs. ResNet); (6) Cloud: more powerful GPUs (A100, V100). Typical speedups: 2-8× with proper optimization.

---

**Question 271:** You have 1M training examples but limited compute. Should you train on all data or sample?

A) Always use all data B) Stratified sampling for initial experiments; scale to full data for final model if resources permit C) Use only 100 examples D) Random sampling without stratification

**Answer: B) Stratified sampling for experiments, full data for final model**

**Explanation:** Balancing data and compute: (1) Experiments: stratified sample (10-20%) maintains class distribution, enables rapid iteration; (2) Learning curves: plot performance vs. data size—if plateaued, full data may not help; (3) Final model: use full data if learning curve still improving and resources permit; (4) Active learning: intelligently select most informative examples. Start with samples for speed; scale up based on evidence. Many problems plateau before using all data.

---

**Question 272:** Your neural network overfits training data. What training strategies reduce overfitting?

A) Train longer B) Dropout, L2 regularization, early stopping, data augmentation, reduce model complexity C) Increase model complexity D) Remove all regularization

**Answer: B) Multiple regularization techniques**

**Explanation:** Overfitting prevention: (1) Dropout: randomly deactivate neurons during training; (2) L2 regularization: penalize large weights; (3) Early stopping: stop when validation performance degrades; (4) Data augmentation: artificially expand training data; (5) Architecture: reduce layers/units; (6) Batch normalization: improves generalization. Combine multiple techniques. Start with early stopping + dropout; add others if needed. Monitor validation performance—gap between training and validation indicates overfitting.

---

**Question 273:** You're training multiple models for hyperparameter tuning. How should you parallelize this?

A) Train sequentially B) Parallel training: independent hyperparameter combinations run simultaneously on multiple GPUs/machines C) Train one model forever D) Don't do hyperparameter tuning

**Answer: B) Parallel hyperparameter search**

**Explanation:** Efficient hyperparameter tuning: (1) Parallelization: train multiple hyperparameter combinations simultaneously; (2) Tools: Ray Tune, Optuna parallel execution; (3) Resources: multiple GPUs, cloud instances; (4) Strategies: random search (easily parallelizable), Bayesian optimization (sequential but can batch), genetic algorithms (population-based); (5) Early stopping: terminate poor-performing trials early. Hyperparameter search is embarrassingly parallel—each trial independent. Parallelization reduces wall-clock time dramatically.

---

**Question 274:** Your training data has temporal structure (time series). How should you handle data splitting?

A) Random split B) Temporal split: train on past, validate on middle period, test on most recent C) Shuffle completely D) Use only most recent data

**Answer: B) Temporal split preserving time order**

**Explanation:** Time series splitting prevents data leakage: (1) Temporal order: train on 2019-2021, validate on 2022, test on 2023; (2) Prevents future information in training; (3) Simulates production (train on past, predict future); (4) Rolling windows: multiple splits testing robustness across time; (5) Gap: optional gap between train/test prevents leakage from autocorrelation. Random splitting leaks future into training, creating optimistic performance estimates. Time-aware splitting essential for temporal data.

---

**Question 275:** You're training a model that needs to run on mobile devices with limited memory. How does this affect training strategy?

A) Train largest possible model B) Train small/efficient model (MobileNet, DistilBERT); use knowledge distillation; quantization-aware training C) Ignore deployment constraints D) Mobile deployment is impossible

**Answer: B) Train efficient model or compress for mobile**

**Explanation:** Mobile-constrained training: (1) Efficient architectures: MobileNet (vision), DistilBERT (NLP) designed for mobile; (2) Knowledge distillation: train large teacher model, distill knowledge into small student model; (3) Quantization-aware training: train model anticipating int8 quantization; (4) Pruning: remove unnecessary weights during training; (5) Architecture search: AutoML finds efficient architectures. Deployment constraints should inform training from start. Retrofitting is harder than designing for constraints.

---

**Question 276:** Your model achieves 99.9% training accuracy but only 60% validation accuracy. What's the problem and solution?

A) Validation set is wrong B) Severe overfitting; solutions: reduce model complexity, add regularization, get more training data C) Model is perfect D) Ignore validation accuracy

**Answer: B) Severe overfitting requiring intervention**

**Explanation:** Massive train-validation gap (99.9% vs. 60%) indicates severe overfitting: (1) Model memorized training data; (2) Solutions: simpler model (fewer parameters), stronger regularization (dropout, L2), more training data, data augmentation, ensemble (reduces variance); (3) Check for data leakage (target information in features). This gap is too large to ignore. Start over with simpler approach; gradually increase complexity while monitoring validation.

---

**Question 277:** You're training an ensemble of 10 models. Should you train them identically or differently?

A) Train all identically B) Train differently: different algorithms, different hyperparameters, different data subsets for diversity C) Train one model only D) All models should be clones

**Answer: B) Train differently to maximize ensemble diversity**

**Explanation:** Ensemble diversity is key to performance: (1) Algorithm diversity: combine logistic regression, random forest, gradient boosting, neural network; (2) Data diversity: bagging (different samples), different features; (3) Hyperparameter diversity: different regularization, learning rates; (4) Initialization: different random seeds. Diverse models make different errors; averaging reduces error. Identical models provide no ensemble benefit—they make same mistakes. Diversity creates robust predictions.

**Question 278:** Your dataset has 1M examples. Should you use mini-batch, batch, or stochastic gradient descent?

A) Batch gradient descent (entire dataset) B) Mini-batch gradient descent (batch sizes 32-256) for balance of stability and efficiency C) Stochastic gradient descent (single example) D) No gradient descent

**Answer: B) Mini-batch gradient descent**

**Explanation:** Gradient descent variants: (1) Batch (full dataset): accurate gradients, slow, doesn't scale; (2) Stochastic (single example): fast, noisy, unstable; (3) Mini-batch (32-256 examples): best balance—reasonable gradient estimates, efficient GPU utilization, noise helps escape local minima. 1M examples requires mini-batch. Typical sizes: 32-128 (smaller for limited memory, larger for stability). Modern deep learning uses mini-batch exclusively.

---

**Question 279:** During training, you save checkpoints every epoch. Which checkpoints should you keep for production?

A) Keep all checkpoints B) Keep best validation performance checkpoint; possibly final checkpoint and few intermediates C) Keep only initial checkpoint D) Delete all checkpoints

**Answer: B) Keep best validation checkpoint and strategic others**

**Explanation:** Checkpoint management: (1) Best validation: model with lowest validation loss (early stopping); (2) Final: last epoch (if validation still improving); (3) Intermediate: checkpoints at key epochs for analysis; (4) Storage: delete intermediate checkpoints after training; (5) Production: deploy best validation checkpoint (best generalization). Training might continue past best validation (overfitting); early stopping checkpoint performs best on unseen data. Don't deploy final checkpoint blindly.

---

**Question 280:** You're training a model with class imbalance (90/10 split). Should you balance the training batches?

A) Keep imbalanced batches B) Balanced batching: sample batches with equal classes; or class weights to emphasize minority C) Use only majority class D) Use only minority class

**Answer: B) Balanced batching or class weights**

**Explanation:** Imbalanced training strategies: (1) Balanced batching: sample batches ensuring equal class representation; (2) Class weights: weight loss function to emphasize minority class; (3) Oversampling: duplicate minority examples; (4) SMOTE: generate synthetic minority examples. Imbalanced batches cause model to ignore minority class. Balancing helps model learn minority patterns. Combine with appropriate metrics (F1-score, not accuracy). Test multiple approaches empirically.

---

## Model Complexity Trade-offs

**Question 281:** You're deciding between a simple logistic regression (10K parameters) and a neural network (1M parameters). Both achieve similar validation performance. Which should you choose?

A) Always neural network (more parameters) B) Logistic regression: simpler, faster, more interpretable, easier to maintain with similar performance C) Neither D) Random model

**Answer: B) Logistic regression for simplicity**

**Explanation:** Occam's Razor in ML: prefer simpler models when performance is similar. Benefits of simpler models: (1) Faster training and inference; (2) Less data required; (3) Easier to interpret; (4) Lower overfitting risk; (5) Easier maintenance; (6) Lower infrastructure costs. Neural networks justified when they significantly outperform simpler alternatives. Don't use complex models for complexity's sake. Start simple; add complexity only when evidence supports it.

---

**Question 282:** Your model has 500 features. Feature importance analysis shows 50 features account for 95% of predictions. Should you remove the other 450 features?

A) Keep all 500 features B) Test simplified model with top 50 features; if performance similar, use simplified model C) Remove all features D) Add more features

**Answer: B) Test simplified model; use if performance maintained**

**Explanation:** Feature selection benefits: (1) Reduces overfitting (fewer parameters); (2) Faster training and inference; (3) Simpler interpretation; (4) Lower storage; (5) Removes noise from irrelevant features. Process: (1) Train on top 50 features; (2) Compare validation performance; (3) If similar, use simplified; if worse, determine acceptable tradeoff. Sometimes redundant features help ensemble diversity or capture rare cases. Empirical testing beats assumptions.

---

**Question 283:** You're building a decision tree and must choose maximum depth. Deeper trees fit training data better but risk overfitting. How do you decide?

A) Maximum depth always B) Cross-validation: test multiple depths, choose depth with best validation performance C) Minimum depth always D) Random depth

**Answer: B) Cross-validation to find optimal depth**

**Explanation:** Depth is critical hyperparameter: (1) Too shallow (depth=3): underfits, poor training and validation; (2) Too deep (depth=20): overfits, perfect training, poor validation; (3) Optimal depth (depth=8): balances bias-variance, best validation. Cross-validation systematically tests depths; plots validation performance vs. depth (validation curve). Choose depth where validation performance peaks. This balances model complexity with generalization.

---

**Question 284:** Your neural network has 10 layers. Training is slow and validation performance isn't better than a 3-layer network. What should you do?

A) Add more layers B) Reduce to 3 layers; deeper isn't always better, especially without sufficient data or proper architecture C) Keep 10 layers regardless of performance D) Use 100 layers

**Answer: B) Reduce to 3 layers**

**Explanation:** Depth doesn't guarantee performance: (1) Deep networks need more data; (2) Require careful initialization and regularization; (3) Slower training; (4) Higher overfitting risk without benefits. If 3 layers achieves similar validation performance, use it—faster, simpler, less prone to overfitting. Add depth only when it improves validation performance. Modern architectures (ResNet, Transformers) enable very deep networks, but require specialized techniques (skip connections, layer normalization).

---

**Question 285:** You're deciding between an interpretable model (logistic regression, 78% accuracy) and a complex ensemble (85% accuracy, hard to interpret). How do you decide?

A) Always choose highest accuracy B) Consider use case: high-stakes decisions need interpretability; low-stakes might prioritize accuracy C) Always choose interpretability D) Flip a coin

**Answer: B) Context-dependent accuracy-interpretability tradeoff**

**Explanation:** Accuracy-interpretability tradeoff requires business judgment: (1) High-stakes (lending, healthcare): 7% accuracy gain may not justify interpretability loss—regulatory, ethical, trust concerns; (2) Low-stakes (content recommendation): 7% accuracy improvement worth complexity; (3) Hybrid: use complex model, add post-hoc explanations (SHAP, LIME); (4) Stakeholder input: involve decision-makers in tradeoff. No universal answer—context determines appropriate balance.

---

**Question 286:** Your random forest has 1,000 trees. Training on 100 trees takes 10 minutes; on 1,000 trees takes 100 minutes. Performance difference is 0.5%. Should you use 1,000 trees?

A) Yes, always more trees B) Use 100 trees: 10× faster training, minimal performance loss (0.5%) C) Use 10,000 trees D) Use 1 tree

**Answer: B) Use 100 trees for efficiency**

**Explanation:** Diminishing returns with ensemble size: (1) Performance improves rapidly with first 50-100 trees; (2) Additional trees provide marginal gains; (3) Training time grows linearly with tree count; (4) Memory and inference time increase. 0.5% improvement rarely justifies 10× training cost. Random forests typically plateau by 100-500 trees. Plot learning curves (performance vs. number of trees) to find point of diminishing returns. Balance performance with practical constraints.

---

**Question 287:** You have a choice: train one very accurate model requiring 48 hours or five decent models in 10 hours each. Which is better?

A) One model is always better B) Five models enable ensemble (often better than single model), faster iteration, robustness C) Never use ensembles D) Don't train any models

**Answer: B) Five models for ensemble benefits**

**Explanation:** Ensemble advantages: (1) Averaging five models often outperforms single model (even if individual models slightly worse); (2) Diversity reduces overfitting; (3) Faster iteration (10 hours vs. 48); (4) Robustness (one model fails, others compensate); (5) Parallel training. Ensemble of decent models frequently beats single excellent model. Exception: when single model already saturates performance limit. Empirically test both approaches.

---

**Question 288:** Your model has batch normalization, dropout, and L2 regularization. Training is slow. Can you remove some regularization to speed up training?

A) Remove all regularization B) Experiment with removing one at a time; monitor validation performance to ensure regularization still sufficient C) Keep all regularization always D) Add more regularization

**Answer: B) Systematically experiment with removal**

**Explanation:** Regularization ablation study: (1) Baseline: current model with all regularization; (2) Remove dropout: test performance; (3) Remove L2: test performance; (4) Remove batch norm: test performance; (5) Compare validation performance and training time. Multiple regularization types can be redundant. Batch norm often provides sufficient regularization alone. Empirically determine minimum regularization maintaining performance. Balance training efficiency with generalization.

---

**Question 289:** You're building a model where false negatives are 10× more costly than false positives. How does this affect model complexity decisions?

A) Complexity doesn't affect false negative rate B) Might justify complex model if it significantly reduces false negatives despite other tradeoffs C) Use simplest possible model regardless of errors D) Ignore error costs

**Answer: B) High false negative cost may justify complexity**

**Explanation:** Error costs inform complexity tradeoffs: (1) High false negative cost (missing disease, missing fraud) justifies investment in accuracy; (2) Complex models reducing false negatives by 5% might justify interpretability loss; (3) Cost-benefit analysis: value of reduced false negatives vs. complexity costs; (4) Alternative: simple model with adjusted threshold to reduce false negatives. Error costs make performance improvements more valuable, potentially justifying complexity that wouldn't be warranted for balanced costs.

---

**Question 290:** You have option to use 100 raw features or engineer 20 sophisticated features. Which approach is better?

A) Always 100 raw features (more is better) B) Test both: 20 engineered features might outperform 100 raw, especially for simple models; deep learning can learn from raw C) Always 20 features D) Use no features

**Answer: B) Test both; depends on algorithm and domain**

**Explanation:** Feature engineering vs. raw features: (1) Traditional ML (linear, tree-based): benefit from engineered features capturing domain knowledge; (2) Deep learning: can learn representations from raw features; (3) 20 good features might outperform 100 mediocre features (less noise); (4) Effort: feature engineering requires domain expertise and time. Test both approaches. Often hybrid works: include both raw and engineered features, let model selection decide. Context and empirical results guide choice.

---

**Question 291:** Your gradient boosting model has 5,000 trees with depth 10. It achieves 99.5% training accuracy but 75% validation accuracy. How should you adjust complexity?

A) Add more trees and depth B) Reduce complexity: fewer trees, shallower depth, stronger regularization C) Perfect model; deploy D) Delete the model

**Answer: B) Reduce complexity to address overfitting**

**Explanation:** Clear overfitting (99.5% vs. 75%): (1) Reduce trees: 5,000 is excessive; try 100-500; (2) Reduce depth: depth 10 might be too deep; try 3-6; (3) Increase learning rate: compensates for fewer trees; (4) Add regularization: min_child_weight, subsample; (5) Early stopping: stop when validation stops improving. Gradient boosting easy to overfit with too many complex trees. Start with moderate complexity; increase only if validation benefits.

---

**Question 292:** You need to choose between a single complex model and an ensemble of simple models. Both achieve similar validation performance. Which factors favor the single complex model?

A) Ensembles are always better B) Single model advantages: simpler deployment, faster inference, easier maintenance, one model to explain C) No advantages exist D) Always use ensembles

**Answer: B) Single model advantages in deployment and operations**

**Explanation:** Single model advantages: (1) Deployment: one artifact to version and deploy; (2) Inference latency: single prediction vs. multiple predictions and aggregation; (3) Maintenance: update one model vs. ensemble coordination; (4) Explainability: explain one model's logic; (5) Infrastructure: less memory, compute. Ensembles justify complexity when performance significantly better. Equal performance favors simpler single-model deployment. Production operations costs often outweigh training-time model selection.

---

**Question 293:** Your model pipeline has 5 sequential models (output of model 1 feeds model 2, etc.). Each model has 95% accuracy. What's the approximate pipeline accuracy?

A) 95% B) ~77% ($0.95^5 \approx 0.77$) – errors compound through pipeline C) 100% D) 50%

**Answer: B) ~77% as errors compound**

**Explanation:** Sequential pipelines multiply error rates: (1) Each model adds errors; (2) Errors propagate (early mistakes affect later stages); (3) 95% accuracy per stage: $0.95 \times 0.95 \times 0.95 \times 0.95 \times 0.95 = 0.77$ (77% overall); (4) Long pipelines amplify errors.

Mitigation: (1) Minimize pipeline length; (2) Improve individual model accuracy; (3) Error correction stages; (4) End-to-end training (when possible). Pipeline complexity should be justified by necessity, not added casually.

---

**Question 294:** You're comparing two architectures: Architecture A has 100K parameters and trains in 1 hour; Architecture B has 10M parameters and trains in 50 hours. Both achieve 90% validation accuracy. Which should you choose?

A) Always Architecture B (more parameters) B) Architecture A: similar performance, 100× fewer parameters, 50× faster training C) Architecture B regardless of efficiency D) Neither

**Answer: B) Architecture A for efficiency**

**Explanation:** Efficiency matters: (1) Fewer parameters: faster inference, less memory, lower overfitting risk; (2) Faster training: rapid iteration, more experiments in same time; (3) Lower compute costs; (4) Easier deployment. 10M parameters justified only if they significantly improve performance. Equal performance makes efficient model superior choice. This principle guides modern architecture research: efficient models (MobileNet, DistilBERT) achieving comparable performance to larger models with fraction of parameters.

---

**Question 295:** Your model's complexity is appropriate for training data size (good bias-variance balance). Stakeholder offers 10× more training data. Should you increase model complexity?

A) Never increase complexity B) Yes, more data supports more complex models; test if increased complexity improves validation performance C) Keep model unchanged always D) Decrease complexity

**Answer: B) More data enables safe complexity increase**

**Explanation:** Data size and model complexity relationship: (1) More data reduces overfitting risk; (2) Enables training more complex models; (3) May improve performance if current model underfits; (4) Test empirically: train more complex model on larger dataset. Learning curves guide decision: if performance still improving with more data, more complex models might help. If plateaued, complexity increase won't help. Additional data is opportunity to explore previously infeasible complexity levels.

---

## Performance Improvement Scenarios

**Question 296:** Your model achieves 75% validation accuracy. You try various improvements: better features (+3%), hyperparameter tuning (+1%), ensemble (+2%), data cleaning (+5%). What's the likely cumulative improvement?

A) +11% (additive) B) Less than +11% due to diminishing returns and overlap C) Exactly +11% D) No improvement possible

**Answer: B) Less than +11% due to diminishing returns**

**Explanation:** Improvements don't simply add: (1) Overlapping impact: better features and data cleaning address similar issues; (2) Diminishing returns: each improvement faces higher baseline; (3) Interactions: some combinations work better together, others don't; (4) Typical reality: expect 6-8% from +11% potential. Improvements compound multiplicatively, not additively. Multiple small improvements together deliver meaningful gains. Track each improvement independently to understand contributions.

---

**Question 297:** Your model performance plateaued at 80% accuracy despite trying multiple algorithms. What should you investigate next?

A) Try more algorithms B) Examine data quality, add features, collect more data, check label quality C) Accept 80% as maximum possible D) Give up

**Answer: B) Focus on data improvements rather than algorithms**

**Explanation:** Performance plateaus often indicate data limitations, not algorithm limitations: (1) Data quality: errors in features or labels; (2) Feature engineering: missing important signals; (3) Data volume: insufficient examples, especially for minority classes; (4) Label quality: inconsistent or incorrect labels. Algorithm swapping delivers diminishing returns after testing few strong options. Data improvements (more, better quality, better features) often provide breakthroughs when algorithm changes don't. "More data beats better algorithms" in many cases.

---

**Question 298:** Your text classification model achieves 82% accuracy with bag-of-words features. What's the most promising improvement approach?

A) Try 100 different algorithms B) Upgrade to contextual embeddings (BERT, RoBERTa) capturing semantic meaning C) Remove all features D) Reduce data

**Answer: B) Upgrade to contextual embeddings**

**Explanation:** NLP architecture evolution provides major improvements: (1) Bag-of-words: ignores word order and context; (2) Word2Vec/GloVe: captures some semantic similarity; (3) BERT/RoBERTa: contextual understanding, transfer learning. Upgrading from bag-of-words to transformers typically yields 5-15% accuracy improvements. This architectural advancement provides more impact than hyperparameter tuning bag-of-words models. Modern NLP starts with transformers, not bag-of-words.

---

**Question 299:** Your model performs well on validation data but poorly in production. What's the most likely cause and solution?

A) Validation data is perfect representation; ignore production issues B) Train-serve distribution mismatch; investigate production data characteristics and retrain on representative data C) Production users are wrong D) Keep model unchanged

**Answer: B) Distribution mismatch; retrain on representative data**

**Explanation:** Validation-production performance gaps indicate: (1) Validation data doesn't represent production population; (2) Data collection differences (sampling bias); (3) Temporal drift (validation from past, production is current); (4) Feature computation differences (training-serving skew). Solutions: (1) Collect production data for validation; (2) A/B test with holdout production traffic; (3) Monitor production performance; (4) Retrain on representative data. Validation should mirror production as closely as possible.

---

**Question 300:** Your image classification model performs well on clear, high-quality images but poorly on blurry, low-light, or occluded images. How do you improve robustness?

A) Train only on clear images B) Data augmentation: add blur, noise, occlusion, lighting variations to training data C) Ignore challenging images D) Filter out difficult images

**Answer: B) Data augmentation with realistic distortions**

**Explanation:** Robustness through training distribution: (1) Data augmentation: artificially add blur, noise, occlusion, lighting variations, rotations; (2) Collect real challenging examples; (3) Synthetic data generation; (4) Adversarial training: train on adversarially perturbed images. Models generalize to conditions seen during training. If training includes only clear images, models fail on challenging conditions. Augmentation exposes model to variations, improving robustness.

**Question 301:** Your neural network training is stuck in a local minimum with sub-optimal performance. What techniques might help escape?

A) Keep training indefinitely with same settings B) Learning rate scheduling, learning rate restarts, different optimizers (Adam, SGD with momentum), different initialization C) Give up immediately D) Use same approach forever

**Answer: B) Learning rate techniques and optimizer changes**

**Explanation:** Escaping local minima strategies: (1) Learning rate scheduling: cosine annealing, warm restarts enable exploration; (2) Optimizer changes: switch between Adam (adaptive) and SGD with momentum (better generalization); (3) Stochastic weight averaging: average multiple checkpoints; (4) Different initialization: restart with different random seeds; (5) Architecture changes: different network structure. Local minima are common in non-convex optimization. Multiple techniques help find better solutions.

---

**Question 302:** Your model achieves 90% accuracy overall but only 60% on a critical customer segment representing 10% of data. How should you address this?

A) Overall accuracy is good; ignore segment B) Weighted loss emphasizing critical segment, oversample segment, or train segment-specific model C) Remove the underperforming segment D) Performance gaps don't matter

**Answer: B) Focus on underperforming segment through weighting or specific modeling**

**Explanation:** Segment-specific performance matters: (1) Weighted loss: assign higher cost to critical segment errors; (2) Oversampling: increase critical segment representation in training; (3) Stratified training: ensure adequate segment representation; (4) Segment-specific model: train specialized model for critical segment; (5) Feature engineering: add segment-specific features. Overall metrics can mask critical segment failures. Business importance determines appropriate interventions.

---

**Question 303:** You train a model and achieve 85% accuracy. After feature engineering, accuracy improves to 88%. Is this improvement statistically significant?

A) Yes, any improvement is significant B) Need statistical testing: confidence intervals, paired t-tests on cross-validation folds C) 3% is always insignificant D) Don't test significance

**Answer: B) Statistical testing required to confirm significance**

**Explanation:** Statistical significance testing: (1) Use cross-validation: train both models on same folds; (2) Paired comparison: test if differences are consistent across folds; (3) Confidence intervals: quantify uncertainty; (4) Significance tests: paired t-test, Wilcoxon signed-rank test; (5) Effect size: is 3% meaningful for business? Single validation accuracy differences might be random variation. Rigorous testing distinguishes real improvements from noise.

---

**Question 304:** Your gradient boosting model has high variance (performance varies significantly across different train/test splits). What reduces variance?

A) Increase variance B) Increase regularization (lower learning rate, fewer trees, higher min_child_weight), ensemble multiple models, more training data C) Remove all regularization D) Variance doesn't matter

**Answer: B) Regularization and ensembling reduce variance**

**Explanation:** Variance reduction techniques: (1) Regularization: constraints limit model flexibility reducing variance; (2) Ensemble: averaging multiple models reduces individual model variance; (3) Bagging: train on bootstrap samples, average predictions; (4) Cross-validation: provides variance estimates; (5) More data: larger datasets reduce variance.

High variance indicates model is sensitive to particular training examples. Regularization and ensembling create more stable predictions.

---

**Question 305:** Your model has good validation performance but stakeholders say predictions "don't make sense" for certain cases. What should you investigate?

A) Ignore stakeholder feedback B) Feature importance analysis, prediction explanations (SHAP), case studies of "nonsensical" predictions C) Stakeholders are always wrong D) Don't investigate

**Answer: B) Investigate through explanations and case analysis**

**Explanation:** Stakeholder concerns warrant investigation: (1) Feature importance: which features drive predictions? (2) SHAP/LIME: explain specific "nonsensical" predictions; (3) Domain validation: do features align with business logic? (4) Edge cases: model might fail on unusual cases; (5) Data quality: predictions reveal data issues. Quantitative metrics don't capture all quality dimensions. Domain expert feedback identifies model failures metrics miss.

---

**Question 306:** You have limited labeled data (1,000 examples) but millions of unlabeled examples. What approach leverages unlabeled data?

A) Use only labeled data B) Semi-supervised learning: self-training, co-training, or pretrain on unlabeled data then fine-tune on labeled C) Delete unlabeled data D) Labels aren't needed

**Answer: B) Semi-supervised learning leverages both**

**Explanation:** Semi-supervised approaches: (1) Self-training: train on labeled data, predict on unlabeled, add high-confidence predictions to training; (2) Co-training: train multiple models on different feature sets, use agreements on unlabeled data; (3) Contrastive learning: pretrain on unlabeled data learning representations, fine-tune on labeled; (4) Consistency regularization: predictions should be consistent across data augmentations. Unlabeled data provides valuable signal despite lacking labels.

---

**Question 307:** Your model performance improves significantly when you increase training time 10×, but production latency requirements are strict. How do you resolve this?

A) Accept poor performance B) Model distillation: train large slow model, distill knowledge into fast model for production C) Deploy slow model anyway D) Give up

**Answer: B) Knowledge distillation to compact model**

**Explanation:** Training-deployment tradeoff: (1) Teacher model: large, slow, accurate model trained with unlimited resources; (2) Student model: small, fast model trained to mimic teacher predictions; (3) Distillation: student learns from teacher's soft predictions and internal representations; (4) Result: student achieves much of teacher's accuracy with fraction of parameters. Separates training complexity (unlimited resources) from deployment constraints (speed, size).

---

**Question 308:** Your team disagrees on performance metric: data scientists prefer AUC-ROC (0.85), business prefers precision at operating point (0.45). How do you resolve this?

A) Force one metric B) Align on business-relevant metric; understand why precision is low despite good AUC; optimize for business metric C) Ignore business requirements D) Use only AUC

**Answer: B) Align on business-relevant metric and optimize accordingly**

**Explanation:** Metric alignment critical: (1) Business metric (precision at operating point) reflects actual deployment value; (2) AUC measures overall discrimination but doesn't reflect operational tradeoffs; (3) Investigate: why is precision low? Threshold choice? Class imbalance? (4) Optimize directly for business metric; (5) Balance: report both but prioritize business metric for decisions. Technical metrics should serve business goals, not replace them.

---

**Question 309:** You implement batch normalization in your neural network and performance improves dramatically. Why does batch normalization help?

A) It only adds complexity B) Stabilizes training by normalizing activations, enables higher learning rates, reduces internal covariate shift, acts as regularization C) Hurts performance D) Has no effect

**Answer: B) Multiple benefits including training stability and regularization**

**Explanation:** Batch normalization benefits: (1) Normalizes layer inputs: reduces internal covariate shift; (2) Enables higher learning rates: more stable gradient flow; (3) Regularization effect: noise from batch statistics acts like dropout; (4) Faster convergence: can train deeper networks; (5) Less sensitive to initialization. Batch norm is standard in modern architectures. Benefits typically outweigh small computational overhead.

---

**Question 310:** Your model performs well on recent data but poorly on older historical data. Stakeholder needs model working across full time range. What do you do?

A) Only use recent data B) Train on data spanning full time range; consider time-based features; potentially weighted toward recent data C) Ignore historical data D) Performance differences don't matter

**Answer: B) Train on full time range with appropriate weighting**

**Explanation:** Temporal generalization: (1) Training data must span deployment time range; (2) Weighted training: weight recent data higher while including historical; (3) Time-based features: capture temporal patterns enabling adaptation; (4) Regular retraining: adapt to evolving patterns; (5) Validation: test across different time periods. Model trained only on recent data fails on historical patterns. Business requirements determine temporal scope needed.

---

## Technical Implementation Challenges

**Question 311:** Your production inference requires processing 1M requests per second. Single model instance handles 100 requests/second. How do you scale?

A) Use one model instance B) Horizontal scaling: deploy 10,000+ model instances behind load balancer; use model serving platforms (TensorFlow Serving, Triton) C) Process requests slowly D) Reduce requests

**Answer: B) Horizontal scaling with model serving infrastructure**

**Explanation:** High-throughput inference: (1) Horizontal scaling: deploy many model instances, distribute load; (2) Load balancing: distribute requests across instances; (3) Model serving: specialized infrastructure (TensorFlow Serving, Triton Inference Server) optimized for ML; (4) Batching: process multiple requests together; (5) GPU acceleration: higher throughput per instance; (6) Caching: cache predictions for repeated inputs. Cloud auto-scaling adjusts capacity to load.

---

**Question 312:** You need to deploy models to thousands of edge devices with intermittent connectivity. How do you handle model updates?

A) Update all devices simultaneously B) Staged rollout: update device batches progressively; fallback to cached models offline; version compatibility C) Never update models D) Require constant connectivity

**Answer: B) Staged rollout with offline capability**

**Explanation:** Edge deployment challenges: (1) Staged rollout: update small percentage, monitor, expand if successful; (2) Offline operation: devices must work without connectivity; (3) Background updates: sync when connected; (4) Version compatibility: handle mixed model versions; (5) Rollback capability: revert problematic updates; (6) Differential updates: only send changed weights. Edge deployment requires resilience to network issues.

---

**Question 313:** Your model requires features computed from multiple services. One service has 50ms latency, creating bottleneck. How do you optimize?

A) Accept slow predictions B) Cache frequently accessed features; precompute features where possible; async feature fetching; simplify feature set C) Don't use features D) Ignore latency

**Answer: B) Caching, precomputation, and async fetching**

**Explanation:** Feature serving optimization: (1) Caching: store frequently used features in Redis/Memcached; (2) Precomputation: calculate features offline for known entities; (3)

Feature store: centralized low-latency feature access; (4) Async fetching: request features in parallel; (5) Feature selection: remove low-value high-latency features; (6) Fallback: default values when features unavailable. Feature engineering must consider serving latency constraints.

---

**Question 314:** You're deploying a new model version. How do you minimize risk during rollout?

A) Deploy to all users immediately B) Canary deployment: route small traffic percentage to new model; monitor; gradually increase; rollback if issues C) No testing needed D) Deploy randomly

**Answer: B) Canary deployment with gradual rollout**

**Explanation:** Safe deployment practices: (1) Canary: route 1-5% traffic to new model initially; (2) Monitoring: track performance, errors, latency; (3) Comparison: A/B test new vs. old model; (4) Gradual increase: 5% → 25% → 50% → 100% if successful; (5) Rollback: quick revert if problems detected; (6) Feature flags: control rollout programmatically. Gradual rollout limits blast radius of failures while enabling production validation.

---

**Question 315:** Your model file is 2GB. Loading it at inference time takes 30 seconds. How do you reduce startup time?

A) Accept 30-second startup B) Model compression (quantization, pruning), lazy loading, keep model in memory, use model serving infrastructure C) Increase model size D) Startup time doesn't matter

**Answer: B) Compression and infrastructure optimization**

**Explanation:** Startup optimization: (1) Model compression: quantization (int8), pruning reduce size; (2) Lazy loading: load components on-demand; (3) Persistent serving: keep model loaded in memory between requests; (4) Serialization: use efficient formats (ONNX); (5) Model serving: infrastructure (TensorFlow Serving) optimizes loading; (6) Warm-up: pre-load models before receiving traffic. 2GB models suggest optimization opportunities.

---

**Question 316:** Your model training requires coordinating data from 5 different databases with different schemas. How do you manage this complexity?

A) Manual coordination each time B) Data integration layer: ETL pipeline creating unified view; feature store; schema mapping; version control C) Ignore data integration D) Use only one database

**Answer: B) Data integration layer with unified access**

**Explanation:** Multi-source integration: (1) ETL pipeline: extract, transform, load into unified schema; (2) Feature store: centralized feature computation and storage; (3) Schema mapping: document transformations between source schemas; (4) Version control: track data versions and transformations; (5) Orchestration: automated pipeline execution; (6) Monitoring: data quality checks. Integration layer abstracts complexity, provides consistent interface.

---

**Question 317:** You need to train models on confidential data that can't leave secure environment, but data scientists need access. How do you enable this?

A) Copy data outside secure environment B) Secure access: VPN/remote access to secure environment; data anonymization for development; synthetic data; federated learning C) No one accesses data D) Ignore security

**Answer: B) Secure access methods preserving confidentiality**

**Explanation:** Secure ML development: (1) Remote access: VPN to secure environment for development; (2) Anonymization: create development dataset with PII removed; (3) Synthetic data: generate statistically similar data for development; (4) Federated learning: train without centralizing data; (5) Secure enclaves: process in isolated secure environment; (6) Audit logging: track all data access. Balance security requirements with development needs.

---

**Question 318:** Your model uses custom preprocessing code. How do you ensure production uses identical preprocessing?

A) Rewrite preprocessing for production B) Package preprocessing with model; use same code in training and serving; containerization; version control C) Hope they match D) Different preprocessing is fine

**Answer: B) Shared preprocessing code between training and serving**

**Explanation:** Training-serving consistency: (1) Shared library: same preprocessing code for training and production; (2) Serialization: save preprocessing pipeline with model; (3) Containerization: Docker packages code and dependencies together; (4) Testing: validate training and serving produce identical outputs; (5) Version control: match code version to model version. Training-serving skew causes silent failures. Shared code prevents divergence.

---

**Question 319:** You're training models on sensitive data. How do you prevent model from memorizing sensitive information?

A) Models never memorize data B) Differential privacy, regularization, membership inference testing, limit model capacity C) Ignore privacy concerns D) All models memorize everything

**Answer: B) Privacy techniques preventing memorization**

**Explanation:** Privacy-preserving training: (1) Differential privacy: adds noise preventing individual memorization; (2) Regularization: limits model capacity to memorize; (3) Membership inference testing: verify if model reveals training set membership; (4) Gradient clipping: bounds per-example gradient contribution; (5) Model capacity: limit parameters relative to data size. Some memorization is inherent in ML; mitigation techniques reduce risk.

---

**Question 320:** Your model deployment requires strict versioning: exactly reproducible results from same model version and inputs. What's required?

A) Reproducibility is impossible B) Deterministic inference: fixed random seeds, deterministic operations, versioned dependencies, frozen models C) Results should vary randomly D) Versioning doesn't matter

**Answer: B) Deterministic inference with version control**

**Explanation:** Reproducibility requirements: (1) Fixed random seeds: eliminate randomness in inference; (2) Deterministic operations: avoid non-deterministic GPU operations; (3) Version control: code, model weights, dependencies all versioned; (4) Frozen models: models don't change post-deployment; (5) Input validation: verify input format consistency; (6) Testing: validate same input produces same output. Regulatory and debugging requirements often demand reproducibility.

---

# Chapter 6.0: Phase V Model Evaluation

You've built your model. Now comes the moment of truth: does it actually work? Not "does it achieve 95% accuracy on training data"—that's meaningless if it fails on new examples. Phase V Model Evaluation provides honest, rigorous assessment of model performance using metrics that align with business objectives and reveal where models succeed and fail (Japkowicz & Shah, 2011).

## 6.1 Performance Metrics Fundamentals

**Accuracy** measures the proportion of correct predictions. For a binary classifier: accuracy = (true positives + true negatives) / total examples. Simple and intuitive, but dangerously misleading for imbalanced datasets. A model predicting "no fraud" for every transaction achieves 99.9% accuracy with 0.1% fraud rate while being completely useless.

**Precision** answers: of examples predicted positive, what proportion are actually positive? Precision = true positives / (true positives + false positives). High precision means few false alarms. A medical test with high precision rarely diagnoses disease in healthy patients.

**Recall** (also called sensitivity or true positive rate) answers: of actual positive examples, what proportion did we correctly identify? Recall = true positives / (true positives + false negatives). High recall means few missed cases. A medical test with high recall catches most disease cases.

**F1-score** harmonically averages precision and recall: F1 = 2 × (precision × recall) / (precision + recall). This balances both metrics, useful when you care equally about false positives and false negatives. F1-score ranges from 0 (worst) to 1 (perfect).

**Specificity** measures true negative rate: of actual negatives, what proportion were correctly identified? Specificity = true negatives / (true negatives + false positives). This complements recall—recall focuses on catching positives, specificity on avoiding false positives (Fawcett, 2006).

**ROC-AUC** (Receiver Operating Characteristic - Area Under Curve) plots true positive rate against false positive rate across all classification thresholds. AUC measures overall discrimination ability regardless of threshold choice. AUC of 1.0 is perfect, 0.5 is random guessing. This metric works well for imbalanced datasets and provides threshold-independent assessment.

For **regression**, different metrics apply. **Mean Absolute Error (MAE)** averages absolute prediction errors: MAE = mean(|predicted - actual|). MAE has same units as target variable and is easy to interpret. **Mean Squared Error (MSE)** squares errors before averaging: MSE = mean((predicted - actual)$^2$). MSE penalizes large errors more heavily than small ones. **Root Mean Squared Error (RMSE)** is the square root of MSE, returning to original units. **R-squared** measures proportion of variance explained, ranging from 0 (model explains nothing) to 1 (perfect fit).

**Case Example: The Metric Revelation**

Zara, evaluating credit default models, initially focused on accuracy. Her best model achieved 94% accuracy—impressive until she examined the confusion matrix. The dataset had 6% default rate. The model achieved 94% accuracy by predicting "no default" for almost everyone, catching only 15% of actual defaults.

Zara switched focus to recall and precision. For credit risk, missing defaults (false negatives) costs much more than false alarms (false positives). She optimized for high recall while maintaining acceptable precision. The revised model achieved 78% accuracy but caught 85% of defaults with 35% precision—much better business outcomes despite lower accuracy.

This illustrates a crucial lesson: choose metrics aligned with business objectives, not those that make models look impressive.

## 6.2 Cross Validation Techniques

**Cross-validation** provides more robust performance estimates than single train-test splits, especially with limited data. It trains and evaluates models multiple times on different data subsets, averaging results (Kohavi, 1995).

**K-fold cross-validation** divides data into k equal subsets (folds). Train k models, each using k-1 folds for training and one fold for validation. Each fold serves as validation set exactly once. Average performance across k folds provides final estimate. Common choices are k=5 or k=10, balancing computational cost with estimate quality.

**Stratified k-fold** maintains class proportions within each fold. For a 90/10 class split, each fold has approximately 90/10 split. This prevents folds with no minority class examples, particularly important for imbalanced datasets.

**Leave-one-out cross-validation (LOOCV)** is extreme k-fold where k equals dataset size. Each example serves as validation set once while all others train the model. This maximizes training data usage but is computationally expensive and can have high variance in performance estimates.

**Time series cross-validation** respects temporal order. Use expanding or sliding windows rather than random folds. Train on months 1-6, validate on month 7. Train on months 1-7, validate on month 8. Continue forward through time. This prevents future information from leaking into training.

**Nested cross-validation** combines hyperparameter tuning with performance estimation. Outer loop assesses performance, inner loop tunes hyperparameters. This prevents overoptimistic estimates from tuning on the same data used for final assessment.

## 6.3 Bias and Variance Analysis

**Bias** measures systematic errors from incorrect assumptions in the learning algorithm. High bias means the model consistently misses patterns in data—underfitting. Simple linear models applied to complex non-linear problems exhibit high bias.

**Variance** measures sensitivity to training data fluctuations. High variance means the model learns noise and random fluctuations rather than true patterns—overfitting. Complex models trained on small datasets exhibit high variance, performing well on training data but poorly on new examples (Geman et al., 1992).

The **bias-variance tradeoff** represents a fundamental tension. Increasing model complexity reduces bias by capturing more patterns but increases variance by learning noise. Decreasing complexity increases bias but reduces variance. Optimal models balance both.

**Learning curves** plot training and validation performance versus training set size. For high bias, both training and validation errors converge to high values—more data doesn't help much. For high variance, training error is low but validation error is much higher—more data helps by reducing overfitting.

**Validation curves** plot training and validation performance versus model complexity (e.g., polynomial degree, tree depth). Training error decreases with complexity. Validation error decreases initially then increases as overfitting dominates. The sweet spot minimizes validation error.

**Diagnosis strategies**: If validation error >> training error, you have high variance—try simpler models, more data, or regularization. If both errors are high, you have high bias—try more complex models, better features, or reduced regularization.

## 6.4 Model Comparison and Selection

Comparing models requires standardized protocols ensuring fair evaluation. Run all models on identical train/test splits, use same cross-validation folds, apply consistent preprocessing, and evaluate on same metrics (Demšar, 2006).

**Statistical significance testing** determines whether performance differences are meaningful or due to random variation. Paired t-tests compare model performance across cross-validation folds. McNemar's test assesses classifier agreement on test examples. Confidence intervals quantify uncertainty in performance estimates.

**Multiple comparison corrections** prevent false discoveries when testing many models. Bonferroni correction adjusts significance thresholds. Friedman test with Nemenyi post-hoc analysis ranks multiple classifiers across datasets. Without corrections, testing 20 models at $p<0.05$ nearly guarantees finding "significant" differences by chance.

**Business criteria** often trump statistical performance. A model with 1% better accuracy requiring 10x more compute might not be worth deploying. A model achieving 95% performance in 1 hour training beats 96% performance requiring 48 hours when rapid iteration matters. Interpretability, maintenance complexity, and operational costs influence final selection.

**Ensemble combination** sometimes beats selecting single best models. If multiple models achieve similar performance with different error patterns, combining them often improves results as we covered in section 5.7.

### Case Example: The Model Selection Matrix

Miguel evaluated five classification models for customer churn prediction: logistic regression, random forest, XGBoost, neural network, and stacked ensemble. Raw performance metrics showed XGBoost best (AUC 0.87), followed closely by neural network (0.86) and stacked ensemble (0.86).

However, business context revealed hidden tradeoffs. XGBoost required daily retraining (4 hours compute) to maintain performance as customer behaviors changed. The neural network needed specialized GPU infrastructure unavailable in production. The stacked ensemble combined logistic regression and random forest, achieving 0.86 AUC with 30-minute training on standard CPU hardware.

Miguel also considered interpretability. Marketing teams needed to understand churn drivers to design retention campaigns. The logistic regression component provided clear feature importance and direction (increasing variable X raises churn risk by Y%). XGBoost offered some interpretability through SHAP values but required more explanation.

The final decision: deploy the stacked ensemble. It delivered competitive performance (0.86 vs. 0.87 for XGBoost), fast training enabling frequent updates, ran on existing infrastructure, and provided interpretability through the logistic regression component. The 1% performance sacrifice was worth operational benefits.

## 6.5 AB Testing Strategies

**A/B testing** compares model performance in production by randomly assigning users to control (existing model) or treatment (new model) groups and measuring business outcomes. This provides ground truth about real-world performance beyond offline metrics (Kohavi et al., 2009).

**Randomization** ensures groups are statistically identical except for model assignment. Random assignment eliminates confounding variables that could bias results. Stratified randomization maintains balance across important segments (geography, user types, devices).

**Sample size** determination requires statistical power analysis. Small experiments might miss meaningful differences. Large experiments waste resources testing obvious improvements. Calculate required sample size based on minimum detectable effect, baseline conversion rate, significance level, and statistical power.

**Duration** balances detecting effects versus business risk. Run tests long enough to capture weekly patterns—users behave differently on weekends versus weekdays. Account for novelty effects where users react to changes initially but behavior normalizes. Typical test durations range from 1-4 weeks depending on traffic volume and effect size.

**Guardrail metrics** protect against harmful unintended consequences. While optimizing for primary metric (e.g., click-through rate), monitor secondary metrics (revenue, user satisfaction, latency) to ensure improvements don't come at unacceptable costs.

**Statistical rigor** prevents false conclusions. Set significance thresholds before running experiments (typically $p<0.05$). Don't peek at results repeatedly—each peek increases false positive rates. Use sequential testing procedures or Bayesian methods if early stopping is essential. Document and report all tests, not just successful ones, to avoid publication bias.

## 6.6 Model Explainability Assessment

**Model explainability** describes the ability to understand why models make specific predictions. This matters for debugging, building trust, meeting regulatory requirements, and extracting insights (Molnar, 2020).

**Intrinsically interpretable models** are transparent by design. Linear regression reveals feature importance through coefficients—increasing feature X by one unit changes prediction by coefficient value. Decision trees show decision paths explicitly. Rule-based models provide if-then logic humans naturally understand.

**Model-agnostic explanations** work with any model type. **SHAP (SHapley Additive exPlanations)** values quantify each feature's contribution to individual predictions using game-theoretic principles. SHAP provides both local explanations (why this specific prediction) and global explanations (overall feature importance). **LIME (Local Interpretable Model-agnostic Explanations)** approximates complex models locally with simple interpretable models around specific predictions.

**Feature importance** ranks features by impact on model performance. Permutation importance measures performance decrease when shuffling each feature. Tree-based models provide built-in feature importance from split frequency and information gain.

**Partial dependence plots** show relationships between features and predictions while averaging out other features. They reveal whether relationships are linear, non-linear, or have interactions.

**Counterfactual explanations** answer "what if" questions: what minimal changes to inputs would flip the prediction? For loan applications, this answers "what would I need to change to get approved?" These actionable explanations help users understand and improve.

**Explainability-performance tradeoffs** exist. Simple models (linear regression, decision trees) are highly interpretable but may underperform. Complex models (deep neural networks, large ensembles) achieve better performance but resist interpretation. Post-hoc explanation methods bridge this gap imperfectly—they approximate black-box behavior rather than revealing true mechanisms.

## 6.7 Edge Case Testing

**Edge cases** are unusual, rare, or extreme scenarios where models may fail despite good average performance. Systematic edge case testing reveals blind spots before production deployment (Breck et al., 2017).

Test **distribution boundaries**. How does the model perform on minimum and maximum values for continuous features? On rare categories for categorical features? On combinations of extreme values?

Evaluate **missing data handling**. What happens when required features are unavailable? Does the model fail gracefully or produce nonsensical predictions? Test with various missing patterns.

Check **adversarial inputs** designed to fool the model. For image classification, adversarially perturbed images look identical to humans but cause misclassification. For text classification, misspellings or character substitutions might evade detection. For fraud detection, fraudsters actively try to bypass systems.

Assess **temporal robustness**. Models trained on historical data may fail when patterns change. Test on data from different time periods, especially recent data not in training sets.

Examine **subgroup performance**. Models might achieve good overall performance while failing on important subgroups. A hiring algorithm might perform well on average while discriminating against protected groups. Break down performance by demographic characteristics, geographic regions, product categories, or other relevant segments.

**Case Example: The Deployment Disaster Averted**

Amina developed a fraud detection model achieving 92% precision and 88% recall on test data—excellent performance. Days before production deployment, she conducted systematic edge case testing.

She discovered the model failed on new merchant categories not present in training data, always predicting "not fraud" for transactions at merchants added in the past month. She found it struggled with international transactions where amount currency conversions introduced rounding errors the model interpreted as anomalies.

Testing revealed poor performance on very high-value transactions (over $10,000) because training data had few such examples—the model was uncertain in this regime. She also found it falsely flagged legitimate transactions during holiday shopping patterns not represented in training data from January-October.

Without edge case testing, these failures would have emerged in production, causing customer friction and fraud losses. Amina expanded training data to include international transactions and holiday patterns, added fallback rules for new merchant categories, and implemented human review for high-value transactions. The refined system handled edge cases gracefully instead of failing spectacularly.

## 6.8 Final Model Selection

After comprehensive evaluation, you must select the model for production deployment. This decision balances multiple competing factors beyond test set performance (Paleyes et al., 2022).

**Performance requirements** define minimum acceptable levels across primary and secondary metrics. Does the model meet business needs? Is performance sufficient for the intended use case? Are all critical guardrail metrics within bounds?

**Operational constraints** limit viable models. What compute resources are available? What latency requirements must be met? How frequently must the model retrain? Can production infrastructure support the model's dependencies?

**Maintenance considerations** affect long-term success. How difficult is the model to update when data distributions change? Can the team maintain and debug it? Is required expertise available?

**Risk assessment** evaluates potential harms. What happens if the model fails? Are there liability or regulatory concerns? Does the model create bias or fairness issues? What monitoring and safeguards are needed?

**Stakeholder alignment** ensures the selected model satisfies key stakeholders. Do business leaders understand the performance-cost tradeoffs? Do engineering teams have infrastructure to support deployment? Do end users find predictions useful and trustworthy?

**Approval processes** formalize model selection. Document model performance, architecture, training details, evaluation results, risk assessment, and operational plans. Present to review committees. Obtain sign-offs from stakeholders. Establish success metrics for production monitoring.

## 6.9 Performance versus Interpretability Tradeoffs

The tension between predictive performance and interpretability forces difficult decisions. Complex models (gradient boosting, deep neural networks) often achieve better accuracy but resist explanation. Simple models (linear regression, decision trees) are transparent but may underperform (Rudin, 2019).

**Regulated industries** often require interpretability. Financial services must explain credit decisions to applicants and regulators. Healthcare needs to justify treatment recommendations to clinicians and patients. Fair lending laws mandate transparency in loan underwriting.

**High-stakes decisions** demand explainability even without regulatory requirements. Hiring, criminal justice, medical diagnosis, and resource allocation decisions affecting people's lives require justification beyond "the algorithm said so."

**Low-stakes applications** tolerate black boxes. Recommending movies or products rarely requires explanation. Users accept opaque predictions when consequences are minimal and opt-out is easy.

**Strategies for managing tradeoffs** include: deploying interpretable models when performance sufficiently meets needs even if not optimal; using complex models with post-hoc explanations when performance improvement justifies additional explanation effort; building hybrid systems where interpretable models handle most cases and complex models trigger human review for edge cases; and investing in interpretable ML research developing models that achieve both goals.

The right balance depends on context. A 2% performance gain might be worth opacity for recommending content but unacceptable for approving loans or diagnosing disease.

## 6.10 What This Means for You

Phase V Model Evaluation provides rigorous, honest assessment of model performance using metrics aligned with business objectives. Accuracy alone misleads, especially for imbalanced datasets. Precision, recall, F1-score, and ROC-AUC provide nuanced performance understanding. Cross-validation delivers robust estimates through multiple train-test splits.

Bias-variance analysis diagnoses underfitting and overfitting, guiding remediation strategies. Model comparison requires standardized protocols and statistical testing to ensure differences are meaningful. A/B testing validates real-world performance beyond offline metrics.

Model explainability ranges from intrinsically interpretable models to post-hoc explanation methods for black boxes. Edge case testing reveals blind spots before production deployment. Final model selection balances performance, operational constraints, maintenance, risk, and stakeholder alignment.

The performance-interpretability tradeoff forces context-dependent decisions based on stakes, regulation, and business requirements. Comprehensive evaluation prevents deploying models that look good on paper but fail in practice.

# 6.11 Practice Questions

## Metric Interpretation Scenarios

**Question 321:** Your binary classifier achieves: Accuracy: 95%, Precision: 50%, Recall: 90%. What does this tell you about the model and data?

A) Model is excellent B) Highly imbalanced data (likely rare positive class); model predicts many false positives but catches most true positives C) Perfect performance D) Model is useless

**Answer: B) Imbalanced data with high false positive rate**

**Explanation:** Metric interpretation: (1) High accuracy (95%) with low precision (50%) indicates class imbalance; (2) Precision 50% means half of positive predictions are wrong (false positives); (3) Recall 90% means model catches most actual positives; (4) Pattern: model is generous with positive predictions (causing false positives) but doesn't miss many true positives. This might be appropriate for disease screening (don't miss cases) but problematic for spam filtering (too many false alarms).

---

**Question 322:** Two models: Model A has AUC-ROC of 0.90; Model B has AUC-ROC of 0.85 but better precision at your operating threshold (0.75 vs 0.60). Which is better?

A) Model A has higher AUC B) Model B better at operating point where you'll use it; optimize for deployment conditions C) AUC is the only metric that matters D) Both are equally good

**Answer: B) Model B better at actual operating point**

**Explanation:** AUC vs. operating point performance: (1) AUC measures overall discrimination across all thresholds; (2) Production uses specific threshold with specific precision/recall tradeoff; (3) Model B performs better at your chosen operating point despite lower overall AUC; (4) Optimize for deployment conditions, not abstract metrics. When choosing between models, evaluate at operating conditions (threshold, business constraints). AUC useful for model selection but operating point performance determines production value.

---

**Question 323:** Your regression model achieves: MAE: 10, RMSE: 25, $R^2$: 0.85. What does the MAE/RMSE difference indicate?

A) Metrics agree perfectly B) Large RMSE relative to MAE indicates presence of large outlier errors C) No outliers present D) Metrics are unrelated

**Answer: B) Large outliers causing RMSE to be much higher than MAE**

**Explanation:** MAE vs. RMSE interpretation: (1) MAE averages absolute errors (all errors weighted equally); (2) RMSE squares errors (large errors weighted heavily); (3) RMSE > 2× MAE indicates outlier predictions; (4) In this case, most predictions within ±10 (MAE) but some predictions off by 50+ (driving RMSE to 25). Investigate large errors: are they data quality issues, model failures on specific segments, or legitimate high-variance predictions?

---

**Question 324:** Your multi-class classifier (5 classes) has overall accuracy of 75%. Examining per-class recall: Class A: 95%, Class B: 85%, Class C: 60%, Class D: 55%, Class E: 25%. What should you investigate?

A) Overall accuracy is acceptable; no investigation needed B) Class E performance is terrible; investigate data distribution, representation, inherent difficulty C) All classes perform equally D) Metrics are fine

**Answer: B) Investigate Class E's poor performance**

**Explanation:** Per-class analysis reveals: (1) Overall metrics mask individual class problems; (2) Class E recall (25%) is unacceptable—missing 75% of E instances; (3) Investigation: Is Class E underrepresented in training? Inherently difficult? Confused with other classes? Data quality issues? (4) Solutions might include: oversampling Class E, collecting more E examples, feature engineering for E discrimination, separate E vs. not-E classifier. Aggregate metrics hide critical failures.

---

**Question 325:** You calculate 95% confidence interval for model accuracy: [0.78, 0.82]. Stakeholder asks what this means. How do you explain?

A) Accuracy is exactly 0.80 B) We're 95% confident true performance is between 78-82%; with different test data, measured accuracy would fall in this range 95% of times C) Confidence intervals are meaningless D) Accuracy is always 78%

**Answer: B) Quantifies uncertainty in performance estimate**

**Explanation:** Confidence interval interpretation: (1) Point estimate (80%) has uncertainty due to limited test data; (2) 95% CI says: if we repeated testing many times with different samples, 95% of intervals would contain true performance; (3) Wider intervals indicate higher uncertainty (smaller test sets); (4) Practical meaning: true performance likely in this range. Confidence intervals quantify measurement uncertainty—essential for informed decisions.

---

**Question 326:** Your fraud detection model shows: True Positive Rate: 85%, False Positive Rate: 3%. Business says 3% false positives creates too many manual reviews. What do you do?

A) Accept poor tradeoff B) Adjust classification threshold to reduce false positives; explain precision-recall tradeoff; optimize for business constraints C) Ignore business feedback D) False positive rate doesn't matter

**Answer: B) Adjust threshold and explain tradeoffs**

**Explanation:** Business-driven threshold selection: (1) Current threshold: 85% TPR, 3% FPR; (2) Increase threshold to reduce FPR: might achieve 1% FPR but at cost of lower TPR (perhaps 70%); (3) Present tradeoff curve: show FPR/TPR at different thresholds; (4) Business decides: is catching 70% of fraud with 1% FPR better than 85% fraud with 3% FPR? (5) Optimization: find threshold matching business constraints. Technical metrics serve business goals.

---

**Question 327:** You evaluate model on test set: Precision: 0.70, Recall: 0.80, F1-Score: 0.75. What does F1-score tell you that precision/recall don't?

A) F1 provides completely new information B) F1 is harmonic mean balancing precision and recall; single metric for models trading off both C) F1 is meaningless D) F1 always equals (precision + recall)/2

**Answer: B) F1 balances precision and recall in single metric**

**Explanation:** F1-score interpretation: (1) Harmonic mean of precision and recall: $2 \times (P \times R)/(P+R)$; (2) Single metric for model comparison when you care about both P and R; (3) Penalizes extreme imbalances (one high, one low); (4) In this case: F1=0.75 is between P=0.70 and R=0.80, closer to lower value. Use when: you care equally about precision and recall, need single metric for model selection. Doesn't replace precision/recall but complements them.

---

**Question 328:** Your model achieves 80% accuracy on training data and 79% on test data. What does this indicate?

A) Severe overfitting B) Good generalization; model performs similarly on unseen data C) Underfitting D) Model failure

**Answer: B) Good generalization with minimal overfitting**

**Explanation:** Train-test gap analysis: (1) Small gap (80% vs 79%) indicates model generalizes well; (2) No significant overfitting—model doesn't memorize training data; (3) Performance level might still be insufficient for business needs, but generalization is healthy; (4) Compare to severe overfitting: 99% train, 70% test (memorization). Small train-test gaps suggest: appropriate model complexity, sufficient regularization, adequate training data. Focus on improving overall performance rather than overfitting concerns.

---

**Question 329:** Your customer churn model has AUC of 0.75. Stakeholder asks, "Is 0.75 good?" How do you respond?

A) "Yes, always good" B) "Context-dependent: compare to baseline, random (0.5), perfect (1.0), and business value; 0.75 shows discriminatory ability" C) "No, always bad" D) "AUC doesn't matter"

**Answer: B) Interpret in context with baselines and business value**

**Explanation:** AUC interpretation: (1) Random classifier: 0.5 (no discrimination); (2) Perfect classifier: 1.0; (3) 0.75 shows moderate discrimination—model can distinguish churners from non-churners; (4) Compare to baseline (simple rules, previous model); (5) Business value: does 0.75 enable profitable interventions? (6) Industry context: some domains achieve 0.95+, others struggle to exceed 0.70. Metrics need context. "Good" depends on business requirements, baselines, and alternatives.

---

**Question 330:** You calculate precision-recall curve showing precision drops from 0.90 to 0.30 as recall increases from 0.20 to 0.90. What does this curve tell you?

A) Model is useless B) Model confident about some positives (high precision at low recall) but predictions degrade as it becomes less selective C) Curve is meaningless D) Model is perfect

**Answer: B) High-confidence predictions are precise; lower-confidence predictions less reliable**

**Explanation:** Precision-recall curve interpretation: (1) High recall requires predicting more positives, including less certain ones; (2) At low recall (0.20), model predicts only most confident positives (90% precision); (3) At high recall (0.90), model casts wider net, including uncertain predictions (30% precision); (4) Curve shape indicates model has reasonable confidence calibration—confident predictions are more accurate. Use case determines operating point: high-stakes decisions operate at high precision (low recall); screening operates at high recall (lower precision acceptable).

---

## Model Comparison Situations

**Question 331:** You compare three models on validation data: Model A: 82%, Model B: 84%, Model C: 83%. Differences are within confidence intervals. How do you choose?

A) Model B has highest accuracy B) Performance essentially equal; choose based on: interpretability, inference speed, maintenance complexity, deployment ease C) Randomly select D) Use all three always

**Answer: B) Differences not significant; use secondary criteria**

**Explanation:** Statistically equivalent performance: (1) Overlapping confidence intervals mean differences might be noise; (2) Secondary criteria matter: speed (Model A faster?), interpretability (Model C explainable?), complexity (Model B harder to maintain?), cost (computational resources); (3) Practical considerations often outweigh marginal statistical differences. When performance is equivalent, operational factors determine optimal choice. Don't over-optimize for small metric differences that might disappear with different test sets.

---

**Question 332:** Model A trained on 10K examples achieves 75% accuracy. Model B trained on 100K examples achieves 77% accuracy. Which is more impressive?

A) Model B has higher accuracy B) Model A achieves competitive performance with 10× less data; consider data efficiency C) They're identical D) Neither is impressive

**Answer: B) Model A's data efficiency is notable**

**Explanation:** Data efficiency matters: (1) Model A achieves 75% with 10K examples shows efficient learning; (2) Model B's 77% with 100K examples is only marginally better despite 10× more data; (3) Suggests Model A architecture or features more efficient; (4) Practical implications: if data is scarce or expensive, Model A preferable; if data abundant, Model B's marginal improvement might justify additional data. Consider learning curves: would Model A with 100K examples surpass Model B? Data efficiency indicates model quality beyond raw performance.

---

**Question 333:** You train identical model architecture with 5 different random seeds. Validation accuracies: 78%, 81%, 79%, 82%, 77%. What does this variation indicate?

A) Model is deterministic B) Random initialization creates variance; model sensitive to initialization; consider ensemble or multiple runs C) Perfect stability D) Variation is meaningless

**Answer: B) Initialization sensitivity; ensemble or averaging helps**

**Explanation:** Random seed variation: (1) 5% range (77-82%) indicates significant initialization sensitivity; (2) Neural networks particularly sensitive to initial weights; (3) Solutions: train multiple models with different seeds and ensemble (average predictions);

report mean ± std of multiple runs; (4) If variation is large relative to improvements you're testing, statistical significance harder to establish. Variation quantifies model stability. High variance suggests ensembling beneficial.

---

**Question 334:** Model A is interpretable (logistic regression) with 80% accuracy. Model B is black box (deep neural network) with 85% accuracy. Regulatory environment requires explainability. How do you compare?

A) Always use Model B (higher accuracy) B) Model A required for regulatory compliance despite lower accuracy; or use Model B with post-hoc explanations if acceptable C) Ignore regulations D) Can't deploy any model

**Answer: B) Regulatory compliance determines choice; consider explanation techniques**

**Explanation:** Regulatory constraints: (1) Explainability requirement might mandate interpretable model (Model A); (2) 5% accuracy sacrifice for compliance is necessary tradeoff; (3) Alternative: use Model B with SHAP/LIME if post-hoc explanations satisfy regulations; (4) Hybrid: Model B for predictions, Model A for explanations and validation; (5) Regulatory environment determines feasibility. Consult legal/compliance teams. Can't choose optimal technical model ignoring business constraints.

---

**Question 335:** Comparing models: Model A has 85% accuracy and 0.5 second latency. Model B has 83% accuracy and 0.05 second latency. Use case requires sub-100ms response. Which is better?

A) Model A has better accuracy B) Model B is only viable option; latency requirement eliminates Model A C) Both work D) Neither works

**Answer: B) Latency requirement eliminates Model A**

**Explanation:** Hard constraints vs. soft optimization: (1) Sub-100ms latency is hard requirement (system won't function above it); (2) Model A's 500ms latency is 5× over limit—not viable regardless of accuracy advantage; (3) Model B meets latency requirement (50ms) with acceptable accuracy; (4) Must optimize within constraints. Hard requirements (latency, memory, regulatory) eliminate options before comparing soft metrics (accuracy differences). Architecture decisions must account for deployment constraints from the start.

---

**Question 336:** You have Model A (simple, interpretable, 75% accuracy) already in production. Model B (complex, 78% accuracy) is candidate replacement. What factors inform replacement decision?

A) 3% improvement always justifies replacement B) Consider: improvement significance, migration cost/risk, maintenance complexity increase, business value of improvement C) Never replace existing models D) Always replace with newer models

**Answer: B) Holistic cost-benefit analysis including migration risk**

**Explanation:** Replacement decisions consider: (1) Statistical significance: is 3% improvement real? (2) Business value: does 3% accuracy translate to meaningful value? (3) Migration costs: development, testing, deployment effort; (4) Migration risk: might introduce new failures; (5) Maintenance: is Model B harder to maintain? (6) Opportunity cost: could effort be better spent elsewhere? Small improvements might not justify complex migrations. Incumbent models have proven stability. New models carry risk. Value must exceed total costs.

---

**Question 337:** Comparing on test set: Model A: Precision=0.80, Recall=0.60. Model B: Precision=0.70, Recall=0.75. Which is better?

A) Model A clearly better B) Depends on business preference: favor precision (Model A) or recall (Model B)? C) Model B clearly better D) Can't compare

**Answer: B) Business preference determines optimal precision-recall tradeoff**

**Explanation:** Precision-recall tradeoff: (1) Model A: fewer false positives (80% precision), misses more true positives (60% recall); (2) Model B: fewer missed positives (75% recall), more false positives (70% precision); (3) Business context determines preference: spam filter prefers precision (don't block legitimate email); disease screening prefers recall (don't miss cases); (4) Calculate F1-scores for single metric; (5) Consider costs of false positives vs. false negatives. No universal "better"—depends on application.

**Question 338:** You compare models using 10-fold cross-validation. Model A mean accuracy: 80% (std: 2%). Model B mean: 81% (std: 5%). Which is better?

A) Model B has higher mean B) Model A more stable (lower variance) despite slightly lower mean; consider variance-bias tradeoff C) Model B obviously superior D) Can't compare

**Answer: B) Model A's stability vs. Model B's marginal performance advantage**

**Explanation:** Mean vs. variance tradeoff: (1) Model B: 81% mean but 5% std suggests performance ranges 76-86% across folds; (2) Model A: 80% mean with 2% std suggests 78-82% range—more predictable; (3) High variance indicates model sensitive to training data; (4) Stability matters for production reliability; (5) Statistical testing: confidence intervals likely overlap given Model B's high variance. Model A's consistent performance might be preferable to Model B's higher but unstable performance. Test statistical significance.

---

**Question 339:** Model A excels on common cases (95% accuracy on 90% of data) but fails on rare cases (20% accuracy on 10% of data). Model B is consistent (80% accuracy across all cases). Which is better?

A) Model A has higher overall accuracy B) Depends on importance of rare cases; if critical, Model B better despite lower overall accuracy C) Model B always better D) Can't compare

**Answer: B) Rare case importance determines choice**

**Explanation:** Segment performance matters: (1) Model A overall: $0.90 \times 0.95 + 0.10 \times 0.20 = 87.5\%$ accuracy; (2) Model B overall: 80% accuracy; (3) If rare cases are critical (high-value customers, severe diseases, critical failures), Model B preferable; (4) If rare cases less important, Model A's overall performance better; (5) Consider segment-specific metrics, not just overall. Some applications require consistent performance across all segments; others tolerate failures on edge cases. Business importance of segments determines optimal choice.

---

**Question 340:** You evaluate models monthly. Model A consistently achieves 82-83% accuracy. Model B varies: 78%, 85%, 80%, 84%, 79%. Which is more reliable for production?

A) Model B sometimes achieves 85% B) Model A's consistency more reliable; Model B's variance concerning for production stability C) Model B better D) Both equally reliable

**Answer: B) Model A's consistency indicates production reliability**

**Explanation:** Production reliability: (1) Model A: predictable performance enables reliable system; (2) Model B: high variance suggests performance depends on uncontrolled factors (data distribution shifts, edge cases); (3) Production systems need predictable behavior; (4) Model B might occasionally outperform but inconsistency problematic; (5) Investigate Model B variance: is it data-dependent? Fixable? Consistent performance reduces operational risk. Variance in production is worse than slightly lower but stable performance.

---

## Validation Strategy Selection

**Question 341:** You have 1,000 examples for training a medical diagnosis model. What validation strategy maximizes data usage while providing reliable estimates?

A) 80/20 train/test split B) K-fold cross-validation (k=10) to use all data for both training and validation; separate held-out test set C) 50/50 split D) Use all data for training

**Answer: B) K-fold cross-validation with held-out test set**

**Explanation:** Limited data validation: (1) Hold out 15-20% for final test set; (2) 10-fold CV on remaining data: each example used once for validation, 9 times for training; (3) Provides robust performance estimate despite limited data; (4) Average metrics across folds reduces variance; (5) Final model trains on all non-test data. Simple splits waste data (small validation sets have high variance). K-fold maximizes efficiency. Medical applications especially need reliable estimates given high stakes.

---

**Question 342:** You're building a stock price predictor using time series data (2015-2023). How should you validate?

A) Random 80/20 split B) Temporal split: train on 2015-2020, validate on 2021-2022, test on 2023; or rolling-window cross-validation C) Shuffle all data randomly D) No validation needed

**Answer: B) Temporal split preserving time order**

**Explanation:** Time series validation prevents data leakage: (1) Temporal split: train on past, validate on middle future, test on recent future; (2) Simulates production: train on history, predict future; (3) Rolling windows: multiple temporal splits test across time periods; (4) Gap: optional buffer between train/test prevents autocorrelation leakage. Random splitting leaks future into training, creating unrealistic performance estimates. Time series must preserve temporal order. Validate on future periods relative to training.

---

**Question 343:** Your dataset has 100,000 examples across 50 customer segments with varying sizes. How should you create train/test splits?

A) Random split B) Stratified split ensuring proportional representation of all 50 segments in train and test sets C) Use only largest segment D) Don't split

**Answer: B) Stratified split maintaining segment proportions**

**Explanation:** Stratified sampling: (1) Ensures train and test sets represent all segments proportionally; (2) Small segments might be absent with simple random split; (3) Stratification maintains population structure; (4) Critical for imbalanced multiclass problems; (5) Enables reliable per-segment evaluation. For 50 segments, stratification ensures adequate representation of all groups. Without stratification, rare segments might appear only in train or only in test, preventing proper evaluation.

---

**Question 344:** You're evaluating a model for deployment in Europe using data from North America. What validation strategy reveals generalization across regions?

A) Standard validation on North American data B) Geographic cross-validation: train on North America, validate on small European pilot data; or evaluate generalization limits C) Ignore geographic differences D) Only use North American data

**Answer: B) Test on target deployment region data**

**Explanation:** Domain shift validation: (1) North American validation doesn't predict European performance; (2) Collect European pilot data for validation; (3) Measure performance gap between source and target domains; (4) Consider domain adaptation techniques if gaps large; (5) Document limitations and assumptions. When deployment domain differs from training domain, validate on target domain. Distribution shift is common cause of production failures. Test deployment population before launch.

---

**Question 345:** Your validation set is very small (50 examples) due to limited labeled data. Validation metrics are unstable across runs. What should you do?

A) Use the 50 examples B) Bootstrap validation: resample validation set, evaluate multiple times, report confidence intervals; or increase validation set at expense of training C) Ignore instability D) Give up validation

**Answer: B) Bootstrap or increase validation set despite training tradeoff**

**Explanation:** Small validation set solutions: (1) Bootstrap: resample with replacement, compute metrics on each sample, report mean and confidence intervals; (2) Increase validation size: allocate more data to validation (e.g., 70/30 instead of 80/20); (3) Cross-validation: use k-fold to increase effective validation size; (4) Accept uncertainty: report confidence intervals. Small validation sets provide unreliable estimates. Reliability often more valuable than maximum training data. Unstable metrics prevent informed decisions.

---

**Question 346:** You're validating a model intended for real-time production deployment. Should validation simulate production conditions?

A) Validation conditions don't matter B) Yes, validate under production-like conditions: latency, concurrent load, data formats, edge cases C) Only offline validation needed D) Production testing unnecessary

**Answer: B) Validate under production conditions**

**Explanation:** Production-realistic validation: (1) Latency: measure inference time under expected load; (2) Concurrency: test with multiple simultaneous requests; (3) Data formats: use production data schemas; (4) Edge cases: test unusual inputs model will encounter; (5) Integration: test with actual upstream/downstream systems. Offline validation measures accuracy but misses operational failures (timeouts, memory issues, format incompatibilities). Production simulation catches issues offline metrics miss.

---

**Question 347:** You have data from 2020-2023 for churn prediction model. How should you validate to ensure model works in 2024?

A) Random split across all years B) Train on 2020-2022, validate on 2023 to test future prediction; or rolling-window validation across time periods C) Train only on 2023 data D) Use only 2020 data

**Answer: B) Temporal validation testing future prediction**

**Explanation:** Future prediction validation: (1) Train on past (2020-2022), validate on recent past (2023) as proxy for future; (2) Tests model's ability to predict forward in time; (3) Rolling windows: multiple temporal splits (train 2020-2021→validate 2022, train 2020-2022→validate 2023); (4) Captures temporal patterns and drift; (5) More realistic than random splits. Production will predict future from past; validation should mirror this. Recent validation performance best estimates future performance.

---

**Question 348:** Your validation shows model performs well on average but fails completely on 5% of examples. How should you investigate?

A) Average performance is sufficient B) Analyze failure cases: error analysis identifying patterns in failures, segment-specific validation C) Ignore failures D) Failures don't matter

**Answer: B) Error analysis and segment-specific validation**

**Explanation:** Failure analysis: (1) Identify failure cases systematically; (2) Look for patterns: specific customer segments, input characteristics, edge cases; (3) Segment-specific metrics: evaluate performance on different subgroups; (4) Root causes: data quality, training set gaps, model limitations; (5) Mitigation: additional training data for failure modes, specialized models, rule-based overrides. Average metrics hide systematic failures. 5% failures might be disproportionately important (high-value customers, critical cases). Understanding failures guides improvements.

---

**Question 349:** You're comparing model versions over time. How do you ensure fair comparison as data evolves?

A) Compare on different data each time B) Maintain hold-out test set unchanged; also evaluate on recent data to measure current performance C) Never update test set D) Random comparison

**Answer: B) Fixed test set for consistency; current data for relevance**

**Explanation:** Longitudinal model comparison: (1) Fixed test set: enables controlled comparison over time; (2) Current performance: evaluate on recent data reflecting production; (3) Both matter: fixed set shows intrinsic model improvements; current set shows practical performance; (4) Document: which test set produced which metrics; (5) Retrain periodically on updated data. Fixed benchmarks enable apples-to-apples comparison. Current data ensures production relevance. Use both for complete picture.

---

**Question 350:** Your binary classifier has imbalanced classes (10% positive). What validation metrics are most informative?

A) Accuracy only B) Precision, recall, F1-score, AUC-ROC, AUC-PR (precision-recall curve area) C) Only accuracy D) No metrics needed

**Answer: B) Multiple metrics beyond accuracy**

**Explanation:** Imbalanced validation metrics: (1) Accuracy misleading (90% accuracy by predicting all negative); (2) Precision: positive prediction reliability; (3) Recall: positive case capture rate; (4) F1-score: balanced precision-recall; (5) AUC-ROC: overall discrimination; (6) AUC-PR: especially informative for imbalanced data; (7) Confusion matrix: see all error types. Report multiple metrics. Different metrics reveal different model characteristics. Accuracy alone hides critical information for imbalanced problems.

---

**Performance Troubleshooting**

**Question 351:** Your model's validation performance suddenly dropped from 85% to 70% without code changes. What are likely causes?

A) Model suddenly broke B) Data drift: input distribution changed; data quality issues; pipeline failure; label distribution shift C) Performance drop is normal D) Ignore the drop

**Answer: B) Investigate data changes and pipeline issues**

**Explanation:** Sudden performance drops suggest: (1) Data distribution shift: input features changed (new product launch, seasonal shift); (2) Data quality: pipeline failure, source system changes; (3) Label shift: target distribution changed; (4) System issues: feature computation errors; (5) Test set contamination: evaluation data problems. Investigate: compare current input distributions to historical; check data quality metrics; validate pipeline health; review system logs. Models don't spontaneously degrade without external changes.

---

**Question 352:** Your model performs well on validation data but poorly on specific customer segment. What debugging steps should you take?

A) Ignore segment performance B) Segment-specific analysis: check training set representation, feature relevance, label quality for that segment C) Overall performance is all that matters D) Delete the segment

**Answer: B) Segment-specific debugging**

**Explanation:** Segment failure debugging: (1) Training representation: is segment underrepresented? Collect more segment examples; (2) Feature relevance: do features predict well for this segment? Engineer segment-specific features; (3) Label quality: are labels accurate for this segment? Audit segment labels; (4) Model assumptions: does model make assumptions that break for segment? (5) Solution: segment-specific model, oversampling, or weighted loss. Segment failures indicate training-test distribution mismatch or missing signal for that segment.

---

**Question 353:** Your model achieves 90% accuracy on validation set but users report many incorrect predictions. What might explain this discrepancy?

A) Users are wrong B) Validation data doesn't represent production distribution; users encounter edge cases underrepresented in validation C) 90% accuracy guarantees success D) Ignore user reports

**Answer: B) Validation-production distribution mismatch**

**Explanation:** Validation-production gaps: (1) Validation data selection bias: doesn't represent production population; (2) Edge cases: users encounter rare scenarios absent from validation; (3) Temporal drift: validation is old, production is current; (4) Context: production has different context (interface, user behavior) affecting inputs; (5) Cherry-picking: users remember errors more than successes. Investigate: collect production data; analyze user-reported errors; compare production input distribution to validation. Validation must represent production use cases.

---

**Question 354:** Your classification model has high validation accuracy but terrible calibration (predicted probabilities don't match actual probabilities). What's the issue and solution?

A) Accuracy is all that matters B) Calibration problem: predicted probabilities unreliable; solutions include Platt scaling, isotonic regression, temperature scaling C) Probabilities don't matter D) No solution exists

**Answer: B) Calibration issue requiring post-processing**

**Explanation:** Calibration problems: (1) Model might rank well (good accuracy/AUC) but predicted probabilities misleading; (2) Consequences: if using probabilities for decisions (not just rankings), poor calibration problematic; (3) Calibration methods: Platt scaling (logistic regression on predictions), isotonic regression (monotonic mapping), temperature scaling (neural networks); (4) Validation: reliability diagrams showing calibration; (5) When it matters: probability-based decisions, ensembling, confidence assessment. Calibration is separate concern from discrimination. Both matter for different use cases.

---

**Question 355:** Your model's precision is good but recall is terrible. You've tried class weighting and oversampling without improvement. What else should you investigate?

A) Give up B) Feature engineering: do features contain signal for minority class? Collect more minority examples; try different algorithms C) Recall can't improve D) Ignore recall

**Answer: B) Feature adequacy and data collection**

**Explanation:** Low recall despite imbalance handling: (1) Features might not contain signal to identify minority class; (2) Minority class might be inherently diverse/difficult to characterize; (3) Training examples insufficient even with oversampling (need real examples, not just synthetic); (4) Algorithm limitations: try different approaches (anomaly detection, one-class SVM); (5) Feature engineering: create features specifically discriminating minority class. Imbalance techniques help but can't create signal that doesn't exist in features.

---

**Question 356:** Your ensemble of 5 models performs worse than the best individual model. What went wrong?

A) Ensembles always improve performance B) Models too similar (lack diversity); correlated errors don't cancel; possible implementation error in aggregation C) Perfect ensemble D) Ensembles never work

**Answer: B) Lack of diversity or implementation issues**

**Explanation:** Ensemble failure causes: (1) No diversity: all models make same errors (trained identically, same features); (2) Overfitting ensemble: learning validation set patterns when combining models; (3) Implementation error: incorrect aggregation, feature mismatch; (4) Model quality: including weak models hurts ensemble. Solutions: increase diversity (different algorithms, features, samples); validate ensemble construction; ensure individual models have reasonable quality. Ensembles require diversity to provide value.

---

**Question 357:** Training loss decreases steadily but validation loss starts increasing after epoch 15. What's happening and what should you do?

A) Keep training indefinitely B) Overfitting starting at epoch 15; use early stopping, save model from epoch 15 C) Validation set is wrong D) Ignore validation loss

**Answer: B) Overfitting; implement early stopping**

**Explanation:** Classic overfitting pattern: (1) Training loss improves (model fits training data better); (2) Validation loss worsens (generalization degrades); (3) Divergence indicates memorization; (4) Solution: early stopping—stop training when validation stops improving; save best validation checkpoint (epoch 15); (5) Prevention: stronger regularization, more training data, data augmentation. Don't train past point where validation performance degrades. Best validation performance is optimal stopping point.

---

**Question 358:** Your model shows excellent validation metrics but fails basic sanity checks (predicts negative values for inherently positive quantities). What's wrong?

A) Metrics are all that matters B) Model learned patterns but violates domain constraints; need output constraints or domain validation C) Sanity checks are unnecessary D) Negative values are fine

**Answer: B) Missing domain constraints and validation**

**Explanation:** Domain validation failures: (1) Metrics measure statistical performance but don't enforce domain logic; (2) Example: predicting negative prices, probabilities > 1.0, impossible combinations; (3) Solutions: output constraints (activation functions ensuring valid ranges), post-processing (clip to valid ranges), domain validation in pipeline; (4) Add sanity checks: validate outputs against domain rules. Statistical metrics necessary but insufficient. Domain validity is separate requirement.

---

**Question 359:** You implement data augmentation for training. Validation performance improves but training loss remains high. Is this concerning?

A) Yes, training loss should be near zero B) No, this is expected with augmentation—training becomes harder but generalization improves C) Something is seriously wrong D) Stop using augmentation

**Answer: B) Expected with augmentation; focus on validation performance**

**Explanation:** Data augmentation effects: (1) Training becomes harder (augmented samples are deliberately perturbed); (2) Training loss stays higher than without augmentation; (3) But validation improves (model learns robust features); (4) This is desired tradeoff: sacrifice training performance for better generalization; (5) Monitor validation metrics; training loss less important. Augmentation intentionally makes training difficult to improve generalization. Don't be alarmed by higher training loss if validation improves.

---

**Question 360:** Your model achieves 95% accuracy but has 50% precision for the critical minority class. How do you improve minority class precision specifically?

A) Overall accuracy is sufficient B) Adjust classification threshold higher for stricter positive predictions; add minority class features; improve label quality C) Precision can't improve D) Ignore minority class

**Answer: B) Threshold adjustment and targeted improvements**

**Explanation:** Minority class precision improvement: (1) Increase threshold: predict positive only with higher confidence (trades recall for precision); (2) Feature engineering: add features specifically discriminating minority class; (3) Label quality: audit minority class labels (errors hurt disproportionately with small class); (4) Cost-sensitive learning: penalize false positives specifically; (5) Better training data: collect more minority examples, especially diverse ones. Overall accuracy can be high while minority class performance poor. Target minority class specifically.

---

**Business-Technical Alignment Challenges**

**Question 361:** Your technically excellent model (92% accuracy) gets rejected by business stakeholders who don't trust predictions. What's missing?

A) Accuracy is all that matters B) Explainability and stakeholder engagement; technical metrics don't build trust alone C) Stakeholders are wrong D) Nothing is missing

**Answer: B) Trust requires explainability and engagement beyond metrics**

**Explanation:** Building trust: (1) Explainability: show why predictions make sense; (2) Validation: demonstrate performance on stakeholder-chosen examples; (3) Transparency: show failure modes and limitations; (4) Engagement: involve stakeholders in development; (5) Gradual adoption: pilot with decision support before full automation. Technical excellence necessary but insufficient. Trust comes from understanding, validation, and experience. Involve business throughout, not just at deployment.

---

**Question 362:** Business wants 99% accuracy. Your model achieves 92% and you estimate 95% is maximum achievable with current data. How do you respond?

A) Promise 99% anyway B) Explain technical limitations, show learning curves, discuss cost of improvement, propose alternative approaches or accept 95% ceiling C) Say nothing D) Fake the metrics

**Answer: B) Honest assessment with evidence and alternatives**

**Explanation:** Managing unrealistic expectations: (1) Learning curves show performance plateaus; (2) Error analysis shows remaining errors are fundamental (ambiguous cases, insufficient signal); (3) Cost-benefit: incremental improvements require exponentially more effort; (4) Alternatives: human-in-loop for uncertain cases, focus on reducing high-impact errors; (5) Reframe: is 95% sufficient for business value? Honesty with evidence builds credibility. Promising impossible targets leads to failure. Help stakeholders understand tradeoffs.

---

**Question 363:** Your model optimizes for F1-score (0.85) but business measures success by revenue impact. F1 doesn't correlate perfectly with revenue. What should you do?

A) F1-score is universal metric B) Align on business metric: A/B test measuring actual revenue; optimize model for business objective C) Ignore revenue D) Stick with F1

**Answer: B) Optimize for actual business metric through A/B testing**

**Explanation:** Metric alignment: (1) Technical metrics (F1, AUC) are proxies for business value; (2) Ultimate measure: business outcomes (revenue, costs, customer satisfaction); (3) A/B test models in production measuring business metrics; (4) Optimize model decisions for business objective, not surrogate metric; (5) Sometimes lower F1 produces higher revenue (precision-recall tradeoffs affect different revenue components differently). Technical metrics guide development; business metrics determine success.

---

**Question 364:** Business wants model deployed in 2 weeks. Proper development, testing, and validation needs 8 weeks. How do you respond?

A) Deploy untested model B) Explain risks of rushing; propose MVP for 2-week timeline with limited scope; full deployment in 8 weeks C) Refuse to deploy anything D) Promise 2 weeks for full deployment

**Answer: B) Risk communication and phased approach**

**Explanation:** Timeline negotiation: (1) Explain risks: rushed deployment causes production failures, reputational damage; (2) Phased approach: limited pilot in 2 weeks (decision support, limited users), full deployment in 8 weeks; (3) Technical debt: shortcuts now cost more later; (4) Evidence: show development steps with timeframes; (5) Business impact: what happens if deployed model fails? Balance urgency with quality. Sometimes business needs justify limited scope fast deployment. Full scope needs adequate time.

---

**Question 365:** Your model reduces processing time by 60% but business wants 80% reduction. You could achieve this by lowering accuracy. Should you?

A) Always prioritize speed B) Quantify accuracy-speed tradeoff; let business decide if 80% speedup with lower accuracy preferable to 60% with current accuracy C) Never sacrifice accuracy D) Don't present options

**Answer: B) Present tradeoff; business decides priorities**

**Explanation:** Multi-objective optimization: (1) Create Pareto frontier: show accuracy vs. speed tradeoffs; (2) Options: 60% faster with current accuracy, 80% faster with X% accuracy loss; (3) Business decision: is speed or accuracy more valuable? (4) Context matters: low-stakes applications tolerate accuracy loss for speed; high-stakes don't; (5) Quantify impact: what does accuracy loss cost? Technical team provides options; business chooses based on priorities.

---

**Question 366:** Business stakeholder says: "I can achieve 85% accuracy with simple rules. Why do we need ML for 87% accuracy?" How do you respond?

A) ML is always better B) Quantify 2% accuracy value; consider maintenance complexity; if rules sufficient, maybe ML isn't justified C) Rules are always inferior D) Ignore the question

**Answer: B) Honest assessment: justify ML value or acknowledge rules suffice**

**Explanation:** ML justification: (1) Quantify: what business value does 2% improvement provide? (2) Maintenance: ML might adapt automatically to changes rules can't handle; (3) Scalability: ML might scale to new scenarios rules don't cover; (4) Honesty: if rules are simpler, cheaper, and sufficiently accurate, maybe they're better choice; (5) Justify complexity: ML should provide clear value beyond simple alternatives. Don't use ML for sake of using ML. Simpler solutions are better when they suffice.

---

**Question 367:** Your model performs well on average but business needs guaranteed minimum performance on all customer segments. Current worst segment: 65% accuracy. How do you address this?

A) Average performance is sufficient B) Segment-specific optimization: train segment-specific models, weighted loss, or performance constraints C) Ignore worst segment D) Worst case doesn't matter

**Answer: B) Segment-specific approaches for minimum guarantees**

**Explanation:** Minimum performance requirements: (1) Worst-case constraints matter for fairness, reliability; (2) Segment-specific models: dedicated model for problematic segment; (3) Weighted loss: penalize errors on weak segments heavily; (4) Constrained optimization: optimize average subject to minimum performance constraints; (5) Data collection: gather more data for weak segments. Business might prefer 80% overall with 70% minimum than 85% overall with 65% minimum. Guarantees require explicit attention to worst cases.

---

**Question 368:** Business wants to understand model confidence for risk management (low confidence predictions get human review). Your model outputs class probabilities but they're poorly calibrated. What do you do?

A) Use uncalibrated probabilities anyway B) Calibrate probabilities (Platt scaling, isotonic regression); validate calibration; establish confidence thresholds C) Confidence doesn't matter D) Remove probability outputs

**Answer: B) Calibrate probabilities for reliable confidence estimates**

**Explanation:** Confidence-based workflows: (1) Calibration ensures probabilities reflect true likelihood; (2) Methods: Platt scaling, isotonic regression, temperature scaling; (3) Validation: reliability diagrams confirm calibration; (4) Thresholds: establish confidence cutoffs for human review based on risk tolerance; (5) Monitoring: track confidence distribution over time. Poor calibration makes confidence-based decisions unreliable. Calibration enables trustworthy confidence estimates for risk management.

---

**Question 369:** Your fraud detection model catches 85% of fraud but generates 10,000 false positives daily requiring manual review. Business says this is unsustainable. What do you do?

A) Accept high false positives B) Increase threshold reducing false positives (accepts lower recall); prioritize high-confidence predictions; improve model precision C) Ignore business constraints D) False positives don't matter

**Answer: B) Optimize for business capacity constraints**

**Explanation:** Capacity-constrained optimization: (1) Current: 85% recall creates 10K daily reviews (unsustainable); (2) Adjust threshold: higher threshold reduces false positives to sustainable level (perhaps 1K/day); (3) Tradeoff: might catch 70% fraud instead of 85%; (4) Business decision: is 70% fraud detection with sustainable review volume better than 85% detection with overwhelming volume? (5) Improve model: feature engineering targeting precision. Real-world constraints determine operational points.

---

**Question 370:** Business wants to deploy your model but legal requires documentation of training data sources, model logic, and validation for regulatory audit. What's needed?

A) Just deploy the model B) Comprehensive documentation: data lineage, model card, validation reports, decision logic explanation C) Documentation is unnecessary D) Ignore legal requirements

**Answer: B) Regulatory documentation for compliance**

**Explanation:** Regulatory documentation: (1) Data lineage: source systems, collection dates, preprocessing steps; (2) Model card: intended use, training data characteristics, performance metrics, limitations; (3) Validation reports: test results, fairness analysis, edge cases; (4) Logic documentation: algorithm choice, feature importance, decision rules; (5) Version control: track changes and updates. Regulated industries require comprehensive documentation. Plan for this from project start. Retrofitting documentation is difficult and incomplete.

---

# Chapter 7.0: Phase VI Model Operationalization

Building a great model is only half the battle. Maybe not even half. Production deployment, monitoring, maintenance, and continuous improvement separate successful AI initiatives from research experiments that never deliver business value. Phase VI Model Operationalization (MLOps) bridges the gap between data science and production systems (Paleyes et al., 2022).

## 7.1 Deployment Strategy Selection

**Batch prediction** generates predictions for large datasets on a schedule—nightly, weekly, monthly. The model processes accumulated data in bulk, stores predictions in a database, and applications retrieve precomputed results. This works when immediate predictions aren't required and prediction volume is predictable. A retail demand forecasting system might run weekly batch jobs predicting next month's sales for all products.

**Real-time prediction** generates predictions on-demand for individual requests with low latency requirements—typically milliseconds to seconds. Each API call triggers model inference immediately. This suits applications where predictions must respond to current context—fraud detection evaluating transactions as they occur, recommendation systems personalizing content for each page load, autonomous vehicles processing sensor data continuously (Sculley et al., 2015).

**Streaming prediction** processes continuous data streams, generating predictions as new data arrives. Similar to real-time but handles high-throughput data flows from IoT sensors, application logs, or user activity streams. The model maintains state and updates predictions incrementally rather than starting fresh for each request.

**Edge deployment** runs models on devices (phones, IoT sensors, vehicles) rather than cloud servers. This enables offline operation, reduces latency by eliminating network round-trips, and protects privacy by processing sensitive data locally. Challenges include limited device compute, memory, and battery constraints requiring model optimization.

**Hybrid approaches** combine strategies. Batch predictions might prefill a cache for common cases while real-time predictions handle exceptions. Edge devices might run lightweight models locally with cloud fallback for complex cases.

**Case Example: The Deployment Architecture Evolution**

Jennifer's team built a personalized email recommendation system for a large e-commerce company. Their initial architecture used batch predictions—overnight jobs generated top 10 product recommendations for each customer, stored them in a database, and morning emails displayed those recommendations.

This worked initially but had limitations. Recommendations didn't reflect customer actions during the day—if someone purchased a recommended item, the email still promoted it. New customers added during the day couldn't receive personalized emails until after the next batch run.

Jennifer transitioned to real-time prediction for customer actions (cart abandonment, product views) that triggered immediate personalized emails. Batch predictions continued for scheduled marketing campaigns where overnight computation was acceptable. This hybrid approach balanced freshness for time-sensitive recommendations with cost efficiency for bulk campaigns.

The system eventually incorporated streaming predictions to update customer profiles continuously based on website activity, enabling increasingly personalized real-time recommendations without overwhelming infrastructure with API calls.

## 7.2 Model Monitoring and Maintenance

**Production monitoring** tracks model performance, system health, and data quality continuously. Models degrade over time as data distributions change. Without monitoring, failures go undetected until business impact becomes obvious (Breck et al., 2017).

**Prediction monitoring** logs all inputs, predictions, and confidence scores. Analyzing prediction distributions over time reveals drift—if today's predictions differ systematically from historical patterns, something changed. Confidence score distributions should remain stable; increasing uncertainty suggests the model is encountering unfamiliar data.

**Performance monitoring** measures prediction accuracy on labeled production data. For supervised learning, collect ground truth labels (actual outcomes) and compare with predictions to calculate ongoing precision, recall, and error rates. This requires establishing feedback loops to obtain labels—for fraud detection, investigation results provide labels; for demand forecasting, actual sales provide labels.

**Data quality monitoring** validates input data distributions, checks for missing values, identifies outliers, and flags schema changes. Statistical tests compare current data distributions to training data distributions. Significant divergence triggers alerts. Feature statistics (mean, median, standard deviation, missing rate) should remain within expected ranges.

**System health monitoring** tracks latency, throughput, error rates, resource utilization, and availability. Machine learning systems have complex dependencies—data pipelines, feature stores, model servers, caching layers. Monitor each component and end-to-end metrics.

**Alerting strategies** balance sensitivity and specificity. Too many false alarms cause alert fatigue where critical issues get ignored. Too few alerts allow problems to persist undetected. Use anomaly detection algorithms to identify unusual patterns rather than static thresholds. Implement graduated severity levels—warnings for minor deviations, critical alerts for major issues requiring immediate action.

## 7.3 Drift Detection and Management

**Drift** describes changes in data distributions or relationships between features and targets that degrade model performance over time. Three types matter: data drift, concept drift, and model drift (Lu et al., 2018).

**Data drift** (also called covariate shift) occurs when input feature distributions change while the relationship between features and target remains constant. Imagine training a credit scoring model on data from stable economic conditions and deploying during a recession—application patterns shift dramatically even though the fundamental relationship between financial indicators and default risk is unchanged. The model receives out-of-distribution inputs it wasn't trained to handle.

**Concept drift** happens when the relationship between features and target changes. Consumer preferences evolve, fraud tactics adapt, competitive dynamics shift. A recommendation model trained on 2020 data might fail in 2025 because user tastes changed. The input distributions might look similar, but the patterns connecting features to outcomes are different.

**Model drift** describes general performance degradation regardless of cause. This catchall term encompasses both data and concept drift plus other factors like software version changes, infrastructure modifications, or accumulated technical debt.

**Detection methods** identify drift statistically. **Population stability index (PSI)** quantifies distribution changes for categorical variables. **Kullback-Leibler divergence** measures distribution differences for continuous variables. **Statistical tests** like Kolmogorov-

Smirnov or chi-squared compare current data to reference distributions. **Performance-based detection** monitors prediction accuracy over time; declining performance indicates drift even if the specific cause is unclear.

**Mitigation strategies** handle detected drift. **Retrain models** on recent data to learn current patterns. Establish retraining frequency based on drift rate—monthly, weekly, or triggered by drift detection. **Online learning** updates models continuously with new data rather than batch retraining. **Ensemble methods** combine models trained on different time periods, weighting recent models more heavily. **Human oversight** involves subject matter experts when drift suggests fundamental pattern changes requiring strategic decisions rather than automatic adaptation.

**Case Example: The Seasonal Surprise**

Marcus managed fraud detection models for a payment processor. Three months after deployment, precision dropped from 45% to 28% while recall remained stable. The model flagged more transactions as fraud, but fewer were actually fraudulent—increasing false positive burden without catching more fraud.

Investigation revealed data drift caused by seasonal shopping patterns. The model trained on January-August data, but November-December holiday shopping behavior differed dramatically. Transaction amounts spiked, new merchants appeared, geographic patterns changed. The model interpreted normal holiday behavior as suspicious because it looked different from training data.

Marcus implemented automated drift detection monitoring input feature distributions weekly. When PSI exceeded thresholds for transaction amount, merchant category, or geographic location, the system triggered retraining on recent data incorporating seasonal patterns. The team also created a rolling retraining schedule, updating models monthly rather than waiting for drift to cause problems.

After implementing these changes, performance stabilized. The model adapted to seasonal patterns automatically, maintaining consistent precision and recall throughout the year.

## 7.4 Model Versioning and Rollback

**Model versioning** tracks model iterations enabling reproduction, comparison, and rollback. Production systems run specific model versions; knowing exactly which version is deployed is essential for debugging and compliance (Polyzotis et al., 2018).

**Semantic versioning** uses MAJOR.MINOR.PATCH numbering. Increment MAJOR for incompatible changes (different features, model architecture), MINOR for backward-compatible additions (new features, retrained models), PATCH for bug fixes and small updates. Version 2.3.1 is the third patch to the second minor release of the second major version.

**Artifacts to version** include model weights/parameters, training code, preprocessing code, feature engineering code, configuration files, hyperparameters, training data snapshots or references, evaluation metrics, and documentation. Tools like MLflow, Weights & Biases, or DVC track these components together ensuring complete reproducibility.

**Deployment pipeline** automates model versioning. When training completes, the pipeline tags code, stores model artifacts, logs metrics, and registers the new version. A governance process decides whether to promote versions to staging and production. Automated tests validate new versions before deployment.

**A/B testing infrastructure** enables gradual rollout. Deploy new versions to small traffic percentages initially (1-5%), monitor performance, expand to larger percentages if successful, roll back if problems emerge. This limits blast radius of failures while enabling rapid iteration.

**Rollback procedures** quickly revert to previous versions when problems arise. Automated rollback triggers on performance degradation, error rate spikes, or latency increases. Manual rollback procedures give operators control for nuanced situations. **Blue-green deployment** maintains two identical production environments, switching traffic instantly between them for zero-downtime rollback. **Canary deployment** gradually shifts traffic from old to new versions, monitoring continuously for issues.

**Model registry** provides centralized tracking of all model versions, their deployment status (staging, production, archived), performance metrics, metadata, and lineage. This creates an audit trail showing which models ran when, enabling regulatory compliance and post-incident investigation.

## 7.5 CICD for ML Pipelines

**Continuous Integration/Continuous Deployment (CI/CD)** for machine learning automates testing, building, and deploying ML systems. Traditional software CI/CD focuses on code; ML CI/CD additionally handles data, models, and infrastructure (Amershi et al., 2019).

**Continuous Integration** automatically tests code changes. When developers commit code, automated pipelines run unit tests (testing individual functions), integration tests (testing component interactions), and data validation tests (checking data schemas and distributions). ML pipelines additionally test data processing, feature engineering, and model training code.

**Continuous Training** automatically retrains models when triggering conditions occur—new data arrival, scheduled intervals, or performance degradation. The pipeline fetches recent data, runs preprocessing, trains models, evaluates performance, and registers new versions if they meet quality thresholds.

**Continuous Deployment** automatically deploys models passing validation checks. After training, automated tests evaluate model performance, bias metrics, and resource requirements. Models meeting criteria deploy to staging for additional validation, then production following approval workflows.

**Testing strategies** for ML systems include: **data tests** validating data quality and schema compliance; **model tests** checking prediction distributions, error rates, and fairness metrics; **infrastructure tests** validating deployment configuration, resource availability, and dependency versions; and **integration tests** ensuring end-to-end workflows function correctly.

**Orchestration platforms** like Apache Airflow, Kubeflow, or cloud-native ML services coordinate complex pipelines with interdependent tasks. They handle scheduling, dependency management, failure recovery, and monitoring across distributed infrastructure.

## 7.6 Scalability and Performance Optimization

**Scalability** enables systems to handle increasing load—more predictions, larger models, higher throughput requirements. Poor scalability creates production bottlenecks limiting business value (Crankshaw et al., 2017).

**Horizontal scaling** adds more machines to distribute load. Model serving can parallelize across multiple servers handling prediction requests independently. Load balancers distribute incoming requests across servers. This scales linearly with server count up to practical limits.

**Vertical scaling** uses more powerful machines—faster CPUs, more memory, bigger GPUs. This helps for models requiring large memory or intensive computation but has hard limits and often costs more per unit capacity than horizontal scaling.

**Model optimization** reduces computational requirements without sacrificing accuracy. **Quantization** converts 32-bit floating point weights to 8-bit integers, reducing model size by 4x and speeding inference. **Pruning** removes unnecessary connections in neural networks, eliminating redundant weights. **Knowledge distillation** trains smaller "student" models to mimic larger "teacher" models, achieving similar performance with less compute.

**Caching strategies** store frequent predictions avoiding repeated computation. For batch predictions, cache results keyed by inputs. For real-time predictions with limited input variability, cache common cases. Cache invalidation strategies determine when to refresh stale predictions.

**Asynchronous processing** decouples request handling from prediction computation. API endpoints queue prediction requests and return immediately; worker processes consume the queue, generate predictions, and store results; clients poll for results or receive callbacks. This prevents timeout issues for slow models while maintaining responsiveness.

**GPU acceleration** dramatically speeds deep learning inference. A single modern GPU can match dozens of CPUs for neural network evaluation. Cloud providers offer GPU instances; model serving platforms support GPU deployment.

**Case Example: The Scaling Challenge**

Wei's image classification model worked beautifully in development—500 millisecond latency on a development laptop. Production requirements demanded processing 1000 images per second with sub-100ms latency per image.

Simple deployment on cloud servers achieved 2 images per second per server—requiring 500 servers to handle peak load at prohibitive cost. Wei implemented aggressive optimization. Model quantization reduced size from 100MB to 25MB and halved inference time without accuracy loss. Pruning eliminated 30% of network weights, further reducing compute.

He deployed the optimized model on GPU instances processing 100 images per second per instance—5 instances handled peak load instead of 500. He added aggressive caching for repeated images (product photos processed multiple times), reducing actual inference load by 40%. Horizontal scaling during traffic spikes automatically added instances to maintain sub-100ms latency.

The optimized system handled 10x initial requirements while costing 1/10th of naive deployment. Systematic optimization made production deployment economically viable.

## 7.7 Model Governance and Documentation

**Model governance** establishes policies, procedures, and accountability for responsible AI development and deployment. This becomes increasingly important as AI systems affect more business decisions and face growing regulatory scrutiny (Raji et al., 2020).

**Approval workflows** formalize model promotion through environments. Models move from development to staging to production only after meeting criteria and obtaining required approvals. Documentation, testing results, and risk assessments inform approval decisions.

**Access controls** restrict who can train, deploy, or modify production models. Role-based permissions prevent unauthorized changes. Audit logs track all modifications, creating accountability.

**Model documentation** captures essential information using model cards or similar frameworks. Describe intended use cases, training data characteristics, performance metrics across different populations, known limitations, ethical considerations, and deployment requirements. This enables stakeholders to assess appropriateness for their contexts (Mitchell et al., 2019).

**Risk assessments** evaluate potential harms before deployment. Consider fairness implications across demographic groups, privacy risks from data usage, security vulnerabilities to adversarial attacks, and potential misuse. Document mitigation strategies for identified risks.

**Regulatory compliance** addresses industry-specific requirements. Financial services face fair lending laws and explainability requirements. Healthcare must comply with HIPAA and clinical validation standards. EU AI Act classifies AI systems by risk and imposes requirements accordingly. Governance processes ensure regulatory compliance.

**Ethical review boards** provide oversight for high-stakes applications. Cross-functional teams including ethicists, domain experts, and impacted community representatives review models before deployment, assess societal implications, and recommend safeguards.

## 7.8 Retraining Strategies and Triggers

**Retraining** updates models with new data maintaining performance as patterns evolve. The challenge is determining when and how to retrain balancing freshness versus stability (Žliobaitė, 2010).

**Scheduled retraining** occurs at regular intervals—daily, weekly, monthly. This simple approach ensures models incorporate recent data systematically. Choose frequency based on drift rate and training costs. Rapidly changing domains need frequent retraining; stable domains tolerate infrequent updates.

**Triggered retraining** responds to specific events—performance drops below thresholds, data drift detection, significant business changes, or data volume accumulation. This reactive approach retrains only when necessary, saving compute costs.

**Online learning** updates models continuously as new data arrives. Each prediction or small batch updates model parameters incrementally. This suits applications with continuous feedback and rapidly changing patterns but requires careful learning rate management to prevent catastrophic forgetting of older patterns.

**Active retraining** selectively incorporates examples where the model is uncertain or performed poorly. Rather than retraining on all new data, focus on informative examples providing maximum learning value. This improves sample efficiency, particularly valuable when labeling is expensive.

**Retraining pipelines** automate the process. Trigger conditions invoke data collection, preprocessing, model training, evaluation, and deployment. Automated validation ensures new models meet performance thresholds before replacing production versions. Rollback procedures activate if retrained models underperform.

## 7.9 Production Incident Response

**Incidents** are production failures requiring immediate attention—model outages, severe performance degradation, incorrect predictions causing business harm, or security breaches. Effective incident response minimizes impact and prevents recurrence (Sculley et al., 2015).

**Detection** relies on comprehensive monitoring. Automated alerts notify on-call engineers when thresholds are exceeded. Integrate ML monitoring into existing incident management systems. Severity levels prioritize response—critical incidents page immediately, warnings generate tickets for investigation during business hours.

**Triage** quickly assesses incident scope and impact. How many users affected? What business processes are disrupted? Are predictions incorrect or system unavailable? This guides response urgency and resource allocation.

**Diagnosis** identifies root causes. Check recent deployments—did a new model version introduce problems? Examine data quality—is input data malformed or drifted? Review system metrics—are infrastructure resources exhausted? Access logs and error messages guide investigation.

**Mitigation** quickly reduces harm while root cause investigation continues. Roll back to previous model version, route traffic to backup systems, implement temporary rules bypassing the model, or adjust prediction thresholds reducing false positive impact.

**Resolution** fixes underlying problems. Address code bugs, retrain models on corrected data, scale infrastructure, or implement missing validations. Verify fixes in staging before production deployment.

**Post-incident review** documents what happened, why, timeline of events, impact assessment, root causes, and action items preventing recurrence. Blameless culture encourages learning from failures rather than punishment, promoting transparency and continuous improvement.

## 7.10 Integration with Existing Systems

**System integration** connects ML models with existing business applications, databases, and workflows. Poor integration causes friction limiting AI value realization (Polyzotis et al., 2018).

**API design** provides interfaces for predictions. RESTful APIs handle synchronous requests returning predictions directly. gRPC offers efficient binary protocols for high-throughput systems. GraphQL enables flexible querying for complex data needs. Batch APIs accept multiple predictions reducing network overhead.

**Authentication and authorization** secure prediction endpoints. API keys, OAuth tokens, or JWT authenticate clients. Role-based access controls authorize specific operations. Rate limiting prevents abuse.

**Data format standardization** ensures consistent inputs and outputs. Define JSON schemas for request and response formats. Validate requests against schemas before processing. Provide clear error messages for malformed requests.

**Feature stores** centralize feature computation providing consistent features for training and serving. This prevents training-serving skew where features are computed differently offline versus online causing model failures. Tools like Feast or Tecton manage feature pipelines, storage, and serving.

**Embedding in applications** varies by use case. Recommendation systems return ranked items displayed in user interfaces. Fraud detection flags suspicious transactions for review. Demand forecasting feeds inventory systems. Design integrations matching user workflows and decision processes.

**Fallback strategies** handle failures gracefully. When the model is unavailable, revert to rule-based logic, return default predictions, or queue requests for later processing. Never fail customer-facing features due to ML system issues.

## 7.11 Taking It Forward

Phase VI Model Operationalization transforms models from experiments into production systems delivering business value. Deployment strategies (batch, real-time, streaming, edge) match system requirements with infrastructure capabilities. Comprehensive monitoring tracks performance, data quality, and system health enabling early problem detection.

Drift detection identifies data and concept changes degrading performance over time. Model versioning and rollback procedures provide safety nets when deployments fail. CI/CD pipelines automate testing, training, and deployment reducing manual effort and errors.

Scalability optimizations through caching, model compression, GPU acceleration, and horizontal scaling enable handling production loads economically. Model governance establishes oversight, documentation, and accountability for responsible AI deployment.

Retraining strategies maintain model performance as patterns evolve through scheduled, triggered, or online learning approaches. Production incident response procedures minimize failure impact through rapid detection, diagnosis, mitigation, and resolution.

Integration with existing systems via APIs, feature stores, and thoughtful application embedding ensures ML capabilities enhance rather than disrupt business workflows. The gap between building models and delivering value closes through disciplined MLOps practices.

# 7.12 Practice Questions

## Deployment Architecture Decisions

**Question 371:** You need to deploy a model that must respond within 100ms to user requests. Which deployment architecture is most appropriate?

A) Batch processing overnight B) Real-time REST API with model loaded in memory; horizontally scaled for load C) Email-based predictions D) Manual predictions

**Answer: B) Real-time REST API with in-memory model**

**Explanation:** Real-time latency requirements necessitate: (1) Synchronous API: REST/gRPC endpoints responding immediately; (2) In-memory models: pre-loaded to avoid disk I/O latency; (3) Horizontal scaling: multiple instances behind load balancer; (4) Low-latency infrastructure: optimized model serving (TensorFlow Serving, Triton); (5) Caching: frequently requested predictions cached. Batch processing can't meet real-time needs. Architecture must match latency requirements from the start.

---

**Question 372:** Your model needs to process millions of predictions daily but has no real-time requirements (results needed within 24 hours). What deployment pattern is most cost-effective?

A) Real-time API with constant resources B) Scheduled batch processing using elastic compute that scales to zero when idle C) Manual processing D) Continuous real-time processing

**Answer: B) Scheduled batch processing with elastic compute**

**Explanation:** Batch processing for non-real-time needs: (1) Scheduled execution: run daily/hourly processing jobs; (2) Elastic compute: spin up resources for batch, shut down when complete (cloud cost optimization); (3) Parallel processing: distribute batch across multiple workers; (4) Output storage: write results to database/data lake; (5) Cost efficiency: pay only for processing time, not 24/7 infrastructure. Real-time APIs waste resources when requests aren't constant. Match architecture to business requirements.

---

**Question 373:** You're deploying models to handle traffic with unpredictable spikes (normal load: 100 requests/second, peak: 10,000 requests/second). How should you architect this?

A) Fixed infrastructure for peak load B) Auto-scaling deployment with horizontal pod autoscaling based on request rate and latency C) Handle only normal load D) Reject spike traffic

**Answer: B) Auto-scaling with request-based scaling**

**Explanation:** Handling variable load: (1) Auto-scaling: automatically add/remove instances based on metrics; (2) Scaling metrics: request rate, CPU utilization, latency; (3) Kubernetes HPA: horizontal pod autoscaling responds to load; (4) Serverless options: AWS Lambda, Cloud Run scale automatically; (5) Load balancing: distribute traffic across instances; (6) Cost optimization: scale down during low traffic. Fixed infrastructure for peak wastes 99× resources during normal periods. Auto-scaling matches capacity to demand.

---

**Question 374:** Your model requires GPU for inference but GPU instances are expensive. Only 10% of requests need GPU (complex cases); 90% can use CPU. How should you architect this?

A) GPU for all requests B) Routing architecture: lightweight CPU model for most requests; GPU model for complex cases identified by CPU model or business rules C) CPU only D) Random routing

**Answer: B) Tiered architecture with intelligent routing**

**Explanation:** Cost-optimized tiered deployment: (1) Fast path: CPU model handles simple cases (90%); (2) Slow path: GPU model handles complex cases (10%); (3) Routing logic: CPU model outputs confidence score, low confidence routes to GPU; (4) Cost savings: 90% reduction in GPU usage; (5) Latency: most requests get fast CPU response. Variants: ensemble (CPU model + GPU model for uncertain cases), cascading (simple rules → CPU model → GPU model). Match compute resources to case complexity.

---

**Question 375:** You need to deploy the same model to cloud (for web traffic) and edge devices (for offline operation). How do you maintain consistency?

A) Develop separate models B) Unified model format (ONNX, TensorFlow Lite) deployable to both environments; shared versioning and testing C) Only deploy to cloud D) Only deploy to edge

**Answer: B) Unified model format with multi-environment support**

**Explanation:** Multi-environment deployment: (1) Standard format: ONNX enables cross-platform deployment; (2) Model optimization: quantization for edge devices (reduced size/memory); (3) Version control: same model version deployed everywhere; (4) Testing: validate predictions match across environments; (5) Update synchronization: coordinated rollout to cloud and edge. Separate models risk divergence. Unified approach ensures consistency while handling environment-specific constraints (edge: limited compute, offline; cloud: scalable, always-on).

---

**Question 376:** Your model deployment needs to handle both individual predictions (real-time API) and bulk predictions (batch files). Should you maintain separate infrastructure?

A) Yes, completely separate systems B) Unified model serving handling both: API endpoint for individual requests, batch endpoint accepting files C) Only support one use case D) No infrastructure needed

**Answer: B) Unified serving infrastructure supporting both patterns**

**Explanation:** Hybrid serving architecture: (1) Shared model: same model version serves both patterns; (2) API endpoint: synchronous predictions for real-time requests; (3) Batch endpoint: asynchronous processing for bulk uploads; (4) Internal batching: API aggregates requests into micro-batches for efficiency; (5) Single deployment: reduces maintenance, ensures consistency. Many organizations need both patterns. Unified infrastructure is simpler than separate systems. Modern serving platforms (TensorFlow Serving, Triton) support both.

---

**Question 377:** You're deploying a recommendation model that personalizes based on user history. History lookup adds 200ms latency. How do you optimize?

A) Accept 200ms latency B) Feature store with pre-computed user features; caching frequent users; async feature updates C) Don't use user history D) Increase latency

**Answer: B) Feature store with caching and precomputation**

**Explanation:** Feature serving optimization: (1) Feature store: centralized system for low-latency feature access; (2) Precomputation: calculate user features offline, store in Redis/DynamoDB; (3) Caching: frequently accessed users in memory cache; (4) Async updates: update features asynchronously, serve slightly stale but fast; (5) Fallback: default features when user history unavailable. 200ms latency often unacceptable for real-time systems. Feature engineering must consider serving constraints. Separate feature computation (expensive, offline) from serving (fast, online).

---

**Question 378:** Your model needs to be deployed across multiple regions for low latency to global users. How do you manage multi-region deployment?

A) Single region deployment B) Multi-region deployment with model replication; global load balancing; consistent versioning; regional fallbacks C) Random region selection D) Don't serve global users

**Answer: B) Multi-region deployment with consistency controls**

**Explanation:** Global deployment architecture: (1) Regional replicas: deploy model to multiple regions (US, EU, Asia); (2) Geo-routing: route users to nearest region for low latency; (3) Consistent versioning: same model version across regions; (4) Synchronized updates: coordinated deployment preventing version skew; (5) Fallback: cross-region failover if regional failure; (6) Monitoring: per-region performance tracking. Multi-region reduces latency but adds complexity. Essential for global applications with strict latency requirements.

---

**Question 379:** Your deployment needs blue-green deployment capability (instant rollback). How do you implement this?

A) Direct replacement B) Parallel environments: blue (current production), green (new version); switch traffic atomically; keep blue running for rollback C) Sequential deployment D) No versioning

**Answer: B) Blue-green deployment with instant traffic switching**

**Explanation:** Blue-green deployment: (1) Blue environment: current production version serving traffic; (2) Green environment: new version deployed, tested with synthetic traffic; (3) Traffic switch: atomically route all traffic from blue to green; (4) Monitoring: watch green performance after switch; (5) Instant rollback: switch back to blue if issues detected; (6) Cleanup: decommission blue once green proven stable. Benefits: zero-downtime deployments, instant rollback, full testing before traffic. Cost: temporary 2× resources. Critical for high-availability systems.

---

**Question 380:** You're deploying a large language model requiring significant memory (40GB). How do you handle deployment constraints?

A) Don't deploy large models B) Model optimization: quantization (int8), model compression; specialized hardware (high-memory instances); model sharding across devices C) Use only small models D) Ignore memory requirements

**Answer: B) Optimization and specialized infrastructure**

**Explanation:** Large model deployment: (1) Quantization: int8/int4 quantization reduces memory 2-4×; (2) Model compression: pruning, distillation reduce size; (3) Infrastructure: high-memory GPU instances (A100, H100); (4) Model parallelism: split model across multiple GPUs; (5) Batching: process multiple requests together for efficiency; (6) Caching: KV-cache for transformer efficiency. Large models challenge deployment but optimization techniques enable production use. Balance model size with performance and cost.

---

**Drift Management Scenarios**

**Question 381:** Your production model's performance has gradually declined from 85% to 78% accuracy over 6 months. What's the likely cause and solution?

A) Model randomly degraded B) Data drift: input distribution changed; solution: retrain on recent data, implement continuous monitoring C) Performance decline is normal D) Ignore the decline

**Answer: B) Data drift requiring retraining**

**Explanation:** Gradual performance degradation indicates drift: (1) Concept drift: relationship between features and target changed; (2) Data drift: input distribution shifted (customer behavior evolved, market changed); (3) Label drift: target distribution changed; (4) Solution: retrain on recent data reflecting current patterns; (5) Prevention: automated monitoring detecting drift, triggered retraining; (6) Schedule: establish retraining cadence (monthly, quarterly). Models trained on historical data become stale. Regular retraining maintains performance.

---

**Question 382:** Your drift detection system alerts that input feature distributions have shifted significantly. However, model performance on recent data remains stable. What should you do?

A) Immediately retrain despite stable performance B) Continue monitoring; data drift without performance impact may not require action; investigate why performance maintained C) Ignore drift detection D) Shut down the model

**Answer: B) Monitor but don't act prematurely; investigate stability**

**Explanation:** Data drift without performance degradation: (1) Not all drift is harmful; (2) Model might be robust to this particular shift; (3) Features experiencing drift might not be critical; (4) Important features might be stable; (5) Continue monitoring: performance might degrade later; (6) Investigate: why is model robust? Can we learn from this? Don't retrain unnecessarily—retraining has costs and risks. Act on drift when it impacts performance or when you have evidence it will.

---

**Question 383:** You detect sudden concept drift (abrupt change in feature-target relationship) due to market disruption. Retraining takes 2 weeks. How do you handle this interim period?

A) Continue with degraded model B) Temporary measures: adjust decision thresholds, add business rules, fall back to simpler reliable logic, or human oversight while retraining C) Shut down service D) Ignore drift

**Answer: B) Temporary mitigations while retraining**

**Explanation:** Emergency drift response: (1) Threshold adjustment: compensate for known shift patterns; (2) Business rules: add rules addressing specific disruption patterns; (3) Human-in-loop: increase human review during uncertain period; (4) Fallback logic: revert to stable simpler model or rules; (5) Accelerate retraining: prioritize getting updated model deployed; (6) Communication: inform stakeholders of temporary measures. Sudden drift requires immediate action. Temporary mitigations bridge gap until proper retraining. Balance service continuity with accuracy.

---

**Question 384:** Your model was trained on pre-pandemic data and deployed during pandemic. Performance dropped dramatically. What's the lesson for future deployments?

A) Pandemic was unpredictable; nothing to learn B) Need drift detection and rapid retraining capabilities; consider ensemble including recent data; plan for non-stationarity C) Can't prepare for disruptions D) Pre-pandemic models are fine

**Answer: B) Build adaptability into system design**

**Explanation:** Lessons from disruption: (1) Non-stationarity: world changes, models must adapt; (2) Drift detection: automated monitoring alerts to shifts; (3) Rapid retraining: infrastructure enabling quick model updates; (4) Recent data emphasis: weight recent patterns more heavily; (5) Ensemble approaches: combine stable long-term with adaptive short-term models; (6) Human oversight: ability to quickly add business logic; (7) Testing: validate on recent data before deployment. Assume distributions will shift. Design for adaptability, not static assumptions.

---

**Question 385:** You monitor feature drift and discover that 2 out of 50 features have shifted significantly. Should you retrain the model?

A) Yes, any drift requires retraining B) Depends on feature importance and performance impact; high-importance features require action; low-importance might not C) Never retrain for feature drift D) Remove those features

**Answer: B) Assess importance before deciding**

**Explanation:** Prioritized drift response: (1) Feature importance: check if drifting features are critical to model; (2) Performance impact: measure actual prediction quality degradation; (3) Risk assessment: high-stakes applications have lower tolerance; (4) Cost-benefit: is retraining cost justified by impact? (5) If high-importance features drift: retrain; (6) If low-importance features drift: continue monitoring. Not all drift requires immediate action. Triage based on impact. Focus resources on consequential drift.

---

**Question 386:** Your credit scoring model shows different performance across time periods: Q1: 82%, Q2: 84%, Q3: 81%, Q4: 78%. What strategy should you implement?

A) Accept variable performance B) Quarterly retraining using rolling window of recent data; adaptive learning rate; ensemble approach C) Train once and never update D) Performance variation is irrelevant

**Answer: B) Regular retraining schedule with recent data**

**Explanation:** Handling temporal performance variation: (1) Quarterly retraining: scheduled updates every quarter; (2) Rolling window: train on recent 12-18 months (balance stability with recency); (3) Validation: holdout recent quarter for validation; (4) Monitoring: track performance between retraining cycles; (5) Triggers: additional retraining if performance drops below threshold; (6) Documentation: track model versions with deployment dates. Regular retraining maintains performance in non-stationary environments. Balance stability (don't retrain too often) with recency (don't get stale).

---

**Question 387:** You're designing a drift detection system. Which signals should you monitor?

A) Only model performance metrics B) Input feature distributions, prediction distributions, performance metrics, data quality metrics, label distributions C) Nothing needs monitoring D) Only infrastructure metrics

**Answer: B) Comprehensive drift monitoring across multiple dimensions**

**Explanation:** Holistic drift detection: (1) Input drift: feature distribution shifts (mean, variance, distribution tests); (2) Prediction drift: model output distribution changes; (3) Performance drift: accuracy, precision, recall degradation; (4) Label drift: target distribution shifts; (5) Data quality: missing values, outliers, schema changes; (6) Covariate drift: feature correlations change. Multiple signals provide early warning. Input drift often precedes performance drift. Comprehensive monitoring enables proactive response before customer impact.

---

**Question 388:** Your model retraining is automated but you discover the retrained model sometimes performs worse than the current production model. How do you prevent deploying inferior models?

A) Always deploy retrained models B) Automated validation: test retrained model on holdout set; deploy only if outperforms current production; A/B test before full rollout C) Never retrain automatically D) Deploy all models immediately

**Answer: B) Validation gate preventing inferior model deployment**

**Explanation:** Safe automated retraining: (1) Validation set: hold-out recent data for testing; (2) Performance threshold: new model must exceed current model by minimum margin; (3) A/B testing: gradual rollout comparing new vs. current; (4) Rollback: automatic revert if A/B test shows degradation; (5) Human review: for borderline cases or when validation uncertain; (6) Monitoring: track post-deployment performance. Automation increases velocity but requires safeguards. Never blindly deploy without validation. Retrained models can be worse due to data quality issues, training instability, or overfitting to recent noise.

---

**Question 389:** Your model drift monitoring shows prediction distribution has shifted significantly but you don't have recent labeled data to evaluate performance. How do you assess if this drift is problematic?

A) Assume drift is always bad B) Indirect indicators: user behavior changes, downstream metrics, business outcomes; expedite labeling sample for validation C) Ignore drift without labeled data D) Shut down model

**Answer: B) Use indirect indicators and accelerate label collection**

**Explanation:** Drift assessment without labels: (1) Business metrics: conversion rates, user engagement, revenue (do they align with expectations?); (2) Downstream impact: are downstream systems affected?; (3) User feedback: increased complaints or support tickets?; (4) Prediction analysis: are predictions reasonable? Edge cases?; (5) Expedited labeling: prioritize getting labels for sample of recent data; (6) A/B test: small population on previous model version for comparison. Indirect signals guide urgency while labels are collected. Don't wait passively—actively gather evidence.

---

**Question 390:** You're deploying a model to production with expectation it will need retraining every 3 months. What infrastructure should you establish upfront?

A) No special infrastructure B) Automated retraining pipeline, drift monitoring, validation framework, rollback capability, model versioning C) Manual retraining only D) Never plan for retraining

**Answer: B) Complete MLOps infrastructure for model lifecycle**

**Explanation:** Operationalization infrastructure: (1) Automated pipeline: scheduled or triggered retraining; (2) Drift detection: monitoring alerting to distribution shifts; (3) Validation: automated testing of retrained models; (4) Versioning: track model versions with metadata; (5) Deployment automation: tested deployment process; (6) Rollback: quick reversion if problems; (7) Monitoring: production performance tracking. Plan for full model lifecycle from initial deployment. Retrofitting operations infrastructure is costly. Build operations capabilities before deploying first model.

---

**Production Issue Resolution**

**Question 391:** Your production model suddenly starts returning errors for 10% of requests. Logs show "KeyError: 'new_feature_column'". What happened and how do you fix it?

A) Model is broken B) Training-serving skew: production data schema changed, added column model wasn't trained on; fix: update feature engineering or retrain C) Random errors D) Ignore errors

**Answer: B) Schema mismatch between training and production**

**Explanation:** Production schema issues: (1) Root cause: upstream data source added new column; (2) Model expects specific feature set from training; (3) Immediate fix: modify feature engineering to handle missing columns (skip new columns, use defaults); (4) Long-term fix: retrain model with new schema or establish schema validation; (5) Prevention:

schema validation at ingestion, feature store managing consistency, contract testing between data and model. Training-serving skew is common production failure. Requires coordination between data engineering and ML teams.

---

**Question 392:** Your model inference latency suddenly increased from 50ms to 500ms. What should you investigate first?

A) Model is broken B) Infrastructure issues: check resource utilization (CPU, memory, network), concurrent load, downstream dependencies, cache performance C) Latency increase is normal D) Nothing to investigate

**Answer: B) Infrastructure and dependency investigation**

**Explanation:** Latency troubleshooting: (1) Resource bottlenecks: CPU/memory maxed out, need scaling; (2) Load spike: traffic exceeded capacity; (3) Downstream services: feature lookup or external API slowdowns; (4) Cache issues: cache misses forcing expensive computations; (5) Model loading: if models reload frequently; (6) Network: connectivity issues. Check monitoring dashboards: request rate, resource utilization, downstream latency. Sudden latency increases usually infrastructure-related, not model-related. Systematic troubleshooting from infrastructure up to model logic.

---

**Question 393:** Production model accuracy dropped suddenly overnight. You verify no code changes. What are likely causes?

A) Model spontaneously broke B) Data pipeline failure, upstream data source issues, feature engineering bug, data quality degradation C) Normal variation D) Ignore drop

**Answer: B) Data pipeline and data quality issues**

**Explanation:** Sudden accuracy drops without code changes indicate data problems: (1) Pipeline failure: features not computing correctly; (2) Upstream changes: data source schema or logic changed; (3) Data quality: increased missing values, outliers, errors; (4) Feature staleness: cached features not updating; (5) Investigation: compare current input distributions to historical, check data quality metrics, validate feature engineering. Most production issues are data issues, not model issues. Validate entire data pipeline, not just model code.

---

**Question 394:** Your model is deployed but starts producing extreme predictions (e.g., predicting negative prices). This wasn't caught in validation. What went wrong and how do you prevent recurrence?

A) Validation was perfect B) Validation didn't include production edge cases; add output validation, domain constraints, comprehensive test cases including edge scenarios C) Extreme predictions are acceptable D) Nothing went wrong

**Answer: B) Insufficient validation coverage; need output validation**

**Explanation:** Production validation gaps: (1) Test coverage: validation didn't include edge cases production encounters; (2) Output validation: add runtime checks ensuring outputs valid (positive prices, probabilities in [0,1]); (3) Domain constraints: encode business logic as validation; (4) Comprehensive test suite: adversarial examples, boundary cases, stress tests; (5) Canary testing: gradual production rollout catching issues before full deployment. Validation is never complete. Production always surprises. Defense-in-depth: multiple validation layers including runtime output validation.

---

**Question 395:** Your fraud detection model starts flagging 50× more transactions as fraudulent than normal, overwhelming investigators. What's your immediate response?

A) Process all 50× flagged transactions B) Emergency threshold adjustment to restore normal alert volume; investigate root cause; temporary human review of threshold changes C) Ignore the spike D) Shut down model

**Answer: B) Immediate mitigation then root cause investigation**

**Explanation:** Production emergency response: (1) Immediate: adjust threshold reducing false positives to sustainable level; (2) Communication: alert stakeholders of temporary adjustment; (3) Investigation: what caused spike? Data drift? Feature bug? Model issue?; (4) Parallel work: team fixes root cause while adjusted threshold keeps operations viable; (5) Validation: test fix thoroughly; (6) Restoration: carefully revert to normal threshold once fixed. Balance immediate operational needs with proper diagnosis and fix. Don't let perfect solution delay necessary mitigation.

---

**Question 396:** Your model is in production but you discover a critical data preprocessing bug. The bug has existed for 2 months. What's your response plan?

A) Ignore historical impact B) Assess impact severity, fix bug, retrain model, deploy quickly; analyze historical predictions for impact; communicate to stakeholders C) Leave bug in place D) Delete all data

**Answer: B) Fix, retrain, assess historical impact, communicate**

**Explanation:** Handling discovered production bugs: (1) Severity assessment: how badly did bug affect predictions?; (2) Immediate fix: correct preprocessing logic; (3) Rapid retraining: retrain model with correct preprocessing; (4) Expedited deployment: fast-track corrected model; (5) Historical analysis: reprocess historical data, assess impact on past decisions; (6) Stakeholder communication: transparent about bug, impact, and correction; (7) Process improvement: how did bug escape testing? Enhance validation. Bugs happen. Rapid response, transparent communication, and learning are critical.

---

**Question 397:** Your model deployment uses a third-party API for feature enrichment. The API goes down, causing prediction failures. How should your system handle this?

A) Fail all predictions when API down B) Graceful degradation: fallback to predictions without enriched features, use cached values, or default values; monitor and alert C) Wait indefinitely for API D) Shut down service

**Answer: B) Graceful degradation with fallbacks**

**Explanation:** Resilient architecture: (1) Fallback strategy: make predictions with available features when enrichment fails; (2) Cache: recently fetched enrichment data serves as temporary fallback; (3) Default values: sensible defaults for missing features; (4) Monitoring: alert when fallback mode active; (5) Timeout: don't wait indefinitely for API; (6) Retry logic: with exponential backoff for transient failures. Production systems depend on unreliable external services. Graceful degradation maintains service continuity with potentially reduced accuracy. Better than complete failure.

---

**Question 398:** You receive alert that model predictions are biased toward one class. You investigate and find training data was imbalanced but validation set was artificially balanced. What went wrong?

A) Nothing wrong B) Validation set didn't represent production data distribution; need realistic validation matching production C) Balanced validation is always correct D) Training data should be balanced

**Answer: B) Validation-production distribution mismatch**

**Explanation:** Validation representativeness: (1) Validation must match production distribution; (2) Artificially balanced validation gave misleading performance estimates; (3) Model optimized for balanced distribution, deployed to imbalanced production; (4) Fix: create validation set matching production distribution; (5) Retrain: optimize for actual deployment conditions; (6) Testing: validate on realistic data before deployment. Validation should simulate production as closely as possible. Artificial validation setups create false confidence. Test on representative data.

---

**Question 399:** Your model container crashes in production with "Out of Memory" errors. The same model runs fine in testing environment. What's different and how do you fix it?

A) Testing was wrong B) Production differences: higher concurrent requests, larger input sizes, memory leaks; fix: increase memory limits, optimize model, add request limits C) Production environment is broken D) Ignore crashes

**Answer: B) Production-specific resource constraints**

**Explanation:** Production vs. testing differences: (1) Concurrency: production handles multiple simultaneous requests; (2) Input variation: production inputs larger/more complex than test cases; (3) Memory leaks: long-running processes accumulate memory; (4) Solutions: increase memory allocation, implement request batching, add input size validation, model optimization (quantization), rate limiting preventing overload; (5) Load testing: stress test before production catching resource issues. Testing environments often don't replicate production conditions. Production-like load testing essential.

---

**Question 400:** Your model accuracy metrics look fine but business reports model recommendations are repetitive and not diverse. What's the issue?

A) Accuracy is all that matters B) Optimization for accuracy alone ignores diversity; need multi-objective optimization including diversity metrics C) Repetitive recommendations are fine D) Business is wrong

**Answer: B) Single metric optimization missing business requirement**

**Explanation:** Multi-dimensional quality: (1) Accuracy alone insufficient for recommendation systems; (2) Need diversity: varied recommendations, novelty, serendipity; (3) User experience: repetitive recommendations hurt engagement despite "accuracy"; (4) Solution: multi-objective optimization balancing accuracy and diversity; (5) Evaluation: metrics beyond accuracy (diversity, coverage, serendipity); (6) A/B testing: user engagement metrics validate recommendations. Technical metrics don't fully capture user experience. Business feedback reveals blind spots. Balance multiple objectives for real-world success.

---

**Monitoring Strategy Selection**

**Question 401:** What should you monitor in production for an ML system?

A) Only model accuracy B) Model performance metrics, prediction distributions, input data quality, system metrics (latency, throughput), business metrics C) Nothing needs monitoring D) Only infrastructure metrics

**Answer: B) Comprehensive monitoring across all system layers**

**Explanation:** Holistic ML monitoring: (1) Model metrics: accuracy, precision, recall on labeled sample; (2) Data quality: missing values, schema changes, outliers; (3) Predictions: distribution shifts, extreme values, confidence scores; (4) System performance: latency, throughput, error rates, resource utilization; (5) Business outcomes: conversion rates, revenue, user satisfaction; (6) Fairness: per-segment performance. Different metrics serve different purposes: data quality catches pipeline issues early, system metrics ensure SLA compliance, business metrics validate value. Multi-layer monitoring enables rapid issue detection and diagnosis.

---

**Question 402:** How frequently should you evaluate model performance on labeled production data?

A) Once at deployment, never again B) Depends on drift velocity and labeling cost; high-drift domains need frequent evaluation (daily/weekly); stable domains less frequent (monthly/quarterly) C) Every second D) Never evaluate production performance

**Answer: B) Frequency matches drift characteristics and business criticality**

**Explanation:** Performance evaluation frequency: (1) High-drift environments: daily/weekly evaluation with continuous labeling; (2) Stable environments: monthly/quarterly sufficient; (3) Labeling cost: balance evaluation frequency with cost of obtaining labels; (4) Sampling: don't need labels for all data, stratified sample sufficient; (5) Triggered evaluation: when drift detected, evaluate immediately; (6) Critical systems: more frequent evaluation regardless of cost. No universal frequency. Match monitoring intensity to business risk and environment stability.

---

**Question 403:** Your production model serves 1M predictions daily. You can only label 1,000 examples per week for monitoring. How should you select which examples to label?

A) Random sampling B) Strategic sampling: high-confidence predictions (validation), low-confidence predictions (edge cases), diverse representation, recent data C) First 1,000 chronologically D) Don't label any

**Answer: B) Strategic sampling maximizing information**

**Explanation:** Intelligent sample selection: (1) Stratified sampling: ensure all prediction ranges and classes represented; (2) Uncertainty sampling: prioritize low-confidence predictions (likely errors); (3) Temporal coverage: recent data most relevant for drift detection; (4) Edge cases: unusual inputs revealing failure modes; (5) Random baseline: some random sampling for unbiased estimates; (6) High-impact: prioritize high-value transactions. Limited labels are precious. Strategic sampling provides more information than pure random sampling. Balance multiple objectives.

---

**Question 404:** What alerting thresholds should you set for production model monitoring?

A) Alert on any change B) Statistical significance: confidence intervals, sustained trends over time, not noise; business impact thresholds C) Never alert D) Random thresholds

**Answer: B) Statistically significant changes with business context**

**Explanation:** Effective alerting: (1) Statistical significance: distinguish real changes from noise (confidence intervals, statistical tests); (2) Sustained trends: alert when degradation persists over time, not transient fluctuations; (3) Business impact: threshold based on business tolerance (accuracy drop of 2% might be critical or acceptable depending on application); (4) Severity levels: warning (investigation needed) vs. critical (immediate action); (5) Avoid alert fatigue: too many false alarms get ignored. Good alerting balances sensitivity (catch real issues) with specificity (avoid false alarms).

---

**Question 405:** Your model monitoring dashboard shows dozens of metrics. Stakeholders complain it's overwhelming. How should you organize monitoring?

A) Show all metrics equally B) Tiered approach: executive dashboard (key business metrics), operational dashboard (system health), debugging dashboard (detailed diagnostics) C) Remove all monitoring D) Add more metrics

**Answer: B) Layered dashboards for different audiences**

**Explanation:** Effective monitoring organization: (1) Executive view: business KPIs, high-level health indicators; (2) Operations view: system performance, alerts, recent changes; (3) Debug view: detailed metrics for troubleshooting; (4) Drill-down: start high-level, drill into details as needed; (5) Audience-specific: different roles need different information; (6) Key metrics: highlight most important metrics, minimize noise. Overwhelming dashboards are ignored. Progressive disclosure: high-level overview with detail available on demand. Match monitoring complexity to audience needs.

---

**Question 406:** You're monitoring prediction latency and see occasional spikes to 5× normal latency. These spikes are rare (0.1% of requests) but concerning. Should you alert?

A) Yes, alert immediately on every spike B) Alert if spikes become frequent or sustained; investigate root cause; consider percentile metrics (p99) rather than max C) Ignore rare spikes D) Shut down system

**Answer: B) Monitor trends; use percentile metrics; investigate without immediate alerts**

**Explanation:** Handling outliers in monitoring: (1) Rare spikes normal in production systems; (2) Percentile metrics: p95, p99 latency more informative than max (robust to outliers); (3) Trend monitoring: alert if spike frequency increases; (4) Investigation: understand causes of spikes (GC pauses, slow dependencies, large inputs); (5) SLA alignment: alert based on service level objectives, not every outlier. Alerting on every anomaly creates alert fatigue. Focus on patterns and trends indicating systematic issues. Balance awareness with noise reduction.

---

**Question 407:** Your monitoring shows model performance varies by time of day (better during day, worse at night). What should you investigate?

A) Time-based performance variation is normal; ignore B) Investigate: different user populations, feature availability differences, data pipeline timing, load variations C) Performance variation doesn't matter D) Shut down model at night

**Answer: B) Investigate causes of temporal patterns**

**Explanation:** Temporal performance investigation: (1) User demographics: different populations at different times; (2) Feature availability: features may have different latency or availability by time; (3) Data freshness: batch feature updates might create gaps; (4) Load effects: system performance under different loads; (5) Upstream dependencies: external services have variable performance; (6) Solution depends on cause: might need separate models for different periods, improve feature availability, or accept variation. Temporal patterns reveal system dependencies. Understanding causes informs whether action needed.

---

**Question 408:** You're designing monitoring for a fraud detection model. What business metrics should you track beyond model accuracy?

A) Only accuracy matters B) Fraud caught (true positives), fraud missed (false negatives), false alarm rate, investigation costs, fraud losses prevented, user friction C) No business metrics needed D) Only technical metrics

**Answer: B) Comprehensive business impact metrics**

**Explanation:** Business-aligned monitoring: (1) Fraud caught: absolute number and percentage of fraud detected; (2) Fraud losses: monetary value of prevented fraud; (3) False alarms: legitimate transactions flagged (investigation cost, user friction); (4) Investigation capacity: number of reviews vs. capacity; (5) User experience: legitimate users incorrectly blocked; (6) Net benefit: fraud prevented minus false alarm costs. Technical metrics (precision, recall) are proxies. Business metrics measure actual value. Both needed: technical for model health, business for ROI validation.

---

**Question 409:** Your production monitoring shows input features have shifted but model performance seems stable. How long should you continue monitoring before taking action?

A) Never take action if performance stable B) Continue elevated monitoring; set review period (2-4 weeks); if performance remains stable, adjust baseline; if degrades, retrain C) Immediately retrain D) Stop monitoring

**Answer: B) Extended monitoring with defined review period**

**Explanation:** Drift response with stable performance: (1) Extended observation: monitor for sustained period confirming stability; (2) Review period: define window (2-4 weeks) balancing patience with risk; (3) Baseline adjustment: if stability confirmed, update drift detection baseline; (4) Heightened vigilance: increase monitoring frequency during period; (5) Prepare retraining: plan retraining if performance degrades; (6) Investigation: understand why model is robust to this drift. Not all drift requires immediate action. Sustained monitoring distinguishes stable adaptation from delayed degradation.

---

**Question 410:** You need to monitor fairness across demographic groups but don't have demographic labels in production data. How do you monitor fairness?

A) Can't monitor without labels B) Proxy methods: infer demographics from correlated features, periodic audit studies, user surveys, A/B testing across regions C) Fairness monitoring is unnecessary D) Ignore fairness

**Answer: B) Indirect monitoring and periodic deep-dive studies**

**Explanation:** Fairness monitoring without direct labels: (1) Proxy features: location, language, behavior patterns may correlate with demographics; (2) Periodic audits: collect demographic data in separate studies for deep analysis; (3) Regional analysis: geography often correlates with demographics; (4) User surveys: optional demographic self-reporting; (5) Disparate impact testing: even without perfect demographic data, analyze for unexplained performance variation; (6) A/B testing: test model variations across segments. Fairness monitoring is harder without labels but not impossible. Multiple indirect approaches provide useful signals.

---

Maintenance Planning Situations

**Question 411:** Your model is deployed to production. How should you plan for ongoing maintenance?

A) No maintenance needed after deployment B) Scheduled retraining (quarterly), monitoring and alerting, incident response process, performance review cycles, drift management C) Maintenance is unnecessary D) Check model once per year

**Answer: B) Comprehensive maintenance plan across multiple dimensions**

**Explanation:** ML system maintenance: (1) Retraining schedule: regular updates (monthly, quarterly) based on drift characteristics; (2) Monitoring: continuous performance and drift tracking; (3) Incident response: on-call rotation, runbooks for common issues; (4) Performance reviews: quarterly business reviews assessing value; (5) Technical debt: regular refactoring, dependency updates; (6) Documentation: keep runbooks, architecture docs current; (7) Capacity planning: scale infrastructure as usage grows. Models require ongoing maintenance like any production system. Plan resources for maintenance from initial deployment.

---

**Question 412:** Your organization has 20 ML models in production. Each requires monthly retraining. How should you structure retraining?

A) Manual retraining for each B) Automated retraining infrastructure: orchestration system, shared pipelines, monitoring, standardized validation C) Stop retraining models D) Retrain randomly

**Answer: B) Centralized automated retraining infrastructure**

**Explanation:** Scaled retraining operations: (1) Orchestration: Airflow, Kubeflow managing scheduled retraining; (2) Shared infrastructure: common training pipeline reusable across models; (3) Automated validation: standardized testing before deployment; (4) Monitoring: track retraining success/failure; (5) Resource management: schedule training jobs efficiently; (6) Standardization: common patterns reducing maintenance burden. Manual retraining doesn't scale. Centralized automation enables managing many models efficiently. Invest in infrastructure enabling scale.

---

**Question 413:** A critical model in production requires a major architecture change (moving from sklearn to deep learning). How do you manage this transition?

A) Replace model immediately B) Parallel deployment: run both models side-by-side; validate new model extensively; gradual traffic migration; maintain rollback capability C) Never change deployed models D) Deploy without testing

**Answer: B) Careful parallel deployment with extensive validation**

**Explanation:** Major model transitions: (1) Parallel running: deploy new model alongside existing, don't replace immediately; (2) Shadow mode: new model makes predictions but doesn't serve users, validate correctness; (3) A/B testing: gradual traffic shift (5% → 25% → 50% → 100%); (4) Extended validation: monitor for several weeks before full migration; (5) Rollback plan: keep old model ready for instant rollback; (6) Business alignment: stakeholder awareness of change and potential risks. Major changes carry high risk. Gradual, validated migration minimizes disruption.

---

**Question 414:** Your team maintains 15 ML models. Different models need retraining at different frequencies (weekly to quarterly). How do you manage this complexity?

A) Retrain all models together B) Model registry with metadata: retraining cadence, last training date, performance metrics; automated scheduling based on model-specific needs C) Ignore retraining frequency differences D) Random retraining

**Answer: B) Model registry with differentiated retraining schedules**

**Explanation:** Model portfolio management: (1) Model registry: central catalog with metadata for each model; (2) Retraining schedules: customized per model based on drift characteristics; (3) Triggers: scheduled (time-based) and event-based (performance drops); (4) Prioritization: criticality-based prioritization when resources constrained; (5) Dependency tracking: retrain dependent models together; (6) Reporting: dashboard showing all model statuses and next retraining. Different models have different needs. Centralized management with model-specific configuration enables efficient operations at scale.

---

**Question 415:** Your model's dependencies (libraries, frameworks) are aging. Security vulnerabilities are being reported. How do you handle dependency updates?

A) Never update dependencies B) Regular dependency updates: test in staging, validate model behavior unchanged, coordinated updates across models C) Ignore security vulnerabilities D) Update immediately without testing

**Answer: B) Planned dependency maintenance with validation**

**Explanation:** Dependency management: (1) Regular updates: quarterly review of dependencies; (2) Security monitoring: track CVEs affecting your stack; (3) Testing: validate model predictions unchanged after updates (numerical stability); (4) Staging validation: test updated environment before production; (5) Coordinated updates: update multiple models using same dependencies together; (6) Rollback capability: quick revert if issues discovered; (7) Documentation: track dependency versions with models. Stale dependencies accumulate security and compatibility risks. Regular maintenance prevents technical debt accumulation.

---

**Question 416:** You discover your model training code has poor documentation and only one person understands it. That person is leaving the company. What should you do?

A) Hope for the best B) Knowledge transfer: documentation sprint, code walkthrough, runbook creation, cross-training other team members C) Documentation is unnecessary D) Hire replacement before transfer

**Answer: B) Urgent knowledge transfer and documentation**

**Explanation:** Knowledge continuity: (1) Documentation sprint: comprehensively document training pipeline, feature engineering, deployment process; (2) Pair programming: current owner works with others on codebase; (3) Runbooks: step-by-step guides for common tasks; (4) Video walkthroughs: recorded explanations of complex components; (5) Code cleanup: improve code clarity while documenting; (6) Test coverage: comprehensive tests serve as documentation; (7) Knowledge distribution: ensure multiple people understand system. Bus factor of one is critical risk. Proactively distribute knowledge. Don't wait for crises.

---

**Question 417:** Your model performance has been stable but you haven't retrained in a year. Stakeholder suggests retraining anyway. Should you?

A) Never retrain if performance is stable B) Evaluate evidence: check for hidden drift, analyze recent data, consider if patterns evolved even without apparent performance degradation C) Retrain annually regardless of need D) Ignore stakeholder suggestion

**Answer: B) Investigate whether retraining beneficial despite stable metrics**

**Explanation:** Proactive maintenance assessment: (1) Hidden drift: current metrics might miss subtle degradation; (2) Data analysis: do recent patterns differ from training data?; (3) Model staleness: even without obvious drift, models benefit from recent data; (4) Feature evolution: are there new features that would improve model?; (5) Cost-benefit: is retraining cost justified by potential improvement?; (6) Validation: retrain and compare to current model; deploy if better. Stability doesn't mean optimal. Evaluate whether improvements possible. Proactive retraining can prevent future issues and capture improvements.

---

**Question 418:** Your model maintenance requires significant manual effort: data extraction, preprocessing, training, validation, deployment. How do you reduce maintenance burden?

A) Accept manual effort B) Automation: end-to-end MLOps pipeline automating all steps; reduce human-in-loop to approval/monitoring C) Hire more people for manual work D) Stop maintenance

**Answer: B) Invest in automation infrastructure**

**Explanation:** Automation for maintainability: (1) End-to-end pipeline: data extraction → preprocessing → training → validation → deployment fully automated; (2) Orchestration: Airflow/Kubeflow managing workflow; (3) Human touchpoints: approvals for deployment, investigation of anomalies; (4) Monitoring: automated alerts reducing need for constant watching; (5) Self-service: enable stakeholders to trigger retraining, view metrics; (6) ROI: automation investment pays off with reduced ongoing effort. Manual processes don't scale and are error-prone. Automation enables maintaining many models efficiently.

---

**Question 419:** You maintain multiple models serving different regions. Each region has slightly different data characteristics. Should you maintain separate models or one global model?

A) Always use one global model B) Depends on performance vs. maintenance tradeoff: test global vs. regional models; choose based on performance gain vs. maintenance complexity C) Always use regional models D) Don't serve multiple regions

**Answer: B) Empirical testing of tradeoffs**

**Explanation:** Global vs. regional model decision: (1) Test both approaches: train global model and region-specific models; (2) Performance comparison: do regional models significantly outperform global?; (3) Maintenance cost: regional models multiply maintenance burden; (4) Hybrid: global model with regional features/adjustments; (5) Transfer learning: pretrain global, fine-tune per region; (6) Business context: some regions might be more critical. Significant performance gain (>5%) might justify regional models; small gain (<2%) probably not worth complexity. Evidence-based decisions.

---

**Question 420:** Your model maintenance runbook is 50 pages long and frequently out of date. Team members struggle to follow it. How do you improve operational documentation?

A) Make runbook longer B) Simplify and maintain: focus on common scenarios, automate documented procedures, living documentation close to code, regular reviews C) Delete all documentation D) Documentation is unnecessary

**Answer: B) Simplification, automation, and maintenance culture**

**Explanation:** Effective operational documentation: (1) Prioritize: focus on common scenarios, emergency procedures; (2) Automation: script what can be automated, document reducing need for runbooks; (3) Living docs: documentation in repository, updated with code changes; (4) Clear structure: quick-reference for common tasks, deep-dives for complex scenarios; (5) Regular reviews: quarterly documentation updates; (6) Examples: concrete examples better than abstract instructions; (7) Ownership: clear owners for each document. Long, stale documentation is worse than no documentation. Keep documentation useful by keeping it current and concise.

# Chapter 8.0: Trustworthy AI and Ethics

AI systems affect people's lives in profound ways—determining who gets loans, who receives medical treatment, whose resume gets seen by recruiters, and who gets flagged by law enforcement. When these systems fail ethically, the consequences can be devastating. Trustworthy AI isn't a nice-to-have feature; it's fundamental to deploying AI responsibly and maintaining the social license to operate (Floridi et al., 2018).

## 8.1 The Five Pillars

The CPMAI framework identifies five pillars of Trustworthy AI: ethical, responsible, transparent, governed, and explainable. These aren't separate concerns—they're interconnected requirements that work together to ensure AI systems serve human values (PMI, 2025).

**Ethical AI** ensures systems align with human values and moral principles. This means considering the impact on all stakeholders, not just optimizing metrics. An AI system that maximizes profit while harming vulnerable populations isn't ethical, even if technically successful.

**Responsible AI** establishes accountability for AI outcomes. Someone must answer for what AI systems do. This requires clear ownership, documented decision-making processes, and mechanisms for addressing harms when they occur. Responsibility can't be diffused across organizations or hidden behind algorithmic opacity.

**Transparent AI** makes processes, data sources, and decision logic visible to appropriate stakeholders. Transparency doesn't mean revealing proprietary algorithms to competitors; it means providing sufficient information for stakeholders to understand how systems work and assess their appropriateness.

**Governed AI** operates within established policies, procedures, and oversight mechanisms. Governance defines who can deploy AI, under what conditions, with what safeguards, and subject to what review. Strong governance prevents rogue deployments and ensures consistent standards.

**Explainable AI** provides understandable rationales for decisions. Stakeholders affected by AI should comprehend why specific outcomes occurred. Explainability enables validation, debugging, improvement, and informed consent (Arrieta et al., 2020).

These pillars reinforce each other. Explainability enables transparency. Governance enforces responsibility. Ethical principles guide all pillars. Weakness in any pillar compromises the entire framework.

## 8.2 Detecting and Mitigating Bias

**Bias** in AI systems produces systematically unfair outcomes for specific groups. Bias can enter through training data reflecting historical discrimination, through algorithm design encoding assumptions, through deployment in contexts different from training environments, or through interaction with human decision-makers (Mehrabi et al., 2021).

**Data bias** occurs when training data doesn't represent the population the system will serve. Historical hiring data overrepresenting men in technical roles trains models that discriminate against women. Criminal justice data reflecting biased policing patterns trains recidivism models that unfairly target minorities. Financial data excluding certain demographics trains credit models that deny them opportunities.

**Measurement bias** arises from how features are defined and captured. Resume screening systems using "cultural fit" as a proxy for performance can encode biased assumptions about what "fit" means. Healthcare systems using healthcare costs as a proxy for health needs systematically underestimate needs of Black patients who historically received less care (Obermeyer et al., 2019).

**Aggregation bias** happens when models trained on aggregate data fail to perform equally across subgroups. A model optimized for average performance might work well for the majority but poorly for minorities. Medical diagnostic systems trained predominantly on data from one demographic often underperform on others.

**Evaluation bias** emerges when test datasets don't match deployment populations or when metrics don't capture fairness concerns. A facial recognition system achieving 95% accuracy overall but only 60% for dark-skinned women has severe bias despite good average performance.

**Mitigation strategies** start with diverse, representative training data. Collect data intentionally covering all relevant populations. When historical data contains bias, techniques like re-sampling, re-weighting, or synthetic data generation can help. Document data limitations explicitly.

**Fairness constraints** during training enforce equality across protected groups. Demographic parity requires similar prediction rates across groups. Equalized odds requires similar true positive and false positive rates. Equal opportunity focuses on equal true positive rates. Different fairness definitions suit different contexts; choosing appropriately requires understanding the specific harm being prevented (Chouldechova, 2017).

**Post-processing adjustments** modify model outputs to achieve fairness criteria. Adjust decision thresholds differently for different groups to equalize outcomes. While sometimes necessary, post-processing doesn't address root causes and may reduce overall performance.

### Case Example: The Hiring Algorithm Fix

A large tech company's resume screening AI showed troubling patterns. Women and applicants with names suggesting minority backgrounds received lower scores than equally qualified white male applicants. Investigation revealed multiple bias sources.

Training data came from ten years of historical hiring decisions during a period when the company predominantly hired white men. The model learned that being male and white correlated with being hired, not because of qualifications but because of past hiring biases.

Feature engineering introduced bias. The system downweighted applicants from women's colleges despite strong academic performance there. It penalized career gaps that disproportionately affected mothers. It favored candidates with hobbies like "competitive programming" that correlated with gender due to cultural factors, not job performance.

The team rebuilt the system with debiased data, removing hiring decision outcomes during biased periods and using only objective performance metrics from employees' first two years. They eliminated features with disparate impact across protected groups. They implemented fairness constraints ensuring similar hiring rates across demographics for equivalent qualifications.

Post-deployment monitoring tracked outcomes by demographic group, flagging disparities for investigation. An ethics review board evaluated model updates before deployment. The revised system achieved both better fairness and better prediction of actual job performance.

## 8.3 Fairness in AI Systems

**Fairness** in AI systems is complex because different stakeholders define fairness differently, and various fairness criteria conflict mathematically. You often can't satisfy all fairness definitions simultaneously (Kleinberg et al., 2017).

**Individual fairness** requires similar individuals to receive similar outcomes. This sounds obvious but poses challenges: Who determines which features are relevant for similarity? Two loan applicants with identical credit scores but different races should receive similar treatment—but what about differences in employment history or debt-to-income ratios?

**Group fairness** requires equal treatment of protected groups defined by characteristics like race, gender, age, or disability. Multiple group fairness definitions exist, each capturing different intuitions about equality.

**Demographic parity** (also called statistical parity) requires equal acceptance rates across groups. If 10% of male applicants are approved for loans, 10% of female applicants should be approved. This prevents disparate impact but ignores group differences in qualifications.

**Equal opportunity** requires equal true positive rates across groups. Among qualified applicants, approval rates should be equal across groups. This ensures qualified individuals have equal chances regardless of group membership.

**Predictive parity** requires equal precision across groups. Among those predicted positive, success rates should be equal across groups. This ensures predictions are equally reliable for all groups.

These criteria conflict. You can't generally achieve demographic parity, equal opportunity, and predictive parity simultaneously when base rates differ across groups. Choosing among them requires understanding the specific context and which harms matter most (Chouldechova, 2017).

**Calibration** requires predicted probabilities to match observed outcomes across groups. If the model predicts 70% probability of loan repayment for a group, 70% should actually repay. Calibration ensures predictions mean the same thing across groups.

**Procedural fairness** focuses on decision-making processes rather than outcomes. Were decisions made using appropriate information? Were individuals given opportunity to contest decisions? Were decision processes consistent and transparent? Procedural fairness matters even when outcomes are unequal if differences reflect legitimate factors.

## 8.4 Privacy Protection Frameworks

**Privacy protection** prevents unauthorized access to personal information while enabling legitimate data use for AI. This balance becomes increasingly difficult as AI systems grow more sophisticated at extracting insights from data (Voigt & Von dem Bussche, 2017).

**GDPR** (General Data Protection Regulation) establishes comprehensive privacy requirements for data about European Union residents. Key principles include purpose limitation (data used only for stated purposes), data minimization (collecting only necessary data), storage limitation (retaining data only as long as needed), and accountability (documenting compliance).

**Consent requirements** under GDPR mandate explicit, informed, freely given consent for data collection and processing. Consent must be specific to each purpose—blanket consent doesn't suffice. Individuals can withdraw consent anytime. For AI systems, this means documenting what data is used, why, and obtaining appropriate consent.

**Right to explanation** gives individuals the right to understand automated decisions significantly affecting them. This directly impacts AI systems making high-stakes decisions about employment, credit, insurance, or legal matters. Organizations must provide meaningful explanations of decision logic and consequences.

**Data subject rights** include access to personal data, correction of inaccuracies, deletion (right to be forgotten), restriction of processing, and data portability. AI systems must support these rights through data tracking, version control, and retraining capabilities when data is deleted.

**Privacy by design** embeds privacy protections into system architecture from inception, not bolted on afterward. This includes minimizing data collection, pseudonymizing data when possible, encrypting data in transit and at rest, and limiting access to need-to-know basis.

**Differential privacy** adds calibrated noise to data or query results preventing inference about individuals while preserving aggregate statistics. This mathematical framework provides provable privacy guarantees but requires careful parameter tuning to balance privacy and utility (Dwork & Roth, 2014).

**Federated learning** trains models without centralizing data. Devices or institutions train local models on their data, share only model updates with a central server, which aggregates updates into a global model. This protects privacy by keeping raw data distributed while enabling collaborative learning.

## 8.5 Regulatory Compliance Landscape

**EU AI Act** classifies AI systems by risk level and imposes requirements accordingly. The regulation, finalized in 2024, represents the world's first comprehensive AI regulatory framework (European Commission, 2024).

**Prohibited practices** ban AI systems with unacceptable risk: social scoring by governments, real-time remote biometric identification in public spaces (with limited exceptions), emotion recognition in workplace and education, and exploitative manipulation. These applications are illegal regardless of safeguards.

**High-risk systems** include AI used in critical infrastructure, education and vocational training, employment, essential services, law enforcement, migration and border control, and justice administration. These systems face strict requirements: risk management systems, high-quality training data, technical documentation, transparency to users, human oversight, and accuracy/robustness/cybersecurity standards.

**Limited-risk systems** like chatbots must disclose their AI nature to users. Transparency requirements ensure users know they're interacting with AI and can make informed decisions.

**Minimal-risk systems** like AI-powered video games face no specific requirements beyond general law. This constitutes the majority of AI applications.

**NIST AI Risk Management Framework** provides voluntary guidance for U.S. organizations developing and deploying AI. The framework identifies four functions: govern (culture and processes for responsible AI), map (understand context and impacts), measure (assess risks and impacts), and manage (prioritize and respond to risks) (NIST, 2023).

**Sector-specific regulations** apply beyond horizontal AI laws. Healthcare AI faces HIPAA and FDA requirements. Financial services AI must comply with fair lending laws and SEC regulations. Automotive AI adheres to vehicle safety standards. Understanding applicable regulations requires domain expertise beyond general AI knowledge.

## 8.6 Ethical Use Case Evaluation

**Ethical evaluation** assesses whether AI applications align with societal values before deployment. Not every technically feasible application is ethically acceptable. Structured evaluation frameworks help organizations make principled decisions (Jobin et al., 2019).

**Stakeholder impact analysis** identifies all parties affected by the AI system—direct users, indirect subjects of decisions, communities, society. For each stakeholder group, assess potential benefits and harms. Facial recognition in airports might benefit travelers (faster processing) but harm privacy and enable surveillance.

**Value alignment** checks whether system objectives align with human values. An AI system optimizing engagement metrics might maximize addiction and misinformation spread—technically successful but ethically problematic. Explicit value specification prevents misalignment.

**Dual-use considerations** recognize technologies designed for beneficial purposes can be repurposed for harm. Deepfake technology enables creative expression but also enables fraud and nonconsensual pornography. Computer vision advances enable medical diagnostics but also enable mass surveillance. Ethical evaluation considers not just intended use but potential misuse.

**Proportionality assessment** weighs benefits against risks. Does the application's value justify its risks? Facial recognition for unlocking phones has limited societal impact; facial recognition for law enforcement has enormous implications. The same technology requires different ethical standards based on stakes.

**Alternatives evaluation** considers whether less risky approaches achieve similar benefits. Can human decision-makers supplemented by simpler tools achieve goals without full automation? Can interpretable models replace black boxes? Ethical evaluation favors approaches minimizing risk while achieving objectives.

**Case Example: The Surveillance Dilemma**

A municipal government considered deploying AI-powered video analytics across public spaces to detect crime patterns and respond faster to incidents. The system promised reduced response times and crime prevention through predictive policing.

Ethical review revealed concerns. The technology enabled mass surveillance without individualized suspicion, threatening privacy. Historical crime data reflected biased policing patterns—the system would perpetuate discrimination by directing police to over-policed neighborhoods. Facial recognition components could track individuals across the city, creating detailed movement profiles.

Stakeholder engagement revealed strong opposition from civil liberties groups and minority communities historically subject to over-policing. Legal analysis found potential conflicts with privacy laws and constitutional protections against unreasonable search.

The review committee recommended rejecting the proposal despite technical feasibility and promised crime reduction. The benefits (faster response, potential crime reduction) didn't justify the harms (privacy loss, discrimination amplification, chilling effects on free assembly). The city instead invested in community policing initiatives addressing root causes of crime without surveillance infrastructure.

This decision illustrates ethical evaluation in practice: rejecting technically sound solutions when ethical costs exceed benefits.

## 8.7 Responsible AI Governance Structures

**Governance structures** establish organizational mechanisms ensuring AI development and deployment follows ethical principles and regulatory requirements. Without governance, good intentions don't translate to consistent practice (Raji et al., 2020).

**AI Ethics Boards** provide oversight for high-stakes or novel AI applications. Membership should include technical experts, ethicists, legal counsel, domain specialists, and impacted community representatives. Boards review proposed AI projects, assess risks, recommend safeguards, and monitor deployed systems.

**Review processes** formalize evaluation at key project stages. Pre-project review assesses ethical acceptability before resource commitment. Design review examines data sources, algorithm choices, and fairness considerations. Pre-deployment review validates testing, documentation, and safeguards. Post-deployment review monitors outcomes and impacts.

**Ethical guidelines** document organizational principles for AI development. These go beyond legal compliance to articulate values and commitments. Guidelines should address fairness, privacy, transparency, accountability, safety, and human oversight. They provide practical guidance translating abstract principles into concrete requirements.

**Impact assessments** systematically evaluate potential consequences before deployment. Algorithmic Impact Assessments document system purpose, data sources, decision logic, affected populations, potential benefits and harms, fairness considerations, and mitigation strategies. These assessments inform decision-making and create accountability.

**Escalation procedures** enable raising concerns about AI systems without retaliation. Employees discovering bias, privacy violations, or safety issues need clear paths to report problems and trigger investigation. Without safe escalation, ethical concerns get buried.

**Training and education** ensure everyone involved in AI development understands ethical responsibilities. Technical teams need training on bias detection, fairness metrics, and privacy protection. Business teams need training on ethical implications and regulatory requirements. Leadership needs training on governance and accountability.

## 8.8 Model Interpretability Techniques

**Model interpretability** enables understanding how models produce predictions. This matters for debugging, building trust, meeting regulatory requirements, and extracting insights. Different techniques provide different types of explanations at different levels of fidelity (Molnar, 2020).

**Feature importance** identifies which inputs most influence predictions. Tree-based models provide built-in feature importance from split frequency and information gain. Permutation importance measures performance decrease when shuffling each feature. Drop-column importance measures performance decrease when removing features entirely.

**Partial dependence plots** show relationships between features and predictions while averaging out other features. These reveal whether relationships are linear, non-linear, or involve interactions. Partial dependence plots work for any model type but can be misleading when features correlate strongly.

**Individual Conditional Expectation (ICE) plots** show prediction changes for individual instances as a feature varies. Unlike partial dependence plots averaging across instances, ICE plots reveal heterogeneity in feature effects. Some customers might respond positively to price increases while others respond negatively—ICE plots reveal these patterns.

**SHAP (SHapley Additive exPlanations)** values quantify each feature's contribution to individual predictions using game-theoretic principles. SHAP satisfies desirable properties: local accuracy (explanations sum to actual prediction), missingness (unused features get zero importance), and consistency (increasing feature contribution increases importance) (Lundberg & Lee, 2017).

**LIME (Local Interpretable Model-agnostic Explanations)** approximates complex models locally with simple interpretable models around specific predictions. LIME perturbs inputs around the instance of interest, observes prediction changes, and fits a local linear model explaining local behavior. This provides human-understandable explanations for individual predictions (Ribeiro et al., 2016).

**Counterfactual explanations** identify minimal changes to inputs that would flip predictions. For loan denials, this answers "what would I need to change to get approved?" Counterfactuals provide actionable feedback helping users understand and potentially address decision drivers.

**Attention mechanisms** in neural networks reveal which inputs receive attention when making predictions. For text classification, attention weights show which words most influenced classification. For image recognition, attention maps highlight relevant image regions. Attention provides intuitive explanations for sequence and vision models.

## 8.9 Building Stakeholder Trust

**Trust** in AI systems develops when stakeholders have confidence that systems work as intended, treat people fairly, and can be held accountable when problems arise. Trust requires both technical performance and social legitimacy (Shneiderman, 2020).

**Transparency about capabilities and limitations** sets appropriate expectations. Overpromising erodes trust when reality disappoints. Clearly communicate what systems can and cannot do, accuracy levels, failure modes, and confidence intervals. Uncertainty quantification helps users understand prediction reliability.

**Stakeholder engagement** throughout development incorporates diverse perspectives and builds buy-in. Engage affected communities early to understand concerns, values, and needs. Test prototypes with representative users to identify usability issues and unexpected impacts. Ongoing engagement after deployment maintains trust and surfaces emerging problems.

**Redress mechanisms** provide channels for challenging decisions and seeking remedy when AI systems cause harm. Appeals processes, human review of contested decisions, and compensation for wrongful harms demonstrate accountability and give stakeholders confidence they won't be helpless against algorithmic errors.

**Auditing and certification** by independent third parties validates ethical compliance. External audits verify fairness claims, test for bias, assess privacy protections, and evaluate transparency. Certification programs establish standards and verify compliance, providing stakeholders with credible assurances.

**Plain language communication** makes AI systems understandable to non-technical stakeholders. Avoid jargon and mathematical formalism when explaining systems to general audiences. Use analogies, examples, and visualizations to convey key concepts. Test communications with target audiences to ensure comprehension.

**Case Example: The Healthcare Trust Recovery**

A hospital deployed an AI system prioritizing emergency department patients by severity. The system showed excellent performance in testing but lost clinician trust within weeks of deployment due to several high-profile cases where it misjudged severity, causing treatment delays.

Investigation revealed the system worked well on average but struggled with edge cases and certain patient populations underrepresented in training data. More problematically, clinicians didn't understand why the system made specific recommendations and lacked confidence to override concerning predictions.

The hospital implemented a trust recovery plan. They added explanations showing which symptoms and vital signs drove each priority assignment. They created a dashboard showing system performance by patient demographic and presenting condition, with alerts when accuracy fell below thresholds for specific subgroups. They established a rapid feedback mechanism where clinicians could flag concerning predictions, triggering immediate review and model updates if patterns emerged.

They conducted training sessions explaining system capabilities, limitations, and appropriate use. They emphasized the system as decision support, not replacement for clinical judgment. They created an ethics committee including clinicians, patients, and ethicists reviewing system performance quarterly.

These changes rebuilt trust. Clinicians understood system reasoning, knew its limitations, felt empowered to override when appropriate, and had confidence problems would be addressed. The system became a valued tool rather than a source of anxiety.

## 8.10 Essential Points to Remember

Trustworthy AI rests on five pillars: ethical alignment with human values, responsible accountability for outcomes, transparent processes and decisions, governed deployment with oversight, and explainable reasoning stakeholders can understand. These pillars reinforce each other and collectively ensure AI systems serve societal interests.

Bias enters AI systems through data, measurement, aggregation, and evaluation, requiring proactive detection and mitigation through diverse data, fairness constraints, and ongoing monitoring. Fairness has multiple definitions that conflict mathematically, requiring context-appropriate choices based on which harms matter most.

Privacy protection frameworks like GDPR and techniques like differential privacy and federated learning enable legitimate data use while protecting individual privacy. Regulatory compliance requires understanding applicable laws from horizontal AI regulations to sector-specific requirements.

Ethical use case evaluation assesses stakeholder impacts, value alignment, dual-use potential, proportionality, and alternatives before deploying AI systems. Some technically feasible applications aren't ethically acceptable regardless of safeguards.

Responsible AI governance establishes ethics boards, review processes, impact assessments, escalation procedures, and training programs ensuring consistent ethical practice. Model interpretability techniques from feature importance to SHAP values enable understanding and explaining AI decisions.

Building stakeholder trust requires transparency about capabilities and limitations, meaningful engagement, effective redress mechanisms, independent auditing, and plain language communication. Trust develops when stakeholders have confidence systems work fairly and can be held accountable.

## 8.11 Practice Questions

### Ethical Dilemma Scenarios

**Question 421:** Your credit scoring model performs better overall using zip code as a feature, but zip code correlates with race. Should you use it?

A) Yes, it improves accuracy B) Ethical and legal analysis required: disparate impact testing, consider proxy discrimination, explore alternatives providing accuracy without discrimination C) Always use all features D) Race correlation doesn't matter

**Answer: B) Careful analysis of discrimination risk and alternatives**

**Explanation:** Proxy discrimination concerns: (1) Zip code correlates with protected attributes (race, ethnicity); (2) Using it might constitute indirect discrimination; (3) Legal analysis: does it create disparate impact? Does ECOA apply?; (4) Technical solutions: fairness constraints, removing correlated features, testing for discrimination; (5) Alternatives: features providing information without discrimination; (6) Documentation: justify decision with evidence. Performance isn't the only consideration. Features enabling discrimination require special scrutiny, even if correlation is indirect.

---

**Question 422:** Your facial recognition model has 95% accuracy on light-skinned faces but only 75% accuracy on dark-skinned faces. What should you do?

A) Deploy with 95% overall accuracy B) Address bias: collect more diverse training data, balance dataset, test fairness rigorously, achieve equitable performance before deployment C) Accuracy difference is acceptable D) Ignore performance gaps

**Answer: B) Address bias through data and evaluation**

**Explanation:** Addressing algorithmic bias: (1) Root cause: training data bias (underrepresentation of dark-skinned individuals); (2) Data collection: actively collect balanced, diverse data; (3) Evaluation: test per-group performance as standard practice; (4) Threshold tuning: per-group thresholds if appropriate; (5) Deployment decision: don't deploy until equitable performance achieved; (6) High-stakes applications: facial recognition for law enforcement, security has serious fairness implications. Performance disparities cause real harm. Fairness is requirement, not nice-to-have.

---

**Question 423:** Your hiring recommendation model can improve efficiency but you discover it penalizes employment gaps, which disproportionately affect women (maternity leave, caregiving). What should you do?

A) Use model anyway for efficiency gains B) Remove or modify employment gap features; test for gender bias; ensure model doesn't encode societal biases; consider human-in-loop C) Employment gaps are legitimate factors D) Efficiency is paramount

**Answer: B) Remove discriminatory features and test for bias**

**Explanation:** Addressing systemic bias in features: (1) Employment gap penalization encodes societal inequities; (2) Protected class impact: disproportionate effect on women is discrimination; (3) Feature removal/modification: remove problematic features or modify to avoid bias; (4) Thorough testing: demographic parity, equal opportunity analysis; (5) Legal compliance: employment law prohibits gender discrimination; (6) Human oversight: especially important for hiring decisions. Efficiency can't justify discrimination. Models can perpetuate and amplify societal biases if not carefully designed.

---

**Question 424:** Your health prediction model could identify high-risk patients for preventive intervention but might also be used to deny insurance coverage. Should you deploy it?

A) Deploy freely; use cases are not your responsibility B) Governance and use restrictions: deploy with contractual limits on use, require ethical use review, monitoring for misuse C) Never deploy any model with misuse potential D) Misuse risk is irrelevant

**Answer: B) Deploy with strong governance and use restrictions**

**Explanation:** Dual-use technology governance: (1) Beneficial use: identifying patients needing preventive care improves health outcomes; (2) Harmful use: insurance discrimination is unethical and often illegal; (3) Use restrictions: contractual agreements limiting use cases; (4) Access control: technical controls preventing misuse; (5) Monitoring: audit logs tracking model use; (6) Ethics review: regular assessment of actual use; (7) Kill switch: ability to revoke access if misuse detected. Technology isn't inherently good or bad—use determines ethics. Deploy responsibly with safeguards.

---

**Question 425:** Your e-commerce recommendation model optimizes for revenue by recommending expensive items. Users with lower incomes are systematically shown more expensive items they can't afford. Is this ethical?

A) Revenue optimization is the goal B) Ethical concern: predatory recommendations; consider user value beyond revenue; personalize appropriately; avoid exploitation C) Business goals trump ethics D) No ethical issues

**Answer: B) Balance revenue with user value and non-exploitation**

**Explanation:** Ethical recommendation systems: (1) Exploitation risk: recommending unaffordable items harms users; (2) User value: recommendations should benefit users, not just company; (3) Personalization ethics: personalize to be helpful, not manipulative; (4) Balance: revenue important but not at user expense; (5) Long-term thinking: exploitative recommendations harm trust and long-term relationships; (6) Alternative optimization: user satisfaction, repeat purchases, balanced with revenue. Pure revenue optimization can lead to harmful outcomes. Consider user welfare in optimization objectives.

---

**Question 426:** Your model makes predictions about individuals' future behavior (crime, health, finances). What ethical responsibilities do you have regarding explanation and transparency?

A) No explanation needed B) Transparency obligations: individuals should understand why predictions made about them; provide explanations; enable appeals/corrections C) Models should be black boxes D) Hide all model logic

**Answer: B) Transparency and accountability for consequential decisions**

**Explanation:** Transparency ethics: (1) Consequential decisions: predictions affecting individuals' lives require explanation; (2) Right to explanation: GDPR and ethical principles support explaining automated decisions; (3) Contestability: individuals should be able to challenge predictions; (4) Accuracy corrections: mechanism for correcting errors in input data; (5) Model card: public documentation of model purpose, limitations, performance; (6) Different audiences: technical explanations for experts, accessible explanations for individuals. Opacity is unethical for high-stakes decisions. Transparency enables accountability.

---

**Question 427:** Your emotion recognition model works by analyzing facial expressions. Critics argue this technology is pseudoscientific and invasion of privacy. How do you respond?

A) Ignore critics B) Scientific validation: is emotion recognition scientifically valid? Privacy safeguards: how is it used? Consent: are subjects informed? Careful consideration before deployment C) Critics are always wrong D) Privacy doesn't matter

**Answer: B) Scientific and ethical scrutiny before deployment**

**Explanation:** Controversial technology assessment: (1) Scientific validity: is underlying science sound? Emotion recognition has questionable validity across cultures; (2) Privacy invasion: facial analysis without consent is ethically problematic; (3) Use case evaluation: some uses more problematic than others; (4) Consent: explicit informed consent from subjects; (5) Regulation: many jurisdictions restricting emotion recognition; (6) Stakeholder engagement: engage critics, understand concerns. Don't deploy questionable technology without thorough analysis. Some technologies shouldn't be deployed regardless of technical feasibility.

---

**Question 428:** Your predictive policing model identifies areas likely to have crimes. Police increase patrols there, leading to more arrests, which reinforces the model's predictions in a feedback loop. What's the ethical concern?

A) No concern; model is accurate B) Feedback loop bias: model creates self-fulfilling prophecy; over-policing harms communities; consider fairness and feedback loops in deployment C) More arrests are always good D) Ignore feedback effects

**Answer: B) Self-fulfilling prophecy and feedback loop bias**

**Explanation:** Feedback loop ethics: (1) Model predictions influence reality: more police → more arrests in predicted areas; (2) Perpetuates bias: even if initial predictions unbiased, feedback creates bias; (3) Community harm: over-policing damages police-community relations; (4) Alternative approaches: focus on root causes, community investment; (5) Fairness auditing: monitor for disparate impact over time; (6) Intervention design: predictions should inform resource allocation, not create automatic increases. Models affecting real-world outcomes can create feedback loops reinforcing predictions. Consider downstream effects of deployment.

---

**Question 429:** Your company wants to use ML to analyze employee emails to detect insider threats. What are the ethical considerations?

A) Employee monitoring has no ethical issues B) Privacy rights: employees have privacy expectations; necessity: is monitoring necessary?; transparency: should employees know?; alternative methods C) Monitor everything without consent D) Privacy doesn't apply to employees

**Answer: B) Balance security needs with privacy rights**

**Explanation:** Employee surveillance ethics: (1) Privacy expectations: even at work, employees have some privacy rights; (2) Proportionality: is email monitoring necessary and proportionate to threat?; (3) Transparency: should employees be informed of monitoring? (often legally required); (4) Scope limitation: monitor only what's necessary, not everything; (5) Human review: avoid fully automated decisions on sensitive findings; (6) Alternative approaches: less invasive security measures; (7) Labor law compliance: many jurisdictions regulate employee monitoring. Security is legitimate but not unlimited. Balance security with privacy and transparency.

---

**Question 430:** Your ML system makes loan decisions. Minority applicants are rejected at higher rates. Investigation shows they have lower credit scores on average. Is this acceptable?

A) Yes, model is just following credit scores B) Further investigation needed: are credit scores themselves biased? Historical discrimination in lending creates score disparities; consider fairness beyond surface neutrality C) Different rejection rates are always discrimination D) Credit scores are perfect

**Answer: B) Investigate whether proxies encode historical discrimination**

**Explanation:** Systemic bias consideration: (1) Credit scores reflect historical discrimination in lending; (2) Using them perpetuates past discrimination; (3) Fairness analysis: disparate impact testing required; (4) Alternative data: consider alternative credit indicators; (5) Human review: especially for borderline cases from disadvantaged groups; (6) Legal compliance: fair lending laws require careful analysis; (7) Societal responsibility: financial institutions have role in addressing historical inequities. "Neutral" algorithms can perpetuate historical biases. Fairness requires examining whether seemingly objective criteria encode discrimination.

---

## Bias Identification and Mitigation

**Question 431:** How do you test if your model exhibits gender bias in hiring recommendations?

A) Gender testing is unnecessary B) Disaggregated evaluation: measure performance separately by gender; test for equal opportunity; analyze feature importance; conduct bias audits C) Overall accuracy is sufficient D) Bias testing is impossible

**Answer: B) Systematic bias testing across demographic groups**

**Explanation:** Gender bias testing: (1) Data collection: need gender labels for evaluation (even if not used for training); (2) Disaggregated metrics: precision, recall, false positive/negative rates by gender; (3) Equal opportunity: do qualified candidates have equal selection rates across genders?; (4) Feature analysis: which features differ by gender? Do they encode bias?; (5) Counterfactual testing: would recommendations change if gender changed?; (6) Audit: ongoing monitoring of selection rates. Bias testing requires intentional measurement. Can't improve what you don't measure. Regular auditing essential for high-stakes applications.

---

**Question 432:** Your training data underrepresents minority groups. What strategies address this?

A) Use biased data anyway B) Data collection: actively gather more minority group data; synthetic augmentation; reweighting/resampling; fairness constraints during training C) Ignore underrepresentation D) Delete minority data

**Answer: B) Multiple strategies to address data imbalance**

**Explanation:** Addressing data bias: (1) Balanced collection: actively recruit diverse data sources; (2) Synthetic data: carefully generate additional minority examples (SMOTE,

GANs); (3) Reweighting: assign higher weights to minority examples during training; (4) Resampling: oversample minority, undersample majority; (5) Transfer learning: pretrain on diverse data, fine-tune on yours; (6) Fairness constraints: explicitly optimize for equitable performance; (7) Evaluation: test per-group performance. Data bias leads to model bias. Multiple techniques can help but real diverse data is best. Invest in representative data collection.

---

**Question 433:** Your model shows demographic parity (equal positive prediction rates) across groups but different groups have different qualification rates. Is this fair?

A) Yes, demographic parity ensures fairness B) Depends on definition of fairness: demographic parity might not be appropriate if base rates truly differ; consider equal opportunity instead C) Different prediction rates are always unfair D) No fairness definition matters

**Answer: B) Fairness definition depends on context and base rates**

**Explanation:** Fairness metrics trade-offs: (1) Demographic parity: equal positive rates across groups (might be unfair if true qualification rates differ); (2) Equal opportunity: equal true positive rates (qualified candidates equally likely to be selected); (3) Equalized odds: equal false positive and false negative rates; (4) Context matters: if groups have different qualification rates due to historical discrimination, which metric is "fair"?; (5) Impossibility theorem: can't simultaneously satisfy all fairness criteria; (6) Stakeholder input: affected communities should inform fairness definition. No universal fairness definition. Choose metric appropriate for use case and values.

---

**Question 434:** You discover your model's feature importance shows gender is highly predictive. You want to build a fair model. Should you simply remove gender?

A) Yes, removing gender ensures fairness B) No, removing protected attributes insufficient: correlated features (proxies) encode gender; need comprehensive fairness approach C) Keep gender in model D) Fairness is impossible

**Answer: B) Removing protected attributes alone doesn't ensure fairness**

**Explanation:** Proxy variables challenge: (1) Many features correlate with protected attributes (height with gender, name with ethnicity); (2) Models can infer protected attributes from proxies; (3) Simply removing doesn't prevent discrimination; (4) Comprehensive approach: remove direct attributes, test for proxies, fairness constraints, adversarial debiasing; (5) Evaluation: test for disparate impact even without protected attributes in model; (6) Monitoring: ongoing fairness auditing. Fairness by unawareness doesn't work. Need proactive fairness measures beyond feature removal.

---

**Question 435:** Your image classification model performs poorly on non-Western subjects because training data was predominantly Western images. How do you address this bias?

A) Western performance is sufficient B) Diversify training data globally; test on diverse populations; partner with international collaborators; culturally aware data collection C) Only serve Western users D) Bias is inevitable

**Answer: B) Global diverse data collection and testing**

**Explanation:** Addressing cultural/geographic bias: (1) Diverse data collection: actively gather data from underrepresented geographies; (2) Local partnerships: collaborate with local organizations for culturally appropriate data; (3) Disaggregated testing: evaluate performance across cultures, regions; (4) Cultural consultation: involve domain experts from diverse backgrounds; (5) Localization: region-specific models if global model inadequate; (6) Ethical deployment: don't deploy where performance inadequate. Western-centric models cause harm in global deployment. Intentional diversity required for equitable AI.

---

**Question 436:** Your model achieves fairness (equal false positive rates) across groups by reducing the majority group's performance. Stakeholders object to "lowering standards." How do you respond?

A) Abandon fairness for performance B) Reframe: not lowering standards but equalizing them; unfairness means one group unfairly advantaged; fairness might require accepting lower overall metrics C) Agree standards are being lowered D) Ignore stakeholder concerns

**Answer: B) Fairness means equalizing, not lowering standards**

**Explanation:** Fairness communication: (1) Reframe narrative: previous system wasn't high-performing, it was unfair to disadvantaged groups; (2) Justice argument: equal treatment is the standard, not average performance; (3) Trade-offs: fairness may require accepting slightly lower overall metrics; (4) Alternative improvements: sometimes can improve disadvantaged group performance without harming majority; (5) Values alignment: organization's values should include fairness; (6) Legal compliance: in many domains, fairness is legally required. Fairness discussions are values discussions. Emphasize justice and equality, not just metrics.

---

**Question 437:** You're evaluating bias mitigation techniques: preprocessing (fair data sampling), in-processing (fairness constraints during training), post-processing (threshold adjustment). Which should you use?

A) Only one technique ever needed B) Depends on constraints: preprocessing when you can modify data; in-processing for algorithm access; post-processing for deployed models; often combine C) Never use bias mitigation D) All techniques are equivalent

**Answer: B) Choose technique based on practical constraints; often combine**

**Explanation:** Bias mitigation strategy selection: (1) Preprocessing: fair sampling, reweighting—use when you control data; (2) In-processing: fairness constraints during training—use when you can modify training; (3) Post-processing: threshold adjustment—use when model already trained or deployed; (4) Combination: often most effective to use multiple techniques; (5) Evaluation: empirically test which techniques work for your use case; (6) Constraints: choose based on what you can modify (data, algorithm, decisions). Different techniques have different strengths. Practical constraints often determine choice. Multi-stage approach typically most effective.

---

**Question 438:** Your NLP model trained on internet text exhibits gender bias (associates "programmer" with male pronouns). How do you debias?

A) Ignore linguistic bias B) Debiasing techniques: counterfactual data augmentation, embedding debiasing, balanced training corpus, bias-aware architectures C) Text inherently biased; nothing can be done D) Gender bias in language is acceptable

**Answer: B) NLP-specific debiasing techniques**

**Explanation:** NLP bias mitigation: (1) Counterfactual augmentation: create gender-swapped versions of training examples; (2) Embedding debiasing: remove gender direction from word embeddings; (3) Balanced corpus: ensure professional terms appear equally with all genders; (4) Bias-aware pretraining: train on curated less-biased text; (5) Fine-tuning: debias pretrained models through targeted fine-tuning; (6) Evaluation: test for stereotypical associations (WinoBias, StereoSet benchmarks). Language models inherit societal biases from training text. Active debiasing required for fair NLP systems.

---

**Question 439:** You're building a model for child custody recommendations. What special fairness considerations apply?

A) Standard fairness metrics suffice B) High-stakes decision requiring: extreme caution with algorithmic decisions, human oversight essential, disparate impact testing, contestability mechanisms C) Algorithms should fully automate decisions D) No special considerations needed

**Answer: B) Highest ethical standards for extremely consequential decisions**

**Explanation:** High-stakes application ethics: (1) Human-in-loop: algorithms should support, not replace, human judges; (2) Transparency: families have right to understand how recommendations made; (3) Appeals: mechanism for challenging algorithmic recommendations; (4) Disparate impact: rigorous testing for bias across demographics; (5) Value alignment: custody decisions involve complex values not reducible to algorithms; (6) Regulation: many jurisdictions restrict algorithmic decision-making in family law. Some decisions too important for full automation. Algorithms can assist but humans must decide on questions involving complex values and individual circumstances.

---

**Question 440:** Your fairness audit reveals your model discriminates but fixing it would require collecting demographic data you don't currently have. What should you do?

A) Accept discrimination to avoid collecting demographic data B) Collect demographic data with appropriate consent, privacy protections, and clear purpose limitation for fairness testing C) Never collect demographic data D) Deploy discriminatory model

**Answer: B) Collect necessary data with proper safeguards**

**Explanation:** Fairness-privacy tradeoff: (1) Demographic data necessary for bias testing and mitigation; (2) Collect with consent: transparent about purpose (fairness evaluation); (3) Purpose limitation: use only for fairness work, not targeting; (4) Privacy protection: segregate data, limited access, encryption; (5) Legal compliance: ensure collection lawful under privacy regulations; (6) Minimize retention: delete after fairness work complete if possible; (7) Documentation: clear policy on demographic data use. Can't achieve fairness without measuring it. Collect demographic data responsibly with appropriate protections.

---

## Compliance Requirement Situations

**Question 441:** Your model is deployed in EU and must comply with GDPR's right to explanation. How do you provide meaningful explanations?

A) GDPR doesn't apply to AI B) Explainability implementation: model-agnostic explanations (SHAP, LIME), documentation (model cards), human-readable explanations for data subjects C) Ignore GDPR D) Black box models can't have explanations

**Answer: B) Implement explainability infrastructure for compliance**

**Explanation:** GDPR compliance: (1) Right to explanation: individuals can request explanation of automated decisions; (2) Technical solutions: SHAP, LIME provide feature importance for predictions; (3) User-facing explanations: translate technical explanations into understandable language; (4) Documentation: model cards explaining model purpose, logic, limitations; (5) Human review: option for human review of automated decisions; (6) Data protection impact assessment: evaluate privacy risks before deployment. GDPR mandates explainability for consequential automated decisions. Plan for this from initial design, not as afterthought.

---

**Question 442:** Your healthcare ML model must comply with HIPAA. What requirements affect model development and deployment?

A) HIPAA doesn't apply to AI B) PHI protection: encryption, access controls, audit logging, business associate agreements, de-identification procedures, breach notification plans C) No special healthcare requirements D) Healthcare data has no restrictions

**Answer: B) Comprehensive HIPAA compliance controls**

**Explanation:** HIPAA compliance requirements: (1) PHI protection: encrypted storage and transmission; (2) Access control: minimum necessary access, authentication; (3) Audit trails: log all PHI access; (4) Business associate agreements: with cloud providers, contractors; (5) De-identification: safe harbor or expert determination methods; (6) Breach notification: plan for security incidents; (7) Technical safeguards: following HIPAA Security Rule. Healthcare data is highly regulated. Compliance required throughout ML lifecycle—data collection, training, deployment, monitoring. Violations have severe penalties.

---

**Question 443:** Your credit model must comply with Fair Lending laws requiring disparate impact testing. How do you conduct this?

A) Fair lending doesn't apply to algorithms B) Statistical testing: adverse impact ratio (four-fifths rule), regression analysis controlling for legitimate factors, regular audits, documentation C) Ignore fair lending requirements D) Algorithms are exempt from lending law

**Answer: B) Rigorous statistical testing and documentation**

**Explanation:** Fair lending compliance: (1) Adverse impact testing: four-fifths rule (protected group acceptance rate ≥ 80% of control group); (2) Legitimate factors: ensure disparities explained by creditworthiness, not protected attributes; (3) Regular audits: quarterly/annual disparate impact analysis; (4) Third-party review: independent auditors validate compliance; (5) Documentation: maintain detailed records of testing, mitigation; (6) Monitoring: ongoing compliance monitoring. Equal Credit Opportunity Act and Fair Housing Act apply to algorithmic lending. Compliance not optional. Rigorous testing and documentation essential.

---

**Question 444:** Your facial recognition system is being deployed for law enforcement. Multiple jurisdictions are considering or have passed bans. What compliance considerations apply?

A) Technology bans don't apply to your system B) Regulatory landscape: check federal, state, local regulations; may be prohibited in some jurisdictions; high accuracy and fairness standards if permitted C) Ignore local regulations D) Facial recognition has no restrictions

**Answer: B) Navigate complex regulatory landscape; some deployments may be prohibited**

**Explanation:** Facial recognition regulation: (1) Jurisdictional bans: some cities/states ban government use of facial recognition; (2) Use case restrictions: law enforcement use more restricted than commercial; (3) Accuracy requirements: some regulations mandate minimum accuracy and equity; (4) Transparency: disclosure requirements for use; (5) Oversight: may require independent audits, review boards; (6) Evolving regulation: rapidly changing landscape. Controversial application requiring careful legal analysis. Some use cases may not be legally or ethically viable. Don't assume deployment is permissible.

---

**Question 445:** Your model uses personal data and must comply with CCPA (California Consumer Privacy Act). What rights must you support?

A) CCPA doesn't apply to AI models B) Consumer rights: right to know what data collected, right to deletion, right to opt-out of sale, data minimization, privacy notices C) No consumer rights for AI D) Only GDPR matters

**Answer: B) Multiple consumer privacy rights under CCPA**

**Explanation:** CCPA compliance: (1) Right to know: disclose what personal data collected and how used; (2) Right to deletion: delete consumer data on request (impacts ability to retrain); (3) Right to opt-out: of data selling (if applicable); (4) Non-discrimination: can't discriminate against consumers exercising rights; (5) Data minimization: collect only necessary data; (6) Privacy notice: inform consumers of data practices; (7) Reasonable security: protect personal data. CCPA gives California residents significant control over personal data. Systems must support data deletion, disclosure, opt-out. Plan compliance into architecture.

---

**Question 446:** Your model deployment requires compliance with sector-specific regulations (FDA for medical devices, NHTSA for automotive). How do you approach regulatory compliance?

A) Self-certification is sufficient B) Early engagement: involve regulatory experts from project start; follow established pathways; comprehensive documentation; pre-submission meetings C) Ignore sector regulations D) AI is unregulated

**Answer: B) Proactive regulatory strategy with expert engagement**

**Explanation:** Sector-specific compliance: (1) Early involvement: include regulatory experts in project team from beginning; (2) Regulatory pathways: understand approval process (FDA 510(k), De Novo, PMA); (3) Documentation: comprehensive validation, testing, risk analysis documentation; (4) Standards: follow industry standards (ISO, IEC); (5) Pre-submission: meet with regulators early to align on approach; (6) Post-market surveillance: ongoing monitoring and reporting requirements. Regulated sectors have established approval processes. Don't treat as afterthought. Regulatory strategy shapes development approach. Retrofitting compliance is costly or impossible.

---

**Question 447:** Your company operates globally and must comply with varying privacy regulations (GDPR, CCPA, LGPD, etc.). How do you manage multi-jurisdiction compliance?

A) Comply only with strictest regulation everywhere B) Hybrid approach: global baseline (strictest requirements) for common controls; jurisdiction-specific features where needed; compliance management system C) Ignore international regulations D) Only comply in home country

**Answer: B) Baseline global standards with jurisdiction-specific additions**

**Explanation:** Multi-jurisdiction compliance: (1) Global baseline: adopt strictest requirements (often GDPR) as floor for all operations; (2) Regional additions: jurisdiction-specific features (e.g., CCPA "Do Not Sell" button); (3) Data residency: respect data localization requirements; (4) Compliance tracking: system tracking requirements by jurisdiction; (5) Legal counsel: local legal expertise for each major market; (6) Scalable architecture: design supporting regional variations. Operating globally requires navigating complex regulatory landscape. Baseline-plus approach balances consistency with compliance. Single standard where possible simplifies operations.

---

**Question 448:** Your model falls into regulatory gray area—no clear regulations yet but regulators are paying attention. How do you proceed?

A) Do whatever you want in gray areas B) Proactive compliance: engage regulators proactively; self-regulate beyond current requirements; anticipate future regulation;

transparent about limitations C) Ignore regulators until rules passed D) Gray areas have no obligations

**Answer: B) Proactive engagement and conservative approach**

**Explanation:** Navigating regulatory uncertainty: (1) Engage regulators: proactive discussions demonstrating responsibility; (2) Industry standards: follow emerging best practices even if not required; (3) Ethics review: internal ethics assessment independent of regulation; (4) Transparent disclosure: clear about capabilities and limitations; (5) Anticipate regulation: design for likely future requirements; (6) Document: comprehensive records of responsible development. Lack of regulation isn't license to ignore ethical concerns. Proactive responsibility builds trust and may influence eventual regulation favorably. Waiting for regulation may require expensive rework.

---

**Question 449:** Your model must maintain audit trail for regulatory compliance. What should you log and retain?

A) No logging needed B) Comprehensive audit trail: model versions, training data versions, predictions with timestamps, input features, model updates, access logs C) Only final predictions D) Logging wastes storage

**Answer: B) Comprehensive logging for compliance and troubleshooting**

**Explanation:** Compliance audit trail: (1) Predictions: every prediction with timestamp, model version, inputs; (2) Model versioning: which model version made which predictions; (3) Training data: data used for each model version; (4) Updates: all model updates, deployments, rollbacks; (5) Access: who accessed what data when; (6) Data lineage: tracking data from source through transformations; (7) Retention: comply with retention requirements (often 3-7 years); (8) Immutability: tamper-evident logs. Audit trails enable regulatory audits, troubleshooting, bias investigations. Plan storage and retention from deployment. Retroactive logging impossible.

---

**Question 450:** Your model is subject to algorithmic accountability bill requiring impact assessments before deployment. What should the assessment include?

A) Impact assessments are unnecessary bureaucracy B) Comprehensive assessment: intended use, potential harms, bias testing, mitigation measures, monitoring plans, stakeholder consultation C) Brief description is sufficient D) Only technical metrics

**Answer: B) Thorough multi-stakeholder impact assessment**

**Explanation:** Algorithmic impact assessment: (1) Purpose and scope: intended use, decisions model will inform; (2) Harm identification: potential negative impacts on individuals, groups; (3) Bias testing: demographic disparate impact analysis; (4) Mitigation: how harms will be prevented or minimized; (5) Monitoring: ongoing oversight and accountability; (6) Stakeholder input: consultation with affected communities; (7) Alternatives: consideration of non-algorithmic approaches; (8) Accountability: who is responsible for outcomes. Impact assessments operationalize responsible AI. Forces systematic consideration of risks before deployment. Stakeholder involvement ensures perspectives beyond technical team.

---

**Question 451:** Your model is too complex for users to understand. How do you improve transparency while maintaining performance?

A) Simplify model drastically, sacrificing performance B) Layered transparency: simple explanations for general users, detailed technical docs for experts, visual explanations, model cards C) Don't explain complex models D) Users don't need to understand

**Answer: B) Multi-level explanations for different audiences**

**Explanation:** Accessible transparency: (1) Audience-appropriate explanations: general users get high-level, experts get technical details; (2) Model cards: standardized documentation of purpose, performance, limitations; (3) Visual explanations: feature importance charts, decision path visualizations; (4) Plain language: avoid jargon in user-facing explanations; (5) Examples: concrete examples illustrating model behavior; (6) Interactive exploration: tools letting users explore model decisions; (7) Graduated disclosure: progressive detail as users want more information. Transparency doesn't mean dumbing down—it means appropriate communication. Match explanation depth to audience needs and technical background.

---

**Question 452:** Stakeholders request "full transparency" including complete algorithm code, training data, and model weights. What are the considerations?

A) Provide everything requested B) Balance transparency with: intellectual property protection, privacy concerns, security risks, competitive advantage; provide meaningful transparency without harmful disclosure C) Never provide any information D) Transparency has no limits

**Answer: B) Balanced transparency respecting multiple constraints**

**Explanation:** Transparency boundaries: (1) Privacy: training data contains personal information—can't release raw data; (2) Security: full model access enables adversarial attacks; (3) IP: proprietary algorithms represent competitive advantage; (4) Meaningful transparency: focus on what helps stakeholders evaluate and use model appropriately; (5) Alternatives: aggregate information, synthetic data, API access instead of full model; (6) Differential privacy: publish statistics without revealing individuals. Transparency is important but not absolute. Balance with legitimate privacy, security, and business concerns. Focus on transparency that serves stakeholder needs.

---

**Question 453:** Your model documentation (model card) is technically accurate but users still don't understand limitations. How do you improve communication?

A) Documentation is complete; user comprehension is not your problem B) User-centered design: user testing of documentation, plain language, concrete examples of limitations, warnings about misuse, FAQ based on actual questions C) More technical jargon D) Shorter documentation

**Answer: B) User-centered documentation design**

**Explanation:** Effective documentation: (1) User testing: test documentation with actual users, iterate based on feedback; (2) Plain language: avoid technical jargon, define necessary terms; (3) Concrete examples: "Model may misclassify low-resolution images" better than "Performance varies with input quality"; (4) Visual aids: charts, diagrams, infographics supplement text; (5) Use cases: explicit examples of appropriate and inappropriate uses; (6) FAQ: address common questions and misconceptions; (7) Warnings: highlight critical limitations prominently. Technical accuracy necessary but insufficient. Documentation must actually communicate. Test with users, iterate based on comprehension.

---

**Question 454:** Your company wants to publish research about your model but is concerned about competitive advantage. How do you balance open science with business interests?

A) Never publish anything B) Strategic disclosure: publish methodology, validation approach, learnings without proprietary details; partial transparency advancing field while protecting IP C) Publish everything including code D) Science and business are incompatible

**Answer: B) Strategic publication balancing openness and protection**

**Explanation:** Selective transparency: (1) Publish approach: share overall methodology, architecture, insights; (2) Protect specifics: exact hyperparameters, training tricks, proprietary data; (3) Reproducibility: provide enough detail for scientific evaluation without full replication; (4) Dataset description: characterize training data without releasing it; (5) Benchmarks: use public datasets for comparisons; (6) Collaboration: partner with academic researchers for independent validation; (7) Timing: publish after product launch, not before. Can contribute to scientific knowledge without compromising business. Strategic disclosure advances field while protecting competitive elements. Many companies successfully balance research publication with IP protection.

---

**Question 455:** Users complain your model explanations are inconsistent—small input changes produce dramatically different explanations. What's the issue?

A) Explanations should vary with inputs B) Explanation instability: explanation method problem, not necessarily model problem; use more robust explanation techniques, ensemble explanations C) Inconsistency is fine D) Don't provide explanations

**Answer: B) Explanation method stability issues**

**Explanation:** Explanation robustness: (1) Problem: some explanation methods (LIME) are locally unstable; (2) Better methods: SHAP generally more stable than LIME; (3) Ensemble explanations: average multiple explanation samples for stability; (4) Consistent narratives: provide high-level consistent explanations supplemented by detailed feature importance; (5) Validation: test explanation stability as part of validation; (6) User education: explain what explanations show (feature importance, not causation). Unstable explanations confuse users and reduce trust. Explanation method choice matters. Test explanation stability, not just model performance.

---

**Question 456:** Your model makes errors that are obvious to humans but the model can't explain why it made mistakes. How do you improve transparency around failures?

A) Hide failures B) Failure transparency: clearly document known failure modes, provide confidence scores, enable user feedback on errors, maintain error database C) Models shouldn't fail D) Don't acknowledge errors

**Answer: B) Transparent communication about limitations and failures**

**Explanation:** Failure mode transparency: (1) Known limitations: explicitly document edge cases where model fails; (2) Confidence scores: provide uncertainty estimates with predictions; (3) Human review: low-confidence predictions flagged for human oversight; (4) Error reporting: easy mechanism for users to report errors; (5) Error analysis: maintain database of failures, analyze patterns; (6) Updates: communicate how failures are being addressed; (7) Honest marketing: don't overstate capabilities. All models fail. Transparency about failures builds trust more than hiding them. Users need to understand where not to rely on model.

---

**Question 457:** Regulators request explanation for a specific prediction your model made six months ago. You don't have the model version or input data from that time. What went wrong?

A) Six-month retention is unreasonable B) Insufficient audit trail: need versioning, logging, data retention strategy for regulatory compliance C) Regulators shouldn't ask for old predictions D) Nothing went wrong

**Answer: B) Missing audit trail and versioning infrastructure**

**Explanation:** Audit trail requirements: (1) Model versioning: track which model version made which predictions; (2) Prediction logging: store predictions with timestamps and model versions; (3) Input retention: retain inputs for predictions (within data retention policies); (4) Version-specific explainability: ability to load old model versions for explanation; (5) Retention policy: define retention periods meeting regulatory needs (often

3-7 years); (6) Storage optimization: balance retention with storage costs. Regulated industries require explaining historical predictions. Plan audit trail and retention from deployment. Retroactive reconstruction impossible.

---

**Question 458:** Your model card says model is for "decision support only" but you discover users treating predictions as final decisions without review. How do you address this?

A) Users' misuse is their responsibility B) Stronger guidance: clearer warnings, technical controls requiring human review, user training, monitor actual usage patterns C) Remove decision support language D) Ignore how users use the model

**Answer: B) Reinforce intended use through multiple channels**

**Explanation:** Preventing misuse: (1) Prominent warnings: clear, repeated warnings about proper use; (2) UI design: interface requiring human input, not one-click acceptance of predictions; (3) Technical controls: system design preventing fully automated decisions; (4) Training: educate users on proper use and limitations; (5) Monitoring: track usage patterns, identify misuse; (6) Enforcement: intervene when misuse detected; (7) Accountability: clear responsibility for final decisions. Can't fully prevent misuse but can make it harder. Multiple reinforcement mechanisms. Monitor actual usage, not just stated usage.

---

**Question 459:** Different stakeholders want different levels of transparency: executives want simple summaries, regulators want detailed documentation, researchers want methodology details. How do you manage this?

A) One document for everyone B) Tiered documentation strategy: executive summaries, regulatory reports, technical papers, public model cards—each tailored to audience C) Only provide minimal documentation D) Documentation is too much work

**Answer: B) Audience-specific documentation suite**

**Explanation:** Multi-audience documentation: (1) Executive summary: business impact, high-level performance, risks; (2) Model card: public-facing standardized documentation; (3) Technical documentation: detailed methodology for researchers, auditors; (4) Regulatory reports: compliance-focused documentation with required elements; (5) User guides: practical guidance for end users; (6) Consistent foundation: all documents derived from same source of truth; (7) Maintenance: update all documents together when model changes. Different stakeholders have different needs. Provide appropriate level of detail for each. Maintain consistency across documents.

---

**Question 460:** Your transparency initiatives (publishing model cards, explanations) are time-consuming. Leadership questions the ROI. How do you justify transparency investment?

A) Transparency has no business value B) Articulate value: regulatory compliance, user trust, risk mitigation, better debugging, competitive advantage from responsible AI C) Abandon transparency D) Transparency is pure cost

**Answer: B) Quantify transparency benefits for business case**

**Explanation:** Transparency business value: (1) Regulatory compliance: avoid fines, enable deployment in regulated markets; (2) Trust and adoption: users more willing to adopt transparent systems; (3) Risk mitigation: early detection of problems through transparency; (4) Debugging: good documentation aids troubleshooting, reduces maintenance costs; (5) Responsible AI brand: differentiation in market increasingly valuing ethics; (6) Talent: attract employees who value responsible AI; (7) Long-term: sustainability and avoiding backlash. Transparency isn't just ethics—it has business value. Quantify benefits in business terms. Lack of transparency carries risks and costs.

---

## Governance Implementation Decisions

**Question 461:** Your organization is deploying multiple AI systems. What governance structure should you establish?

A) No governance needed B) Multi-level governance: AI ethics board, review processes for high-risk applications, policies and standards, monitoring and audit functions C) One person decides everything D) Governance slows innovation

**Answer: B) Comprehensive governance framework across organization**

**Explanation:** AI governance structure: (1) Ethics board: diverse stakeholders reviewing high-risk AI applications; (2) Risk tiering: different review processes based on AI risk level; (3) Policies: clear policies on data use, fairness, transparency, accountability; (4) Standards: technical standards for development, testing, deployment; (5) Review processes: mandatory reviews at project milestones; (6) Monitoring: ongoing compliance monitoring; (7) Accountability: clear responsibility for outcomes; (8) Appeals: mechanism for challenging decisions. Governance isn't bureaucracy—it's risk management. Structure ensures consistent responsible AI practices across organization.

---

**Question 462:** Who should be on your AI ethics board?

A) Only AI engineers B) Diverse representation: technologists, ethicists, legal, domain experts, affected community representatives, business leaders C) Only executives D) Ethics boards are unnecessary

**Answer: B) Multi-stakeholder diverse membership**

**Explanation:** Ethics board composition: (1) Technical experts: understand what's technically feasible; (2) Ethicists: provide ethical frameworks and analysis; (3) Legal: understand regulatory landscape; (4) Domain experts: bring application-specific expertise; (5) Community representatives: perspectives from potentially affected groups; (6) Business: understand business context and constraints; (7) Independent members: external perspectives; (8) Diversity: gender, race, age, background diversity essential. Ethics requires multiple perspectives. Technical expertise alone insufficient. Affected communities must have voice. Diversity prevents blind spots.

---

**Question 463:** Your company wants to implement AI risk tiering (high, medium, low risk) to focus governance on highest-risk applications. What criteria determine risk level?

A) All AI is equally risky B) Multiple factors: stakes of decisions (life, liberty, livelihood), potential for discrimination, scale of impact, reversibility, human oversight C) Random assignment D) Risk doesn't vary

**Answer: B) Multi-dimensional risk assessment framework**

**Explanation:** AI risk factors: (1) Stakes: affects life, freedom, livelihood = high risk; (2) Discrimination potential: risk of disparate impact on protected groups; (3) Scale: millions affected vs. dozens; (4) Reversibility: can decisions be appealed/reversed?; (5) Autonomy:

human oversight present or fully automated?; (6) Transparency: can decisions be explained?; (7) Proven technology: mature vs. experimental; (8) Domain sensitivity: healthcare, criminal justice = higher risk. Comprehensive risk scoring guides appropriate governance rigor. High-risk AI requires stricter review than low-risk. Clear criteria ensure consistent application.

---

**Question 464:** Your AI ethics review process takes 4-6 weeks. Product teams complain it slows development. How do you balance rigor with speed?

A) Eliminate ethics review B) Streamline without sacrificing rigor: risk-based review (high-risk gets full review, low-risk expedited), pre-approval for common patterns, parallel review C) Speed is more important than ethics D) Reviews should be slow

**Answer: B) Efficient risk-proportionate review process**

**Explanation:** Efficient governance: (1) Risk-tiering: low-risk AI gets lightweight review, high-risk gets thorough review; (2) Pre-approval: common low-risk patterns pre-approved (e.g., content recommendation); (3) Parallel work: ethics review concurrent with development, not sequential; (4) Self-assessment: teams complete structured self-assessment before formal review; (5) Clear criteria: explicit criteria speed decision-making; (6) Standing reviews: regular scheduled reviews vs. ad-hoc; (7) Resources: adequate staffing for timely reviews. Governance doesn't require slowness. Efficient processes enable both speed and rigor. Optimize process flow while maintaining substantive review.

---

**Question 465:** Your AI governance policy requires fairness testing but provides no technical guidance on how to do it. Teams struggle with implementation. What's missing?

A) Policy is sufficient B) Implementation support: technical guidelines, tools and templates, training, expert consultation, examples of good fairness testing C) Teams should figure it out D) Fairness testing is too hard

**Answer: B) Practical implementation guidance supporting policy**

**Explanation:** Operationalizing governance: (1) Technical guidelines: step-by-step fairness testing methodology; (2) Tools: recommended libraries, scripts, frameworks; (3)

Templates: fairness testing reports, documentation templates; (4) Training: workshops teaching fairness testing; (5) Expert support: internal experts teams can consult; (6) Examples: case studies of good fairness testing; (7) Integration: fairness testing in CI/CD pipeline. Policy without implementation guidance is ineffective. Teams need practical tools and support. Governance succeeds when made easy to comply with.

---

**Question 466:** Your company operates across multiple business units with different AI maturity levels. Should governance be centralized or decentralized?

A) Fully centralized always B) Hybrid model: centralized core policies and standards; decentralized implementation with business unit flexibility; shared services and expertise C) Fully decentralized always D) No governance structure needed

**Answer: B) Hybrid centralized-decentralized governance**

**Explanation:** Governance at scale: (1) Centralized: core principles, minimum standards, high-risk review board; (2) Decentralized: business units implement within their contexts; (3) Shared services: central AI ethics team provides consulting, tools, training; (4) Communities of practice: cross-BU knowledge sharing; (5) Escalation: BUs handle low-risk, escalate high-risk to central board; (6) Flexibility: adapt to BU needs while maintaining standards; (7) Consistency: central monitoring ensures consistent standards application. Pure centralization can be bureaucratic; pure decentralization creates inconsistency. Hybrid balances consistency with agility.

---

**Question 467:** Your AI governance requires documentation (model cards, impact assessments) but teams view it as paperwork burden. How do you improve compliance?

A) Mandate without support B) Make documentation valuable: integrate into workflow, auto-generate from code/metadata where possible, show usefulness (debugging, compliance), recognition for good docs C) Drop documentation requirements D) Punish non-compliance

**Answer: B) Reduce friction and demonstrate value**

**Explanation:** Documentation adoption: (1) Tooling: automated generation from code, metadata, logs; (2) Templates: structured templates reducing writing effort; (3) Integration:

documentation in development workflow, not separate; (4) Value demonstration: show how documentation helps debugging, compliance, knowledge transfer; (5) Recognition: reward good documentation in performance reviews; (6) Training: make documentation easy by teaching best practices; (7) Examples: showcase good documentation. Documentation feels like burden when manual and low-value. Automation and demonstrated value increase compliance. Make it easy and useful.

---

**Question 468:** Your AI governance board rejected a proposed AI application. The product team is upset and wants to proceed anyway. How do you handle this?

A) Let product team override B) Governance authority: board decisions binding; appeal process available; escalation to executive leadership if needed; clear accountability C) Ethics board has no authority D) Avoid conflict

**Answer: B) Clear authority with structured appeal process**

**Explanation:** Governance enforcement: (1) Clear authority: governance decisions are binding, not advisory; (2) Appeal process: teams can appeal with new information or modified approach; (3) Executive alignment: leadership supports governance authority; (4) Documentation: clear rationale for rejection helps teams understand; (5) Collaboration: governance team works with product to find acceptable alternatives; (6) Accountability: proceeding without approval has consequences; (7) Learning: rejections feed into guidance helping other teams. Governance without enforcement is meaningless. Authority must be real and supported by leadership. Appeals ensure fairness without undermining governance.

---

**Question 469:** Your company is acquiring another company with different AI governance standards. How do you harmonize governance across the merged organization?

A) Force immediate compliance with your standards B) Phased harmonization: assess both approaches, create unified standard incorporating best of both, transition plan respecting ongoing projects C) Maintain separate standards indefinitely D) Ignore governance during merger

**Answer: B) Thoughtful integration of governance practices**

**Explanation:** Governance merger: (1) Assessment: understand both organizations' governance strengths and weaknesses; (2) Best practices: unified standards incorporating best elements from both; (3) Transition plan: phased adoption giving teams time to adapt; (4) Quick wins: immediately adopt uncontroversial improvements from either side; (5) Respect ongoing: don't disrupt projects mid-flight with new requirements; (6) Communication: explain rationale for unified standards; (7) Training: ensure merged organization understands new governance. Mergers require sensitivity. Neither organization should feel their governance was "wrong." Unified standards stronger than either alone.

---

**Question 470:** Your AI governance focuses on technical fairness metrics but ignores broader societal impacts. What's missing?

A) Technical metrics are sufficient B) Holistic assessment: include social impact analysis, stakeholder engagement, environmental considerations, labor impacts, long-term societal effects C) Societal impacts are not relevant D) Only code quality matters

**Answer: B) Broader impact assessment beyond technical metrics**

**Explanation:** Comprehensive governance: (1) Social impact: how does AI affect society, communities, democracy?; (2) Environmental: energy consumption, carbon footprint; (3) Labor: displacement, working conditions, skill requirements; (4) Economic: wealth concentration, market competition; (5) Stakeholder input: affected communities, workers, civil society; (6) Long-term: anticipate long-term societal shifts; (7) Alternatives: consider non-AI solutions. Technical fairness metrics necessary but insufficient. AI affects society broadly. Governance should consider full range of impacts. Engage diverse perspectives to understand implications beyond technical team's view.

---

# Chapter 9.0: Data Governance and Management

AI projects fail more often from organizational dysfunction than technical limitations. You can have brilliant data scientists, powerful algorithms, and massive compute resources, but without effective data governance, clear team structures, and disciplined project management, AI initiatives collapse under their own complexity. This chapter addresses the organizational and managerial foundations successful AI projects require (Dong & Rekatsinas, 2018).

## 9.1 Data Governance Frameworks

**Data governance** establishes policies, procedures, standards, and roles ensuring data quality, security, availability, and usability across the organization. Without governance, data becomes fragmented, inconsistent, insecure, and unreliable—undermining AI initiatives built on that data (Khatri & Brown, 2010).

**Data ownership** assigns accountability for data quality, security, and appropriate use. **Data stewards** are individuals or teams responsible for specific datasets—ensuring accuracy, managing access, documenting definitions, and resolving quality issues. Clear ownership prevents the tragedy of the commons where everyone uses data but nobody maintains it.

**Data policies** document standards for data collection, storage, access, sharing, retention, and disposal. Policies cover classification schemes (public, internal, confidential, restricted), security requirements by classification level, acceptable use restrictions, and compliance requirements. Policies must be specific enough to guide decisions but flexible enough to accommodate legitimate variations.

**Data standards** specify formats, naming conventions, metadata requirements, and quality thresholds. Standardization enables integration and reduces errors from inconsistent representation. Standards should cover data types (how to represent dates, currencies, addresses), field definitions (what does "revenue" include), and codes (product categories, geographic regions).

**Data cataloging** creates searchable inventories of organizational data assets. Catalogs document what data exists, where it lives, who owns it, what it contains, quality levels, and access procedures. This prevents teams from recreating data that already exists and enables discovering relevant datasets for AI projects.

**Master Data Management (MDM)** creates single authoritative sources for critical business entities like customers, products, suppliers, and employees. Without MDM, the same customer might appear differently across systems, making integration difficult and analytics unreliable. MDM reconciles discrepancies and maintains consistent identifiers.

**Metadata management** tracks data about data—schemas, lineage, quality metrics, usage patterns, and business context. Good metadata enables understanding data provenance (where it came from), transformations applied, quality issues, and appropriate uses. This becomes especially critical for AI where understanding training data characteristics is essential.

## 9.2 Data Quality Management

**Data quality** measures fitness for use along dimensions including accuracy, completeness, consistency, timeliness, validity, and uniqueness. Poor data quality generates unreliable AI models regardless of sophisticated algorithms (Batini & Scannapieco, 2016).

**Accuracy** measures whether data correctly represents reality. Inaccurate data arises from measurement errors, data entry mistakes, failed integrations, or outdated information. Accuracy assessment requires comparing data against ground truth sources, validating samples, and investigating anomalies.

**Completeness** assesses whether required data elements are present. Missing data can occur randomly or systematically. Random missingness (technical glitches, occasional user skips) is less problematic than systematic missingness (certain demographics declining to provide information) which introduces bias. Measure completeness at field level and record level.

**Consistency** evaluates whether data agrees across systems and over time. Inconsistency suggests integration failures, process problems, or data quality issues. Compare data about the same entities across systems to identify discrepancies requiring reconciliation.

**Timeliness** determines whether data is current enough for intended use. Yesterday's stock prices don't help today's trading decisions. Last year's customer preferences don't predict current behavior. Define timeliness requirements based on business needs and measure data age against those requirements.

**Validity** checks whether data conforms to defined formats, ranges, and business rules. Validation rules detect impossible values (negative ages, future birthdates), format violations (malformed email addresses), and referential integrity problems (references to non-existent entities). Implement validation at data entry, during integration, and before AI training.

**Uniqueness** ensures entities appear once without duplicates. Duplicate records corrupt analytics and AI training. Identify duplicates through exact matching (identical records) and fuzzy matching (similar but not identical records likely representing the same entity). Develop deduplication logic carefully—overly aggressive matching creates false positives while too conservative matching misses duplicates.

**Quality improvement processes** include **data profiling** (analyzing data distributions and patterns), **data cleansing** (correcting errors and inconsistencies), **data enrichment** (augmenting with additional information), and **continuous monitoring** (ongoing quality measurement with alerting).

**Case Example: The Data Quality Crisis**

A financial services firm built customer lifetime value prediction models achieving impressive test set performance. Production deployment revealed disaster—predictions were wildly inaccurate, damaging business decisions.

Investigation uncovered systemic data quality problems. Customer addresses were 40% outdated because update processes failed. Income estimates were self-reported during account opening and never updated, becoming increasingly inaccurate over time. Transaction categorizations were inconsistent across product lines due to different coding schemes.

Account linking across systems was broken—customers with checking accounts, credit cards, and mortgages appeared as separate entities. The model couldn't learn cross-product patterns because it didn't know products belonged to the same customer. Timestamps had inconsistent time zones, making temporal analysis unreliable.

The firm implemented a comprehensive data quality program. They established data stewardship for each critical dataset with clear quality targets. They built automated data profiling identifying quality issues weekly. They created data quality dashboards making issues visible to business stakeholders who could prioritize remediation.

They fixed broken integration processes, standardized data formats, and implemented validation rules preventing future quality degradation. They invested in master data management creating authoritative customer records linking accounts correctly. Model performance improved dramatically once training data quality reached acceptable levels.

This experience taught them an expensive lesson: data quality isn't a one-time cleanup project; it requires ongoing governance, monitoring, and investment.

## 9.3 Privacy and Security Considerations

**Privacy protection** prevents unauthorized access to sensitive personal information while **security** protects data integrity and availability. Both are essential for trustworthy AI systems (Sweeney, 2002).

**Data classification** categorizes data by sensitivity: public (no harm from disclosure), internal (moderate harm), confidential (serious harm), restricted (severe harm or regulatory violations). Classification drives security controls—access restrictions, encryption requirements, audit logging, and retention policies.

**Access control** implements least privilege principles—users access only data necessary for their roles. Role-based access control (RBAC) grants permissions based on job functions. Attribute-based access control (ABAC) makes decisions based on user attributes, data attributes, and environmental context. Strong authentication (multi-factor) prevents unauthorized access even with compromised credentials.

**Encryption** protects data at rest (in databases and files) and in transit (during network transmission). Modern encryption standards (AES-256) make data unreadable without encryption keys. Key management is critical—encryption is worthless if keys are compromised. Use hardware security modules or key management services for production systems.

**Data masking** replaces sensitive fields with realistic but fake values for non-production use. Development and testing environments shouldn't contain real personal data. Masking techniques include substitution (replacing names with fake names), shuffling (randomly redistributing values), and nullification (blanking fields). Maintain referential integrity and statistical properties while protecting privacy.

**Audit logging** records data access and modifications, enabling detection of unauthorized activity and supporting incident investigation. Log who accessed what data, when, from where, and what they did. Retain logs securely for compliance periods. Monitor logs for suspicious patterns—unusual access volumes, access to sensitive data by unauthorized users, or access from unusual locations.

**Incident response planning** prepares for inevitable security breaches. Define detection mechanisms, escalation procedures, containment strategies, investigation processes, remediation steps, and communication plans. Practice through tabletop exercises and update plans based on lessons learned. Speed matters—delayed detection and response amplifies breach impacts.

**Privacy impact assessments** evaluate privacy risks before deploying AI systems. Document data collection purposes, legal bases, data types, retention periods, sharing arrangements, security measures, and individual rights implementations. Identify risks and mitigation strategies. Update assessments when systems change significantly.

## 9.4 Change Management for AI

**Change management** addresses the human side of AI adoption. Technical success doesn't guarantee organizational adoption. Users resist change that threatens job security, disrupts workflows, or requires new skills. Effective change management builds support and enables smooth transitions (Kotter, 2012).

**Stakeholder analysis** identifies everyone affected by AI initiatives—direct users, displaced workers, managers whose decision authority changes, customers experiencing different interactions, and executives accountable for outcomes. For each stakeholder group, assess attitudes (support, neutral, resistance), influence (high, medium, low), and concerns.

**Communication strategy** keeps stakeholders informed throughout AI projects. Explain why AI is being deployed, what problems it solves, how it works at appropriate detail levels, what will change, and what won't change. Address concerns proactively. Tailor messages to different audiences—executives care about business impact, users care about workflow changes, technical teams care about architecture.

**Training programs** prepare users for AI-augmented roles. Training covers how systems work, how to interpret outputs, when to override recommendations, and how to escalate issues. Hands-on practice with realistic scenarios builds confidence. Ongoing training addresses questions emerging post-deployment.

**Pilot programs** test AI systems with limited user groups before full deployment. Pilots surface usability issues, identify training gaps, validate benefits, and create internal champions who become advocates during broader rollout. Choose pilot groups representing diverse user populations and workflows.

**Feedback mechanisms** enable users to report problems, suggest improvements, and ask questions. User feedback identifies issues technical testing misses and surfaces feature requests guiding development priorities. Responsive feedback handling demonstrates that user input matters, building trust and engagement.

**Success metrics** demonstrate value to stakeholders. Track efficiency gains, quality improvements, error reductions, revenue increases, or cost savings attributable to AI adoption. Celebrate wins and communicate them widely. Success builds momentum for further adoption.

## 9.5 Team Composition and Roles

**Effective AI teams** combine diverse skills and perspectives. No single individual possesses all necessary expertise. Thoughtfully designed team structures enable collaboration while maintaining clear accountabilities (Kim et al., 2018).

**Data scientists** develop models, select algorithms, tune hyperparameters, and evaluate performance. They need statistics, machine learning, and programming skills plus domain knowledge contextualizing technical decisions. Data scientists bridge business problems and technical solutions.

**Machine learning engineers** operationalize models, building production pipelines, optimizing inference performance, implementing monitoring, and maintaining deployments. They need software engineering skills, MLOps expertise, and infrastructure knowledge. ML engineers bridge data science and operations.

**Data engineers** build data pipelines, manage data infrastructure, ensure data quality, and enable data access. They need database expertise, ETL development skills, and distributed systems knowledge. Data engineers provide the foundation enabling AI initiatives.

**Business analysts** translate business needs into technical requirements, evaluate business impact, and communicate results to stakeholders. They need domain expertise, analytical skills, and communication abilities. Business analysts ensure technical work serves business objectives.

**Product managers** define AI product strategy, prioritize features, manage roadmaps, and coordinate cross-functional teams. They need business acumen, technical literacy, and stakeholder management skills. Product managers ensure AI initiatives align with organizational strategy.

**Domain experts** provide subject matter expertise guiding model development, evaluating results, and ensuring practical applicability. A medical AI needs clinician input. A financial AI needs trader input. Domain expertise prevents technically sound but practically useless solutions.

**Ethics and compliance specialists** assess ethical implications, ensure regulatory compliance, evaluate fairness, and manage risk. They need ethics expertise, legal knowledge, and understanding of AI's societal impacts. These specialists protect organizations and individuals from AI harms.

Team structure depends on organization size and maturity. Small organizations might combine roles—a data scientist handles engineering tasks, a business analyst manages products. Large organizations specialize roles. Matrix structures where individuals report both to functional managers and project managers balance deep expertise with project focus.

## 9.6 Agile Approaches for AI

**Agile methodologies** adapted from software development enable iterative AI development. Traditional waterfall approaches poorly suit AI's inherent uncertainty—you often don't know if approaches will work until you try them. Agile embraces uncertainty through short iterations, frequent feedback, and continuous adaptation (Saltz & Dewar, 2019).

**Sprints** are short development cycles (typically 1-2 weeks) producing potentially deployable increments. Each sprint includes planning (defining work), execution (developing features), review (demonstrating results), and retrospective (improving processes). Short cycles enable rapid learning and course correction.

**User stories** capture requirements from user perspectives: "As a customer service agent, I want fraud risk scores for incoming calls so I can prioritize potentially fraudulent accounts." User stories focus on user needs and value rather than technical implementation, keeping teams oriented toward business outcomes.

**Backlog management** prioritizes work based on business value, technical risk, and dependencies. High-priority items get addressed first. The backlog evolves as new information emerges—priorities shift, requirements clarify, and technical challenges surface. Regular backlog grooming ensures teams work on what matters most.

**Daily standups** are brief team synchronizations (15 minutes) where members share progress, plans, and blockers. Standups enable coordination, surface impediments, and build team cohesion. Keep them focused and time-boxed to maintain efficiency.

**Continuous integration** automatically tests and validates code changes. For AI, this includes data validation, model training, performance evaluation, and bias testing. Automated pipelines catch problems early before they compound.

**MVP (Minimum Viable Product)** philosophy starts with the simplest approach potentially delivering value, then iterates based on feedback. Rather than building perfect models before deployment, deploy simple models quickly, learn from production data, and improve iteratively. This accelerates time-to-value and reduces wasted effort on features users don't need.

Agile for AI faces challenges. Model training can take days or weeks, incompatible with daily standups focused on completion. Experimentation might not produce deployable artifacts each sprint. Data dependencies create bottlenecks. Successful teams adapt Agile principles rather than applying software development practices literally.

## 9.7 Risk Management for AI

**AI-specific risks** go beyond traditional project risks. AI systems can fail in subtle ways, perpetuate biases, invade privacy, enable misuse, or create unintended consequences. Structured risk management identifies, assesses, and mitigates these risks (Amodei et al., 2016).

**Technical risks** include model failures (poor accuracy, high false positive/negative rates), brittle behaviors (failures on edge cases), adversarial vulnerabilities (malicious manipulation), concept drift (performance degradation over time), and scalability limits (inability to handle production loads). Mitigation strategies include comprehensive testing, adversarial testing, monitoring, retraining triggers, and architecture reviews.

**Ethical risks** include bias and discrimination, privacy violations, lack of transparency, unintended consequences, and misuse enabling harm. Mitigation requires fairness testing, privacy protections, explainability, impact assessments, and use case restrictions.

**Operational risks** include integration failures, data pipeline breaks, system outages, insufficient monitoring, and poor incident response. Mitigation involves robust engineering, redundancy, monitoring, alerting, and practiced incident procedures.

**Business risks** include misaligned objectives, insufficient value delivery, stakeholder resistance, reputational damage, and competitive disadvantage. Mitigation requires clear value propositions, stakeholder engagement, change management, and contingency planning.

**Regulatory risks** include compliance violations, inadequate documentation, insufficient human oversight, and failure to honor data subject rights. Mitigation involves legal review, compliance frameworks, documentation processes, and governance structures.

**Risk assessment** evaluates likelihood and impact for identified risks. High likelihood, high impact risks demand immediate attention. Low likelihood, low impact risks may be accepted. Risk matrices visualize assessments and guide prioritization.

**Risk mitigation** reduces likelihood or impact. Preventive controls reduce probability—data validation prevents quality issues, fairness testing reduces bias. Detective controls enable early detection—monitoring alerts on performance degradation. Corrective controls reduce impact—rollback procedures limit damage from bad deployments. Risk transfer (insurance, contracts) or acceptance (for low-impact risks) complete the risk response toolkit.

**Case Example: The Risk That Materialized**

An insurance company deployed an AI system recommending claim approvals. Risk assessment identified several concerns: model might discriminate against protected groups, claim adjusters might rubber-stamp AI recommendations without review, system might fail on unusual claims.

The team implemented mitigation strategies. They tested for disparate impact across demographics. They designed UI requiring explicit approval rather than passive acceptance. They implemented human review for claims above certain amounts or falling outside model confidence ranges. They established monitoring for performance drift and bias.

Six months post-deployment, monitoring flagged increasing disparate impact against elderly claimants. Investigation revealed concept drift—medical treatment patterns changed due to new therapies not represented in training data. The model learned associations between age and claim complexity that were outdated, leading to unfair treatment.

Because the team anticipated this risk and implemented monitoring, they detected the problem within weeks rather than months or years. They retrained the model on recent data, validated fairness metrics improved, and deployed the update. Rapid detection and response prevented significant harm and regulatory consequences.

This illustrates effective risk management—not preventing all problems (impossible) but anticipating potential failures, implementing detection, and responding quickly when they occur.

## 9.8 Resource Allocation Decisions

**Resource allocation** determines which AI projects receive funding, staffing, and infrastructure—critical decisions since most organizations can't pursue every opportunity. Thoughtful allocation maximizes value from limited resources (PMI, 2025).

**Business value assessment** estimates each project's potential impact. Consider revenue increases, cost reductions, risk mitigation, competitive advantage, strategic alignment, and customer satisfaction improvements. Quantify benefits where possible but recognize some benefits resist quantification.

**Technical feasibility** evaluates whether projects are achievable given current capabilities, data availability, and known techniques. High-risk technical approaches might fail despite best efforts. Balance high-risk, high-reward innovations with lower-risk, incremental improvements.

**Resource requirements** estimate costs across talent, compute infrastructure, data acquisition, tools and platforms, and ongoing operational expenses. Many organizations dramatically underestimate AI costs, leading to underfunded projects that fail or deliver suboptimal results.

**Timeline and urgency** factor into prioritization. Some opportunities have time windows—competitive threats, regulatory changes, seasonal patterns. Opportunities available indefinitely rank lower than time-sensitive opportunities of comparable value.

**Dependencies and sequencing** affect scheduling. Some projects must complete before others begin—master data management before customer analytics, data governance before AI deployment. Identify these dependencies during planning.

**Portfolio balancing** ensures healthy mixes across multiple dimensions: short-term wins and long-term bets, incremental improvements and transformational changes, low-risk and high-risk initiatives, different business units and functions. Excessive concentration creates vulnerability to single points of failure.

**Scoring models** systematically evaluate projects across criteria with weighted scores guiding prioritization. While imperfect, scoring models make evaluation criteria explicit and consistent. Example criteria: strategic alignment (20% weight), business value (30%), technical feasibility (20%), resource efficiency (15%), risk (15%). Projects are scored on each criterion, weighted scores sum to total scores, and projects are ranked by total.

**Stage-gate processes** release funding incrementally based on demonstrated progress. Initial funding supports business understanding and data exploration. Proof-of-concept funding enables model development on limited data. Full deployment funding requires demonstrating value through pilots. Stage gates reduce waste on projects that would ultimately fail while enabling promising projects to proceed.

## 9.9 Cross Functional Collaboration

**Collaboration** between diverse functions—business, technical, data, operations—determines AI success more than individual excellence. AI projects fail when teams work in silos, miscommunicate, or optimize for conflicting objectives (Davenport & Harris, 2017).

**Common language** prevents misunderstandings. Technical teams use jargon incomprehensible to business stakeholders. Business teams use domain language technical teams don't understand. Establish shared vocabulary defining key terms clearly. Create glossaries, run joint training, and use analogies bridging knowledge gaps.

**Joint workshops** bring together diverse perspectives. Business understanding workshops identify problems worth solving. Data workshops assess data availability and quality. Architecture workshops design systems meeting business and technical requirements. Collaborative design prevents misalignment caught late in projects.

**Embedded team members** place individuals from one function within another team. A business analyst embedded in a data science team ensures technical work serves business needs. A data scientist embedded in a business unit understands operational realities shaping requirements. Embedding builds relationships and mutual understanding.

**Regular checkpoints** maintain alignment throughout projects. Weekly updates keep stakeholders informed. Sprint reviews demonstrate progress and gather feedback. Executive briefings at project milestones secure continued support and address emerging concerns. Structured communication prevents surprises.

**Shared success metrics** align incentives across functions. If data scientists optimize for model accuracy while business teams optimize for user adoption, dysfunction results.

Define success jointly—model performance and business impact both matter. Make everyone accountable for shared outcomes rather than functional metrics alone.

**Conflict resolution processes** address inevitable disagreements. Technical teams might recommend approaches business teams consider too risky. Business teams might demand impossible timelines. Create escalation paths, establish decision-making authority, and document rationales for controversial decisions. Don't let conflicts fester.

## 9.10 Documentation and Knowledge Management

**Documentation** captures decisions, rationales, technical details, and lessons learned—enabling continuity when team members change, supporting debugging when problems arise, facilitating audits demonstrating compliance, and accelerating onboarding of new team members (PMI, 2025).

**Project documentation** includes business requirements, technical specifications, architecture diagrams, data source documentation, model development notebooks, evaluation results, deployment procedures, and operating manuals. Organize documentation logically, keep it current, and make it searchable.

**Data documentation** describes datasets thoroughly—source systems, collection methods, field definitions, quality assessment, known limitations, update frequency, and access procedures. Good data documentation prevents misuse and enables understanding model behavior.

**Model documentation** following model card frameworks captures intended use cases, training data characteristics, performance metrics across subpopulations, known limitations, ethical considerations, update history, and monitoring procedures. This creates accountability and enables informed deployment decisions.

**Decision logs** record why choices were made—algorithm selection rationale, hyperparameter tuning decisions, fairness metric choices, deployment strategy selection. When revisiting projects months later, decision logs explain "why we did it this way" preventing uninformed changes.

**Knowledge bases** organize documentation systematically—wikis, document management systems, or specialized platforms. Implement search functionality, access controls, and version control. Encourage contributions from entire teams, not just designated documenters.

**Lessons learned** capture what worked, what didn't, and what would be done differently. Conduct post-project retrospectives documenting successes, failures, unexpected challenges, and recommendations. Review lessons learned when planning new projects, avoiding repeated mistakes.

**Knowledge transfer** prepares for team transitions. When team members leave, their tacit knowledge leaves too unless explicitly captured. Schedule knowledge transfer sessions, record them, create documentation checklists, and implement shadowing periods for new team members. Don't let key knowledge reside in single individuals.

## 9.11 Your Takeaways

Data governance establishes ownership, policies, standards, cataloging, and master data management ensuring data quality and usability for AI. Without governance, data becomes fragmented and unreliable, undermining AI initiatives.

Data quality management assesses and improves accuracy, completeness, consistency, timeliness, validity, and uniqueness through profiling, cleansing, enrichment, and monitoring. Poor quality data generates unreliable models regardless of sophisticated algorithms.

Privacy and security protections including classification, access control, encryption, masking, audit logging, and incident response prevent unauthorized access while enabling legitimate use. Privacy impact assessments evaluate risks before deployment.

Change management through stakeholder analysis, communication, training, pilots, feedback mechanisms, and success metrics addresses the human side of AI adoption, building support and enabling smooth transitions.

Effective AI teams combine data scientists, ML engineers, data engineers, business analysts, product managers, domain experts, and ethics specialists. Team structures balance expertise depth with cross-functional collaboration.

Agile approaches adapted for AI through sprints, user stories, backlog management, and MVP philosophy enable iterative development embracing AI's inherent uncertainty. Successful teams adapt Agile principles rather than applying software practices literally.

Risk management identifies, assesses, and mitigates technical, ethical, operational, business, and regulatory risks through comprehensive testing, monitoring, governance structures, and incident response procedures.

Resource allocation through business value assessment, technical feasibility, cost estimation, timeline considerations, and portfolio balancing maximizes value from limited resources. Stage-gate processes release funding incrementally based on demonstrated progress.

Cross-functional collaboration through common language, joint workshops, embedded team members, regular checkpoints, and shared success metrics aligns diverse perspectives and prevents siloed work.

Documentation and knowledge management through project documentation, data documentation, model cards, decision logs, knowledge bases, lessons learned, and knowledge transfer enable continuity and support long-term success.

# 9.12 Practice Questions

## Team Structure Scenarios

**Question 471:** You're building an AI team from scratch. What roles are essential?

A) Only data scientists needed B) Diverse team: data scientists, ML engineers, data engineers, domain experts, ethics specialists, product managers C) Only software engineers D) One person can do everything

**Answer: B) Cross-functional team with complementary skills**

**Explanation:** Essential AI team roles: (1) Data scientists: modeling, experimentation, analysis; (2) ML engineers: production deployment, MLOps; (3) Data engineers: pipelines, infrastructure, data quality; (4) Domain experts: business context, feature engineering; (5) Ethics specialists: fairness, transparency, responsible AI; (6) Product managers: requirements, prioritization, stakeholder management; (7) Software engineers: system integration. AI projects require diverse skills. No single role has all needed expertise. Cross-functional collaboration essential for success.

---

**Question 472:** Your data scientists build great models but struggle with production deployment. ML engineers complain models aren't production-ready. What's the structural issue?

A) Data scientists are incompetent B) Lack of collaboration: need shared responsibility, clear handoff processes, embedded ML engineers with data science teams C) ML engineers should build all models D) No structural issue

**Answer: B) Collaboration gap requiring structural solutions**

**Explanation:** Bridging research-production gap: (1) Embedded structure: ML engineers embedded in data science teams, not separate; (2) Shared goals: both teams accountable for production success; (3) Production standards: data scientists follow production coding standards; (4) Collaboration tools: shared repositories, documentation, environments; (5) Early involvement: ML engineers involved from project start, not just at deployment; (6) Skill development: data scientists learn production skills, ML engineers learn modeling. Silos create handoff failures. Cross-functional collaboration from start prevents deployment challenges.

---

**Question 473:** Your AI team is centralized (reports to Chief Data Officer) but serves multiple business units with different priorities. Business units complain about slow response. What structure might work better?

A) Keep fully centralized B) Hybrid model: federated data scientists embedded in business units; centralized platform team providing tools/infrastructure; center of excellence for standards C) Fully decentralize everything D) No structure needed

**Answer: B) Federated model balancing business responsiveness with consistency**

**Explanation:** Organizational models: (1) Federated data scientists: embedded in business units, understand domain, respond quickly; (2) Platform team: centralized team building shared ML infrastructure, tools; (3) Center of excellence: sets standards, best practices, governance; (4) Dotted-line reporting: business unit data scientists report to business but have connection to central team; (5) Communities of practice: cross-BU knowledge sharing; (6) Shared resources: expensive resources (compute, tools) centrally managed. Balances speed (federated) with consistency and efficiency (centralized). Common in mature organizations.

---

**Question 474:** Your organization is hiring data scientists but most candidates lack domain knowledge in your industry (healthcare). Should you hire data scientists with healthcare expertise or strong technical skills with no domain knowledge?

A) Only hire candidates with both B) Hire for technical skills, partner data scientists with domain experts; or hire domain experts and train in ML; depends on availability C) Only hire healthcare experts D) Domain knowledge is irrelevant

**Answer: B) Balance technical and domain skills through teaming or training**

**Explanation:** Domain knowledge acquisition: (1) Hire technical + pair: hire strong data scientists, pair with domain experts; (2) Hire domain + train: hire domain experts (physicians, nurses), train in ML; (3) Hybrid teams: combinations of both; (4) Context matters: regulated industries often prefer domain experts trained in ML for compliance; (5) Timeline: training takes time—urgent projects may need existing expertise; (6) Career development: provide domain training to data scientists. Perfect candidates (technical + domain) are rare and expensive. Build capability through teaming or training.

---

**Question 475:** Your AI projects frequently fail in production due to data quality issues. Data engineering team is overwhelmed. What structural change would help?

A) Hire more data scientists B) Strengthen data engineering: increase data engineering capacity, embed data engineers in AI teams, data quality as shared responsibility C) Ignore data quality D) Replace data engineers

**Answer: B) Elevate data engineering importance and capacity**

**Explanation:** Data infrastructure investment: (1) Capacity: data engineers often outnumbered by data scientists—reverse ratio in data-intensive orgs; (2) Embedded: data engineers work directly with data scientists on projects; (3) Shared responsibility: data quality is everyone's job, not just data engineering; (4) Tools: invest in data quality monitoring, pipeline automation; (5) Priority: recognize data engineering as foundational, not support; (6) Career development: make data engineering attractive career with clear advancement. AI projects fail more often on data than algorithms. Data infrastructure is foundation. Structural recognition and investment essential.

---

**Question 476:** Your company has 5 different teams building ML models using different tools, frameworks, and practices. What governance structure would improve consistency?

A) Let teams use whatever they want B) Platform team approach: centralized team providing standard tools, frameworks, best practices; teams use common platform while retaining autonomy C) Force everyone to use identical tools D) No standardization needed

**Answer: B) Platform team with standards and shared services**

**Explanation:** Standardization without rigidity: (1) Platform team: builds common ML infrastructure (feature stores, model registry, deployment); (2) Standard tooling: recommended tools and frameworks (with flexibility for special cases); (3) Best practices: shared templates, patterns, documentation; (4) Self-service: platform enables teams to be productive independently; (5) Flexibility: standards with escape hatches for justified exceptions; (6) Inner sourcing: teams contribute to shared platform. Standardization brings efficiency (shared tools, knowledge transfer, simplified maintenance). But overly rigid standardization stifles innovation. Platform approach balances both.

---

**Question 477:** Your AI team consists of recent graduates with strong technical skills but little business experience. Projects often solve wrong problems. What's the structural solution?

A) Replace team with experienced people B) Mentorship and pairing: senior business-savvy leaders guide team; cross-functional embedding; rotations to business units C) Technical skills are sufficient D) No solution needed

**Answer: B) Development through mentorship and business exposure**

**Explanation:** Developing business acumen: (1) Senior mentorship: experienced leaders guide junior team on problem framing; (2) Product managers: strong PMs help translate business needs; (3) Business rotations: data scientists spend time in business units; (4) Cross-functional projects: mixed teams with business stakeholders; (5) Customer exposure: data scientists meet customers, see how products used; (6) Training: business training programs; (7) Hiring: balance junior with business-experienced hires. Junior teams are cost-effective but need development. Structured exposure to business builds necessary context. Don't expect immediate business sophistication.

---

**Question 478:** Your ML engineers are excellent at deployment but don't understand modeling choices data scientists make. Data scientists don't understand production constraints. How do you bridge this gap?

A) Keep them separate B) Cross-training and rotation: ML engineers learn modeling, data scientists learn production; job rotation; pair programming; shared on-call C) They don't need to understand each other D) Hire people who do everything

**Answer: B) Intentional skill cross-pollination**

**Explanation:** T-shaped skills development: (1) Cross-training: ML engineers attend modeling training; data scientists learn deployment; (2) Rotations: temporary role swaps building empathy; (3) Pair programming: work together on projects; (4) Shared on-call: data scientists on-call for production builds ownership; (5) Documentation: each side documents their work for other; (6) Collaboration rituals: regular working sessions together. Specialization is efficient but creates silos. T-shaped professionals (depth in one area, breadth in others) collaborate better. Invest in broadening skills strategically.

---

**Question 479:** Your organization wants to scale AI from 5 models to 50 models. Current team structure is ad-hoc. What changes are needed?

A) Keep ad-hoc structure B) Formal structure: specialized roles, defined processes, shared infrastructure, governance, portfolio management, dedicated operations team C) Just hire more people D) Scaling doesn't require changes

**Answer: B) Formalize structure and processes for scale**

**Explanation:** Scaling AI operations: (1) Specialized roles: differentiate data scientists, ML engineers, data engineers clearly; (2) Platform team: dedicated team for shared infrastructure; (3) MLOps team: focused on deployment, monitoring, maintenance; (4) Governance: formal processes for review, approval, compliance; (5) Portfolio management: prioritization across projects, resource allocation; (6) Standards: coding standards, documentation requirements, architectural patterns; (7) Automation: CI/CD, automated testing, monitoring. Ad-hoc works for small scale. Fifty models require structure, processes, and dedicated operations. Formalization enables scale.

---

**Question 480:** Your AI team is all male. How should you address diversity?

A) Diversity doesn't matter for technical work B) Intentional diversity initiatives: inclusive hiring practices, address barriers, mentorship programs, inclusive culture, measure progress C) Hiring based solely on merit means homogenous teams D) Ignore demographics

**Answer: B) Proactive diversity building with systemic approach**

**Explanation:** Building diverse teams: (1) Pipeline: recruit from diverse sources (HBCUs, women-in-tech programs, career changers); (2) Bias reduction: structured interviews, diverse interview panels, blind resume reviews; (3) Inclusive environment: ensure culture welcomes diverse perspectives; (4) Mentorship: support for underrepresented team members; (5) Retention: address why people leave; (6) Leadership commitment: diversity as organizational priority; (7) Metrics: track and report on diversity. Homogenous teams have blind spots. Diversity improves decision-making, catches biases, expands perspective. Requires intentional effort—won't happen accidentally. "Meritocracy" without bias awareness perpetuates homogeneity.

## Governance Policy Decisions

**Question 481:** Your organization needs data governance policies for AI projects. What should they cover?

A) Policies aren't needed B) Comprehensive coverage: data access control, privacy requirements, quality standards, retention policies, usage restrictions, security requirements, compliance C) Only focus on security D) One-page policy is sufficient

**Answer: B) Multi-dimensional data governance framework**

**Explanation:** Data governance scope: (1) Access control: who can access what data, approval processes; (2) Privacy: handling of PII, consent requirements, anonymization standards; (3) Quality: data quality standards, validation requirements; (4) Lifecycle: retention periods, archival, deletion policies; (5) Usage: permitted and prohibited uses, purpose limitation; (6) Security: encryption, transmission, storage requirements; (7) Compliance: GDPR, HIPAA, sector-specific regulations; (8) Documentation: data lineage, metadata requirements. Data governance is foundation for responsible AI. Comprehensive policies prevent problems. Living policies—update as regulations and practices evolve.

---

**Question 482:** Should your data governance policies apply equally to all data or vary by data sensitivity?

A) One policy for all data B) Risk-based tiers: higher requirements for sensitive data (PII, PHI, financial); lighter governance for non-sensitive data; clear classification scheme C) No governance for any data D) Random application

**Answer: B) Risk-based graduated governance**

**Explanation:** Tiered data governance: (1) Classification: clear taxonomy (public, internal, confidential, restricted); (2) Sensitivity factors: PII, trade secrets, regulated data, competitive intelligence; (3) Differential controls: more sensitive = stricter access, audit, retention; (4) Efficiency: don't burden non-sensitive data with unnecessary controls; (5) Clear criteria: objective classification criteria prevent confusion; (6) Labeling: data clearly labeled with classification; (7) Regular review: classifications updated as needed. One-size-fits-all governance is inefficient (over-controls non-sensitive) or ineffective (under-controls sensitive). Risk-based approach optimizes rigor and efficiency.

---

**Question 483:** Your data governance policy prohibits using customer data for training models without consent, but getting consent is difficult. Should you relax the policy?

A) Yes, relax policy for business convenience B) No, find compliant solutions: synthetic data, federated learning, anonymization, transparent consent mechanisms, or accept limitations C) Ignore policy D) Consent is unnecessary

**Answer: B) Maintain privacy standards; find compliant approaches**

**Explanation:** Privacy-preserving alternatives: (1) Synthetic data: generate realistic data without personal information; (2) Federated learning: train on distributed data without centralizing; (3) Differential privacy: add noise protecting individuals while enabling learning; (4) Anonymization: proper de-identification allowing use; (5) Consent infrastructure: invest in making consent easy and transparent; (6) Aggregate insights: learn from aggregate patterns, not individual data; (7) Business model: some use cases not viable without consent—accept limitation. Policies protecting individuals shouldn't be compromised for convenience. Innovation within constraints drives better solutions. Privacy-respecting ML is possible and often legally required.

---

**Question 484:** Different departments interpret your data governance policies differently, creating inconsistency. How do you improve policy clarity?

A) Policies are clear enough B) Implementation support: detailed guidance documents, decision trees, examples, FAQ, interpretation authority, training C) More ambiguous policies D) No interpretation needed

**Answer: B) Supplement policy with practical implementation guidance**

**Explanation:** Operationalizing policy: (1) Guidance documents: detailed how-to guides with examples; (2) Decision trees: flowcharts guiding policy application; (3) Examples: concrete scenarios illustrating policy; (4) FAQ: address common questions and edge cases; (5) Authority: designated team/person for authoritative interpretations; (6) Training: workshops teaching policy application; (7) Templates: standardized forms, checklists implementing policy; (8) Feedback loop: refine policy based on implementation challenges. High-level policies need practical implementation guidance. Ambiguity creates inconsistency. Invest in making compliance clear and easy.

**Question 485:** Your governance policy requires approval for all AI projects, creating bottleneck. How do you maintain governance while improving speed?

A) Remove all approvals B) Risk-based approval: low-risk projects fast-tracked or self-certified; high-risk projects require full review; clear risk criteria C) Approve everything instantly without review D) Governance and speed are incompatible

**Answer: B) Differentiated approval based on risk**

**Explanation:** Efficient governance processes: (1) Risk tiers: clear criteria differentiating high/medium/low risk; (2) Self-certification: low-risk projects self-certify compliance with checklist; (3) Fast track: medium-risk gets expedited review; (4) Full review: only high-risk gets comprehensive review; (5) Clear criteria: objective risk scoring preventing gaming; (6) Escalation: easy path to escalate if uncertain; (7) Resources: adequate review capacity for expected volume. Blanket requirements create bottlenecks. Risk-based approach focuses rigor where needed. Fast-track low-risk projects. Most AI isn't high-risk—reserve intensive governance for applications warranting it.

---

**Question 486:** Your data retention policy says delete data after 2 years, but ML models need retraining on historical data beyond 2 years. How do you resolve this conflict?

A) Ignore retention policy B) Policy refinement: exception for model training data with special protections; or anonymization allowing longer retention; document rationale C) Delete training data anyway D) Retention policies don't apply to ML

**Answer: B) Refine policy addressing ML-specific needs with safeguards**

**Explanation:** Policy adaptation: (1) ML exception: training data retention beyond general limits with business justification; (2) Enhanced protection: stricter access controls for long-retained training data; (3) Anonymization: de-identify data for longer retention; (4) Purpose limitation: long retention only for model training, not other purposes; (5) Periodic review: assess if still needed; (6) Documentation: clearly document retention rationale; (7) Privacy by design: minimal data retention for production predictions even if training data kept longer. Policies should enable business while protecting privacy. Blanket rules may need refinement for ML use cases. Accommodate legitimate needs with appropriate safeguards.

---

**Question 487:** Your organization allows third-party vendors to access customer data for analytics. How should governance address vendor data access?

A) Vendors can access anything B) Vendor governance: data processing agreements, access controls, audit rights, purpose limitation, security requirements, liability provisions C) No vendor access allowed ever D) Vendors are trusted by default

**Answer: B) Comprehensive vendor data governance**

**Explanation:** Third-party data management: (1) Contracts: data processing agreements specifying use, protection, retention; (2) Minimal access: vendors get only necessary data; (3) Purpose limitation: use only for specified purposes; (4) Security requirements: vendors must meet security standards; (5) Audit rights: ability to audit vendor practices; (6) Breach notification: vendors must report incidents; (7) Liability: clear responsibility for data breaches; (8) Vendor assessment: evaluate vendor practices before granting access; (9) Periodic review: reassess vendor access regularly. Vendors extend your data risk. Governance must extend to third parties. Contracts and controls protect your data in vendor hands.

---

**Question 488:** Your governance policy was written three years ago and doesn't address recent AI developments (LLMs, generative AI). How do you keep policies current?

A) Never update policies B) Regular policy review: scheduled reviews (annually/bi-annually), technology monitoring, stakeholder feedback, version control, communication of changes C) Policies are timeless D) Write new policy for each technology

**Answer: B) Living policy with scheduled updates**

**Explanation:** Policy maintenance: (1) Scheduled review: annual or bi-annual comprehensive review; (2) Ad-hoc updates: address urgent gaps as they emerge; (3) Technology monitoring: track AI developments requiring policy updates; (4) Stakeholder input: gather feedback on policy effectiveness; (5) Version control: track policy changes over time; (6) Communication: clearly communicate updates to all stakeholders; (7) Training: retrain staff on updated policies; (8) Sunset clauses: explicitly review and reaffirm or retire policies periodically. Technology and regulation evolve. Policies must evolve too. Regular review prevents obsolescence. Treat policy as living document.

---

**Question 489:** Your governance policy requires human review for certain AI decisions but doesn't specify what "meaningful human review" means. Teams interpret differently. What clarification is needed?

A) Current policy is adequate B) Define meaningful review: reviewer must have relevant information, expertise, time, authority to override; not rubber-stamping C) Human review means any human looks at it D) No clarification needed

**Answer: B) Specify requirements for effective human oversight**

**Explanation:** Meaningful human review elements: (1) Information: human has all information AI used plus context; (2) Expertise: reviewer has domain expertise to evaluate decision; (3) Time: adequate time to review, not rushed; (4) Authority: ability to override AI decision; (5) Accountability: reviewer responsible for final decision; (6) Not automation: review isn't just accepting AI recommendations automatically; (7) Documentation: review process documented; (8) Training: reviewers trained on AI limitations. "Human in the loop" is meaningless if human is rubber stamp. Specify requirements ensuring genuine human judgment. Prevents human-in-the-loop theater.

---

**Question 490:** Should data governance policies be public or internal-only?

A) Always secret B) Hybrid: high-level principles public (transparency, accountability); implementation details internal; balance transparency with security C) Always fully public D) Policies aren't anyone's business

**Answer: B) Tiered transparency with public principles**

**Explanation:** Policy transparency: (1) Public principles: high-level commitments to fairness, privacy, transparency; (2) Internal details: specific procedures, technical implementations kept internal; (3) Stakeholder access: more detailed information for regulators, auditors; (4) Security: don't publish information enabling attacks; (5) Competitive: protect proprietary practices; (6) Accountability: public commitments enable accountability; (7) Trust: transparency builds stakeholder trust. Full transparency might expose security vulnerabilities or competitive information. Full secrecy prevents accountability. Publish high-level commitments; keep sensitive implementation details internal.

## Risk Mitigation Strategies

**Question 491:** What are the major risk categories for AI projects that need mitigation strategies?

A) Only technical risks matter B) Multiple risk dimensions: technical (model failure), ethical (bias), regulatory (compliance), reputational (public backlash), operational (deployment issues), business (ROI failure) C) Risk is not manageable D) AI has no risks

**Answer: B) Holistic risk assessment across dimensions**

**Explanation:** AI risk taxonomy: (1) Technical: model accuracy, robustness, reliability; (2) Ethical: bias, fairness, privacy violations; (3) Regulatory: non-compliance with laws and regulations; (4) Reputational: negative publicity, trust loss; (5) Operational: system failures, security breaches; (6) Business: project failure, insufficient ROI; (7) Societal: broader negative impacts. Comprehensive risk management addresses all dimensions. Technical excellence doesn't eliminate ethical or regulatory risk. Each risk type requires specific mitigation strategies. Mature organizations assess and manage all risk categories.

---

**Question 492:** Your AI project could pose significant bias risk. What mitigation strategies should you implement upfront?

A) Address bias after deployment B) Preventive measures: diverse team, diverse training data, disaggregated evaluation, fairness constraints, external review, monitoring plan C) Ignore bias risk D) Bias is inevitable

**Answer: B) Multi-layered bias prevention from project start**

**Explanation:** Proactive bias mitigation: (1) Diverse team: multiple perspectives in design and development; (2) Data audit: assess training data for representation; (3) Balanced data: actively gather diverse, representative data; (4) Disaggregated testing: evaluate performance across demographic groups; (5) Fairness constraints: incorporate fairness requirements in training; (6) External review: have external experts assess for bias; (7) Monitoring: plan for ongoing bias monitoring post-deployment; (8) Documentation: document bias assessment and mitigation. Prevention is more effective than remediation. Address bias throughout project lifecycle. Multiple defense layers provide robust protection.

**Question 493:** How do you mitigate risk of regulatory non-compliance for an AI project?

A) Ignore regulations until problems arise B) Proactive compliance: early legal review, privacy-by-design, regular compliance checks, documentation, staying current on regulations C) Regulations don't apply to AI D) Compliance is impossible

**Answer: B) Build compliance into project from start**

**Explanation:** Compliance risk mitigation: (1) Early engagement: involve legal/compliance from project initiation; (2) Regulatory mapping: identify applicable regulations (GDPR, HIPAA, ECOA, etc.); (3) Privacy by design: build in privacy protections from architecture; (4) Regular checkpoints: compliance review at project milestones; (5) Documentation: maintain comprehensive compliance documentation; (6) Monitoring: track regulatory changes affecting project; (7) Training: team trained on relevant compliance requirements; (8) External expertise: consult regulatory experts when needed. Retrofitting compliance is costly or impossible. Build in from beginning. Legal and compliance are partners, not obstacles.

---

**Question 494:** Your AI project has potential for significant public criticism if it goes wrong (e.g., facial recognition). How do you mitigate reputational risk?

A) Ignore public opinion B) Reputation management: stakeholder engagement, transparent communication, ethics review, pilot testing, prepared response plans C) Public opinion doesn't matter D) Reputational risk is unmanageable

**Answer: B) Proactive reputation risk management**

**Explanation:** Reputational risk mitigation: (1) Stakeholder engagement: early dialogue with affected communities, civil society; (2) Transparency: public communication about safeguards and limitations; (3) Ethics review: rigorous ethical assessment before deployment; (4) Limited pilot: small-scale deployment with monitoring before full launch; (5) Prepared communications: crisis communication plan if issues arise; (6) Exit strategy: ability to shut down if problems can't be resolved; (7) Advisory board: external advisors providing guidance and credibility. Controversial AI applications need extra scrutiny. Engage critics early rather than defensively after problems. Some projects shouldn't proceed despite technical feasibility.

---

**Question 495:** What operational risks should you mitigate before deploying an AI model to production?

A) Just deploy and fix problems later B) Operational risk mitigation: testing (load, edge cases), monitoring infrastructure, rollback capability, incident response plans, capacity planning C) No operational preparation needed D) Production issues are inevitable and unpreventable

**Answer: B) Comprehensive operational readiness**

**Explanation:** Production risk mitigation: (1) Testing: load testing, edge case testing, integration testing; (2) Monitoring: comprehensive monitoring of performance, latency, errors; (3) Rollback: quick rollback to previous version if issues arise; (4) Incident response: runbooks for common issues, on-call rotation; (5) Capacity: adequate infrastructure for expected load plus buffer; (6) Redundancy: failover capabilities for high-availability; (7) Gradual rollout: canary deployment catching issues at small scale; (8) Data validation: input validation preventing bad data from breaking model. Production failures harm users and reputation. Thorough preparation prevents avoidable problems. Test, monitor, and prepare for failure recovery.

---

**Question 496:** Your AI project depends on a third-party data source that could become unavailable. How do you mitigate this dependency risk?

A) Assume data source will always be available B) Dependency risk mitigation: alternative data sources, cached data, fallback models, contractual guarantees, monitoring C) Single data source is fine D) Can't mitigate dependency risk

**Answer: B) Reduce dependency through redundancy and fallbacks**

**Explanation:** Third-party dependency mitigation: (1) Alternative sources: identify backup data providers; (2) Data caching: cache recent data for temporary outages; (3) Fallback model: simpler model not requiring third-party data; (4) Contracts: SLA agreements with penalties for downtime; (5) Monitoring: alert on data source issues; (6) Graceful degradation: system continues operating with reduced capability when data unavailable; (7) Evaluation: assess criticality of dependency—can you eliminate it? External dependencies create risk. Redundancy and fallbacks maintain service continuity. Balance data value against dependency risk.

---

**Question 497:** How do you mitigate the risk that your AI model becomes obsolete as the world changes?

A) Models never become obsolete B) Obsolescence mitigation: drift monitoring, scheduled retraining, architecture supporting updates, documentation enabling future maintenance C) Deploy once and forget D) Obsolescence is inevitable and unmanageable

**Answer: B) Design for change from the start**

**Explanation:** Maintainability and adaptability: (1) Drift detection: automated monitoring for performance degradation; (2) Retraining pipeline: infrastructure enabling regular model updates; (3) Modular architecture: easy to update components without complete rebuild; (4) Documentation: comprehensive documentation enabling future teams to maintain; (5) Regular reviews: scheduled reassessment of model relevance; (6) Feature store: easy to add new features as world evolves; (7) Version control: ability to update incrementally, not just rebuild. World is non-stationary. Models must evolve or become obsolete. Design for maintainability and updates. Technical debt accelerates obsolescence.

---

**Question 498:** Your AI project has tight deadlines. Stakeholders suggest cutting corners on testing to meet timeline. How do you balance speed and risk?

A) Cut all testing to meet deadline B) Risk-informed shortcuts: prioritize critical testing, defer nice-to-have features, negotiate timeline, communicate risks clearly to stakeholders C) Never take any shortcuts D) Ignore stakeholder timeline

**Answer: B) Informed tradeoffs with stakeholder awareness**

**Explanation:** Managing speed-quality tradeoffs: (1) Risk prioritization: identify must-have vs. nice-to-have testing; (2) Core coverage: never skip critical tests (fairness, security, core functionality); (3) Phased deployment: launch with limited scope, expand after more testing; (4) Technical debt: document shortcuts taken for future remediation; (5) Risk communication: clearly communicate risks of reduced testing to stakeholders; (6) Empowered to push back: organization supports necessary timeline negotiations.

Sometimes deadlines are negotiable; sometimes they're hard. When forced to cut corners, do so strategically with stakeholder awareness of risks. Never cut safety-critical testing.

---

**Question 499:** How do you mitigate risk that your AI model perpetuates or amplifies historical biases present in training data?

A) Historical data is objective; no risk B) Bias mitigation: acknowledge historical bias exists, balance datasets, fairness constraints, debiasing techniques, diverse evaluation C) Historical patterns should always be replicated D) Bias amplification is inevitable

**Answer: B) Actively counter historical bias through multiple techniques**

**Explanation:** Historical bias mitigation: (1) Recognition: acknowledge that historical data reflects past discrimination; (2) Data balancing: oversample underrepresented groups; (3) Debiasing: preprocessing, in-processing, post-processing techniques; (4) Fairness constraints: explicitly optimize for fairness, not just accuracy; (5) Feature selection: remove or modify features encoding bias; (6) Evaluation: test if model amplifies historical disparities; (7) Human oversight: human review for decisions perpetuating bias. Machine learning learns from past—including past injustices. Faithful reproduction of historical patterns can be harmful. Intentionally counter historical bias when it reflects discrimination.

---

**Question 500:** What contingency plans should you have for AI project failure?

A) Failure is not an option; no planning needed B) Failure planning: fallback to non-AI solution, incremental approach allowing early termination, lessons learned process, reputational damage control C) Never consider failure scenarios D) AI projects never fail

**Answer: B) Realistic failure planning and contingencies**

**Explanation:** Failure contingency: (1) Fallback: maintain non-AI solution as backup; (2) Kill criteria: pre-defined conditions triggering project termination; (3) Incremental: phased approach allowing early exit if not working; (4) Pivot planning: alternative approaches if initial doesn't work; (5) Lessons learned: systematic capture of insights from failures; (6) Communication: prepared messaging if project fails; (7) Resource reallocation: plan for reallocating resources from failed projects. Not all AI projects succeed. Failure should be acceptable option when recognized early. Pre-planning reduces cost of failure. Culture punishing all failures encourages hiding problems and continuing failed projects.

---

## Collaboration Challenge Resolution

**Question 501:** Data scientists and business stakeholders speak different languages, leading to miscommunication. How do you bridge this gap?

A) Expect business to learn technical language B) Translation mechanisms: product managers as translators, visualization over equations, business-focused metrics, regular collaboration rituals C) Expect data scientists to become business experts D) Communication isn't important

**Answer: B) Structural bridges between technical and business perspectives**

**Explanation:** Enabling cross-functional communication: (1) Product managers: skilled at translating between technical and business; (2) Visual communication: charts, demos, prototypes more accessible than equations; (3) Business metrics: discuss in terms of revenue, cost, user impact, not just accuracy; (4) Regular interaction: standing meetings, co-location, shared goals; (5) Translation training: train data scientists in business communication; (6) Business training: give stakeholders AI literacy; (7) Documentation: business requirements and technical specs both maintained. Communication gaps cause project failures. Invest in translation mechanisms. Both sides should stretch toward middle.

---

**Question 502:** Your data science team and engineering team have different tooling preferences and workflows, creating friction. How do you resolve this?

A) Force everyone to use same tools B) Common ground: shared tools where necessary (version control, deployment), flexibility where possible, focus on interfaces not implementations C) No coordination needed D) Tools don't matter

**Answer: B) Standardize interfaces; allow flexibility in implementation**

**Explanation:** Managing tool diversity: (1) Critical commonality: shared version control, deployment pipelines, monitoring; (2) Interface standards: APIs, data formats enabling interoperability; (3) Flexibility: teams choose tools for their domain where no interaction needed; (4) Evaluation: test before standardizing—different tools suited for different tasks;

(5) Migration paths: smooth transitions when standardizing; (6) Documentation: clear documentation of all tools and integrations. Total uniformity is rigid; total freedom is chaotic. Standardize what enables collaboration (interfaces, critical infrastructure); allow flexibility elsewhere. Focus on interoperability.

---

**Question 503:** Business stakeholders pressure AI team for unrealistic timelines. How do you manage expectations?

A) Commit to whatever stakeholders want B) Realistic planning: educate on AI project timelines, break projects into phases with early wins, transparent progress reporting, push back on impossible demands C) AI teams just need to work harder D) Ignore stakeholder input

**Answer: B) Education and transparent communication**

**Explanation:** Managing AI project expectations: (1) Education: help stakeholders understand AI development realities; (2) Historical data: show actual timelines from past projects; (3) Phased approach: deliver value incrementally rather than big-bang; (4) Early prototypes: quick POCs demonstrating feasibility; (5) Risk transparency: clearly communicate risks of rushing; (6) Negotiation: find compromises (reduced scope for faster delivery); (7) Organizational backing: leadership supports realistic timelines. Unrealistic expectations lead to cutting corners, poor quality, burnout. Education and transparency help stakeholders understand constraints. Sometimes must deliver bad news on timelines—better early than late.

---

**Question 504:** Your AI team operates in isolation and surprises business with models not meeting actual needs. How do you improve alignment?

A) Keep AI team isolated B) Collaborative development: embed stakeholders in team, frequent demos and feedback, user research, agile methodology with short iterations C) Business should accept whatever AI builds D) Alignment isn't possible

**Answer: B) Frequent feedback loops and stakeholder involvement**

**Explanation:** Continuous alignment: (1) Embedded stakeholders: business representatives participate in team meetings; (2) Frequent demos: biweekly demos of work in progress; (3)

User research: AI team directly interacts with end users; (4) Agile sprints: short cycles with feedback; (5) Acceptance criteria: clear upfront agreement on success criteria; (6) Prototypes: quick prototypes validating approach before full development; (7) Co-design: collaborative problem definition and solution design. Waterfall (build in isolation, reveal at end) fails frequently for AI. Agile collaboration with frequent feedback prevents building wrong solution. Stakeholder involvement throughout, not just at end.

---

**Question 505:** Different teams in your organization are solving similar ML problems independently, duplicating effort. How do you improve knowledge sharing?

A) Let teams work independently B) Knowledge sharing mechanisms: communities of practice, shared model registry, internal tech talks, documentation hub, inner source contributions C) Duplication is fine D) Centralize everything

**Answer: B) Structured knowledge sharing without over-centralization**

**Explanation:** Cross-team learning: (1) Communities of practice: regular meetings of practitioners across teams; (2) Model registry: shared catalog of models with metadata, use cases; (3) Tech talks: internal presentations sharing learnings; (4) Documentation hub: centralized documentation repository; (5) Inner source: teams contribute to shared libraries; (6) Mentorship: experienced teams mentor newer teams; (7) Cross-team rotations: temporary team swaps for knowledge transfer. Duplication wastes resources; over-centralization is slow. Enable discovery and reuse through knowledge sharing infrastructure. Voluntary adoption often more effective than mandated sharing.

---

**Question 506:** Your data engineering team is constantly interrupted by ad-hoc data requests from data scientists, preventing them from working on infrastructure improvements. How do you resolve this?

A) Data engineering should handle all requests immediately B) Request management: self-service data access tools, prioritization framework, dedicated office hours, roadmap for infrastructure improvements C) Ignore infrastructure improvements D) No data access for data scientists

**Answer: B) Balance responsive support with strategic work**

**Explanation:** Managing operational demands vs. strategic work: (1) Self-service: invest in tools enabling data scientists to access data independently; (2) Request queue: structured process for requests with prioritization; (3) Office hours: dedicated support times for ad-hoc questions; (4) Protected time: dedicated time for infrastructure work without interruptions; (5) Roadmap: transparent roadmap of infrastructure work; (6) Urgency triage: distinguish truly urgent from can-wait requests; (7) Automation: automate common requests. Operational demands can crowd out strategic work. Structural solutions (self-service, dedicated time) prevent this. Balance enables both responsive support and platform improvements.

---

**Question 507:** Your organization has acquired another company. Their AI team uses completely different tools, practices, and culture. How do you integrate the teams?

A) Force immediate conformity to your practices B) Gradual integration: learn from both sides, identify best practices from each, unified standards incorporating both, cultural sensitivity C) Keep teams permanently separate D) Fire one team

**Answer: B) Thoughtful integration preserving best of both**

**Explanation:** Post-merger integration: (1) Discovery: understand both teams' strengths and approaches; (2) Best practices: unified standards incorporating best from each; (3) Gradual transition: phased adoption of common practices; (4) Cultural sensitivity: respect both cultures, create new unified culture; (5) Communication: transparent about integration plans and rationale; (6) Quick wins: early successes building trust and momentum; (7) Retain talent: ensure key people from both sides feel valued. Mergers are challenging. Neither side should feel they "lost." Learn from both, create something better than either alone. Rushed integration causes resentment and attrition.

---

**Question 508:** Legal department blocks many AI initiatives citing compliance concerns. AI team views legal as obstacle. How do you improve this relationship?

A) AI team should bypass legal B) Partnership building: educate legal on AI, involve legal early in projects, collaborative problem-solving, legal embedded in AI team C) Legal should approve everything instantly D) Adversarial relationship is fine

**Answer: B) Transform legal from gatekeeper to partner**

**Explanation:** Legal-AI collaboration: (1) Education: teach legal team about AI capabilities and limitations; (2) Early involvement: engage legal at project start, not at end; (3) Embedded: legal representative participates in AI team meetings; (4) Problem-solving: collaborative finding of compliant solutions, not just blocking; (5) Risk framework: agreed-upon risk assessment framework; (6) Trust building: legal sees AI team as responsible, AI team sees legal as partner; (7) External expertise: bring in AI law experts to educate both sides. Legal as afterthought leads to blocked projects. Legal as early partner finds compliant paths forward. Both sides share goal of responsible innovation.

---

**Question 509:** Your distributed AI team spans multiple time zones, making collaboration difficult. How do you manage remote collaboration?

A) Require everyone in same time zone B) Asynchronous work: documentation culture, overlap hours for synchronous meetings, tools enabling async collaboration, clear communication protocols C) No accommodation needed D) Distributed teams can't work

**Answer: B) Design for asynchronous collaboration**

**Explanation:** Distributed team practices: (1) Documentation: comprehensive written communication, not just verbal; (2) Overlap hours: find common hours for critical meetings; (3) Asynchronous tools: collaborative documents, recorded demos, threaded discussions; (4) Clear protocols: response time expectations, communication channels; (5) Rotate meeting times: share burden of inconvenient meeting times; (6) Regular synchronous: some real-time interaction for building relationships; (7) Intentional communication: overcommunicate in writing. Distributed teams are increasingly common. Asynchronous collaboration enables productivity across time zones. Requires intentional practice and discipline.

---

**Question 510:** Your organization's executives don't understand AI, leading to unrealistic expectations and poor decisions about AI investments. How do you address this?

A) Executives don't need to understand AI B) Executive education: AI literacy programs for leadership, regular updates on AI realities, external experts, site visits to successful AI organizations C) Technical details are unnecessary for executives D) Ignore executive knowledge gaps

**Answer: B) Systematic executive AI education**

**Explanation:** Leadership AI literacy: (1) Customized training: AI fundamentals tailored for business leaders; (2) Real examples: concrete use cases and realistic timelines; (3) External perspectives: bring in experts, industry peers; (4) Site visits: see mature AI operations at other companies; (5) Regular updates: quarterly AI state-of-the-art briefings; (6) Experimentation: hands-on experience with AI tools; (7) Advisor network: access to AI advisors for guidance. Executive understanding is critical for realistic AI strategy. Without it, unrealistic expectations, poor investment decisions, and strategic missteps. Investment in leadership education has high ROI. Can't develop effective AI strategy without basic understanding.

---

## Resource Allocation Problems

**Question 511:** You have limited budget for AI initiatives. Five projects proposed, all seem valuable. How do you prioritize?

A) Fund all projects inadequately B) Prioritization framework: business value, technical feasibility, strategic alignment, resource requirements, sequencing dependencies C) Random selection D) Fund nothing

**Answer: B) Systematic prioritization based on multiple criteria**

**Explanation:** AI portfolio prioritization: (1) Business value: expected ROI, revenue impact, cost savings; (2) Strategic importance: alignment with company strategy; (3) Technical feasibility: likelihood of success given current capabilities; (4) Resource requirements: team, data, infrastructure needs; (5) Risk: probability and impact of failure; (6) Dependencies: foundation for future projects?; (7) Timing: market windows, competitive pressure; (8) Stakeholder input: diverse perspectives on priorities. Limited resources require hard choices. Systematic framework enables defensible decisions. Fund fewer projects adequately rather than many poorly. Consider sequencing—some projects enable others.

---

**Question 512:** Your AI team is constantly pulled in multiple directions by different business units competing for their time. How do you manage competing demands?

A) Try to satisfy everyone simultaneously B) Portfolio management: governance board prioritizing work, transparent capacity planning, formal request process, dedicated allocations C) Work on whatever shouts loudest D) Ignore all requests

**Answer: B) Formal governance and capacity management**

**Explanation:** Demand management: (1) Governance board: cross-functional leadership prioritizing AI investments; (2) Capacity planning: transparent view of team capacity and commitments; (3) Request process: structured intake preventing ad-hoc requests; (4) Prioritization: clear criteria for prioritizing work; (5) Dedicated allocations: some capacity reserved for each major business unit; (6) Roadmap: transparent roadmap of committed work; (7) Saying no: ability to decline low-priority work. Unmanaged demand leads to team thrashing, nothing completed. Formal governance enables focus. Transparency on capacity and commitments prevents overpromising.

---

**Question 513:** Your organization wants to hire senior AI talent but can't compete with tech giants on compensation. How do you attract talent with limited budget?

A) Accept you can't compete for talent B) Non-financial attractions: interesting problems, impact, autonomy, learning opportunities, culture, flexibility, mission C) Only hire junior talent D) Ignore talent constraints

**Answer: B) Emphasize non-monetary value proposition**

**Explanation:** Talent attraction beyond compensation: (1) Interesting problems: challenging, meaningful work; (2) Impact: work that matters, clear connection to outcomes; (3) Autonomy: freedom to choose approaches, minimal bureaucracy; (4) Learning: exposure to diverse problems, latest technologies; (5) Culture: collaborative, supportive environment; (6) Flexibility: remote work, flexible hours; (7) Mission: purposeful work aligned with values; (8) Growth: career development opportunities. Compensation matters but isn't everything. Many talent value impact, learning, culture, flexibility. Play to your strengths (meaningful mission, interesting problems, autonomy). Some talent willing to accept lower pay for right opportunity.

---

**Question 514:** Your AI projects need expensive GPUs but budget for compute infrastructure is limited. How do you optimize resource allocation?

A) Buy minimal infrastructure B) Strategic allocation: cloud for elastic needs, shared GPU pools, job scheduling, resource monitoring, optimize model efficiency C) Everyone gets their own GPUs D) Don't use GPUs

**Answer: B) Efficient shared infrastructure with optimization**

**Explanation:** Compute resource optimization: (1) Hybrid cloud: owned hardware for baseline, cloud for spikes; (2) Shared pools: GPU clusters shared across teams with scheduling; (3) Job orchestration: queue management maximizing utilization; (4) Monitoring: track usage, identify waste; (5) Efficiency: model optimization reducing compute needs (quantization, distillation); (6) Right-sizing: match compute to workload; (7) Scheduling: off-peak usage for non-urgent work; (8) Prioritization: critical work gets priority access. Compute is expensive. Shared infrastructure with good orchestration more efficient than dedicated resources per team. Monitor and optimize continuously.

---

**Question 515:** You need to allocate data scientists between maintaining existing models and developing new ones. How do you balance maintenance vs. innovation?

A) All effort on new development B) Balanced allocation: dedicated maintenance capacity (e.g., 20-30% of team), rotation through maintenance, operations team for routine tasks C) All effort on maintenance D) Ignore maintenance

**Answer: B) Planned allocation ensuring both maintenance and innovation**

**Explanation:** Maintenance-innovation balance: (1) Dedicated capacity: allocate percentage of team to maintenance; (2) Rotation: rotate team members through maintenance duty; (3) Operations team: dedicated MLOps team handling routine operational tasks; (4) Automation: automate retraining, monitoring, reducing manual maintenance; (5) Sunsetting: retire low-value models reducing maintenance burden; (6) Build for maintainability: invest in maintainable architecture reducing future cost; (7) Guard innovation time: protect time for new development. Pure innovation creates unmaintainable portfolio; pure maintenance prevents growth. Explicit allocation ensures both. Automation and maintainable design reduce maintenance burden.

---

**Question 516:** Your organization is deciding between building internal AI capabilities vs. buying external solutions. How do you make this decision?

A) Always build B) Build vs. buy analysis: strategic importance, competitive differentiation, available solutions, internal capabilities, total cost of ownership C) Always buy D) Random choice

**Answer: B) Strategic build-vs-buy analysis**

**Explanation:** Build vs. buy factors: (1) Strategic importance: competitive differentiator = build; commodity = buy; (2) Available solutions: good external solution = buy; gap in market = build; (3) Internal capability: have skills = build; lack skills = buy (or build capability); (4) Customization: unique requirements = build; standard needs = buy; (5) TCO: consider all costs including maintenance, not just initial; (6) Speed: buy often faster to deploy; (7) Lock-in: vendor dependence risk; (8) IP: build retains intellectual property. No universal answer. Core competitive capabilities often built. Standard capabilities often bought. Hybrid common (buy foundation, customize). Honest capability assessment critical.

---

**Question 517:** Your data science team wants to experiment with latest ML techniques but stakeholders want them focused on delivering business value. How do you balance exploration and exploitation?

A) Only business-driven work B) Allocated exploration time: 10-20% time for experimentation, innovation sprints, internal research projects with business case C) Only exploration D) No balance needed

**Answer: B) Structured allocation for both delivery and exploration**

**Explanation:** Innovation-delivery balance: (1) Percentage time: Google's "20% time" model for exploration; (2) Innovation sprints: dedicated periods for experimentation; (3) Research projects: small-scale research with potential business impact; (4) Conferences: time for learning through conferences, papers; (5) Business case: exploration tied to potential business value, not purely academic; (6) Sharing: internal talks sharing learnings; (7) Sunset: not all exploration continues. Pure delivery creates stagnation; pure exploration delivers no value. Explicit time allocation enables both. Exploration today is delivery tomorrow. Experimentation attracts and retains talent.

---

**Question 518:** Your organization has limited labeled data. Should you invest budget in labeling more data or improving algorithms to work with less data?

A) Only focus on algorithms B) Depends on returns: which has higher expected improvement? Often more good data beats better algorithms C) Only focus on labeling D) Neither investment is useful

**Answer: B) Evaluate marginal returns on each investment**

**Explanation:** Data vs. algorithms tradeoff: (1) Learning curves: does performance still improve with more data?; (2) Current data quality: is existing data high quality?; (3) Diminishing returns: algorithm improvements plateauing?; (4) Cost comparison: relative cost of labeling vs. algorithm development; (5) Data leverage: can labeled data enable multiple models?; (6) Time: labeling might be faster than algorithm research; (7) Empirical testing: test with sample of additional data vs. algorithm improvements. "Data beats algorithms" is common. If learning curve still rising, more data likely better investment than algorithm refinement. But if learning curve plateaued, algorithm improvements or better features may help. Empirically test when possible.

---

**Question 519:** Multiple teams want to use your organization's feature store, but maintaining it requires dedicated resources. Who should fund feature store development?

A) Whoever builds it funds it B) Shared infrastructure funding: central budget or cost-sharing model across benefiting teams; platform team responsible C) No funding needed D) Each team builds their own

**Answer: B) Treat as shared infrastructure with central or distributed funding**

**Explanation:** Shared service funding models: (1) Central funding: platform team funded centrally as infrastructure; (2) Chargeback: teams pay for usage proportionally; (3) Contribution: teams contribute engineer time to platform; (4) Hybrid: central funding with team contributions; (5) Business case: quantify value across organization justifying investment; (6) Governance: shared governance ensuring platform serves all users. Feature stores benefit all teams but require investment. Central funding treats as core infrastructure. Chargeback aligns incentives but adds complexity. Common pattern: central funding for platform team, team contributions for specific features. Shared benefits justify shared funding.

**Question 520:** Your AI initiatives compete with other tech investments (cloud migration, security upgrades) for limited IT budget. How do you make the case for AI investment?

A) AI is always top priority B) Business case: quantify AI value (revenue, cost savings, competitive advantage), compare to other investments, start with high-ROI projects C) AI doesn't need justification D) Accept no AI investment

**Answer: B) Compete on business value, not technology hype**

**Explanation:** Justifying AI investment: (1) Quantified value: expected revenue increase, cost reduction, efficiency gains; (2) Competitive necessity: capabilities competitors have or are building; (3) Strategic enablement: capabilities enabling business strategy; (4) Risk mitigation: risks of not investing (falling behind); (5) Comparison: how does ROI compare to other investments?; (6) Track record: show value from pilot projects; (7) Incrementalism: start with high-ROI projects, expand from wins. AI competes with all investments. Technology hype doesn't justify investment—business value does. Demonstrate value through pilots. Build track record of delivering ROI. Success creates funding for more ambitious projects. Start where you can show clear value.

# Chapter 10: Comprehensive Practice Exams & Advanced Scenarios

## Mixed-Difficulty Questions Across All Phases

**Instructions:** This practice exam covers all CRISP-ML(Q) phases and related topics. Questions are arranged in random order to simulate a real certification exam. Allow 90 minutes for completion. Passing score: 70% (35/50 correct).

---

**Question 521 [EASY - Business Understanding]:** A retail company wants to "use AI to increase sales." What's the first critical step?

A) Start collecting data immediately B) Hire more data scientists C) Define specific, measurable business objectives (e.g., "increase conversion rate by 10% within 6 months") D) Build a recommendation system

**Answer: C) Define specific, measurable business objectives**

**Explanation:** Vague goals like "increase sales" are unmeasurable and don't provide direction. CRISP-ML(Q) Phase I requires translating business desires into specific objectives with success criteria. "Increase conversion rate by 10% within 6 months" provides clear target, metric, and timeline. This enables appropriate solution design and success measurement. Starting with data collection or solution implementation without clear objectives leads to misaligned efforts.

---

**Question 522 [MEDIUM - Data Understanding]:** During EDA, you discover that customer_age has values: mean=35, median=28, mode=25. What does this distribution suggest?

A) Perfectly normal distribution B) Right-skewed distribution with older outliers pulling the mean up C) Left-skewed distribution D) Bimodal distribution

**Answer: B) Right-skewed distribution with older outliers**

**Explanation:** When mean > median > mode, the distribution is right-skewed (positively skewed) with a long tail of high values. The mean (35) is pulled upward by older customers, while the median (28) and mode (25) indicate most customers are younger. This pattern is common in age distributions where most customers are young but some very old customers increase the average. Understanding distribution shape informs feature engineering and modeling choices.

---

**Question 523 [HARD - Model Development]:** You're building a model for loan approval. Legal requires explanations for all denials. You have three options: Logistic Regression (78% accuracy, fully interpretable), Gradient Boosting (85% accuracy, SHAP explanations available), Deep Neural Network (87% accuracy, difficult to explain). Which should you choose?

A) Deep Neural Network (highest accuracy) B) Logistic Regression (fully interpretable) C) Gradient Boosting (balance of performance and explainability) D) All are equally suitable

**Answer: C) Gradient Boosting with SHAP explanations**

**Explanation:** This requires balancing competing priorities: performance, explainability, and legal compliance. Logistic Regression is interpretable but sacrifices significant performance (9% less than best). Deep Neural Network has best performance but explanation difficulty creates legal risk. Gradient Boosting offers strong compromise: 85% accuracy (7% improvement over logistic regression, only 2% less than neural network) with SHAP providing feature-level explanations meeting legal requirements. In regulated lending, this balance is optimal—significant performance improvement without sacrificing explainability. Legal compliance is non-negotiable; optimize within constraints.

---

**Question 524 [EASY - Data Preparation]:** You have a categorical feature "country" with 195 unique values. One-hot encoding would create 195 columns. What's a better approach for this high-cardinality feature?

A) One-hot encode all 195 countries B) Target encoding, frequency encoding, or grouping rare countries into "Other" C) Delete the country feature D) Convert to numeric by alphabetical order

**Answer: B) Target encoding, frequency encoding, or grouping**

**Explanation:** High-cardinality categorical features (many unique values) create problems with one-hot encoding: dimensionality explosion, sparse data, overfitting risk. Better approaches: (1) Target encoding: replace countries with average target value (use cross-fold to prevent leakage); (2) Frequency encoding: replace with occurrence frequency; (3) Grouping: combine rare countries into "Other" category, one-hot encode remaining; (4) Embeddings: learn dense representations. Choice depends on cardinality level and model type. Target encoding typically works well for 50+ categories.

---

**Question 525 [MEDIUM - Model Evaluation]:** Your model shows: Training Accuracy: 98%, Validation Accuracy: 72%. What's the problem and solution?

A) Model is underfitting; increase complexity B) Model is overfitting; add regularization, reduce complexity, or get more training data C) Model is perfect; deploy immediately D) Validation set is wrong

**Answer: B) Model is overfitting; needs regularization**

**Explanation:** Large gap between training (98%) and validation (72%) indicates severe overfitting—model memorized training data but doesn't generalize. The 26-point gap is concerning. Solutions: (1) Regularization: L1/L2 penalties, dropout for neural networks; (2) Reduce complexity: fewer features, smaller model; (3) More training data: dilutes impact of memorization; (4) Early stopping: stop training before overfitting worsens; (5) Data augmentation: artificially expand training data. Never deploy models with this level of overfitting—validation performance predicts production performance, not training performance.

---

**Question 526 [HARD - Deployment]:** Your model must handle 10,000 requests/second with <50ms latency. Current model: 100ms latency, 90% accuracy. You can optimize to 40ms latency but accuracy drops to 85%. What should you do?

A) Keep current model; latency requirements don't matter B) Optimize to meet latency requirement; 5% accuracy loss acceptable to meet hard constraint C) Deploy despite not meeting latency requirement D) Abandon the project

**Answer: B) Optimize to meet latency requirement**

**Explanation:** Hard constraints (latency, memory, compliance) eliminate options that don't meet them. 100ms latency is 2× over the 50ms requirement—system won't function. The 40ms optimized model meets requirements with acceptable accuracy tradeoff (85% vs 90%). Strategies enabling this: model compression (quantization, pruning), architecture optimization (MobileNet-style efficient architectures), infrastructure optimization (GPU acceleration, caching). Requirements hierarchy: (1) Hard constraints must be met; (2) Optimize soft metrics within constraints. A 90% accurate model that can't meet latency is worthless; an 85% accurate model that works is valuable.

---

**Question 527 [EASY - Ethics & Fairness]:** Your hiring model shows different pass rates for men (60%) vs women (40%). What's your first action?

A) Deploy the model; statistical differences are normal B) Investigate the disparity through disaggregated evaluation and bias testing C) Adjust thresholds to equalize pass rates regardless of qualification differences D) Delete gender from training data and ignore the issue

**Answer: B) Investigate the disparity through bias testing**

**Explanation:** Different outcomes by protected class warrant investigation, not immediate action or dismissal. Investigation steps: (1) Measure: confirm statistical significance of difference; (2) Root cause: do differences reflect true qualification differences or bias?; (3) Features: which features drive disparity? Are they proxies for gender?; (4) Historical bias: does training data reflect past discrimination?; (5) Fairness definition: which fairness metric is appropriate? Investigation determines whether disparity is problematic discrimination or reflects legitimate differences. Simply removing gender doesn't prevent bias (proxies remain). Blindly equalizing rates might introduce reverse discrimination. Thoughtful investigation precedes action.

---

**Question 528 [MEDIUM - Data Governance]:** Your team needs access to customer PII for model training but privacy regulations require data minimization. What's the best approach?

A) Collect all available PII; more data is better B) Collect only PII necessary for defined purpose; anonymize when possible; document necessity C) Never use PII for anything D) PII restrictions don't apply to internal use

**Answer: B) Data minimization with necessity documentation**

**Explanation:** GDPR and privacy-by-design principles require data minimization: collect only data necessary for specified purposes. Process: (1) Define purpose: what exactly is model doing?; (2) Necessity analysis: which PII fields are truly necessary?; (3) Alternatives: can you use aggregated, anonymized, or synthetic data?; (4) Proportionality: is PII use proportionate to benefit?; (5) Documentation: document why each PII field is necessary; (6) Protections: extra safeguards for PII (access control, encryption); (7) Retention limits: delete when no longer needed. Compliance doesn't mean no PII use—it means justified, minimal PII use with appropriate protections.

---

**Question 529 [HARD - Drift Management]:** Production monitoring shows: input distributions stable, model performance stable, but business metrics (conversion rate) declining. What's likely happening?

A) No problem; model performance is stable B) Distribution shift in competitor behavior or market conditions that model doesn't capture C) Monitoring is broken D) Business metrics are unrelated to model

**Answer: B) External environment change not reflected in model**

**Explanation:** This scenario illustrates that stable model metrics don't guarantee business value. Possible explanations: (1) Competitor actions: competitors adapted strategies, changing customer behavior; (2) Market shifts: economic conditions, seasonal changes not in model; (3) Missing features: model doesn't capture relevant changing factors; (4) Lagging indicators: model features updated slowly while market changes quickly. Investigation: (1) Business analysis: what changed in market?; (2) Feature engineering: add features capturing market dynamics; (3) Frequent retraining: capture recent patterns; (4) External data: incorporate competitor, economic data. Technical metrics (accuracy) measure model behavior; business metrics measure business value. Both matter.

---

**Question 530 [EASY - Project Management]:** Your AI project has data scientists, ML engineers, and data engineers working in silos with poor communication. What structural change would help most?

A) Hire more people in each silo B) Create cross-functional teams with all roles working together on shared goals C) Keep silos but add more meetings D) Eliminate ML engineers

**Answer: B) Cross-functional teams with shared goals**

**Explanation:** Silos create handoff failures, miscommunication, and finger-pointing. Cross-functional teams solve this through: (1) Shared goals: entire team accountable for production success, not just their phase; (2) Continuous collaboration: work together throughout project, not just handoffs; (3) T-shaped skills: encourage learning across disciplines; (4) Embedded roles: ML engineers embedded in data science teams; (5) Co-location: physical or virtual co-location; (6) Shared tools: common platforms, repositories. More meetings don't fix structural problems. Reorganizing into cross-functional teams with end-to-end responsibility creates accountability and reduces coordination overhead.

---

**Question 531 [MEDIUM - Model Selection]:** You're building a customer churn prediction model. Dataset: 100,000 customers, 50 features, 5% churn rate. Which algorithm is likely most appropriate as a starting point?

A) Linear regression (continuous target) B) Gradient boosting (handles imbalance, non-linear relationships, structured data) C) Convolutional neural network (image data) D) K-means clustering (unsupervised)

**Answer: B) Gradient boosting for structured imbalanced classification**

**Explanation:** Problem characteristics: (1) Binary classification (churn: yes/no); (2) Structured/tabular data (customer features); (3) Moderate size (100K records); (4) Imbalanced (5% churn); (5) Likely non-linear relationships. Algorithm selection: Linear regression is for continuous targets. CNNs are for images. K-means is unsupervised. Gradient boosting (XGBoost, LightGBM, CatBoost) excels at: structured data, handling imbalance (via class weights), capturing non-linear relationships, providing feature importance, delivering strong performance with moderate tuning. Alternative: Random Forest also suitable. Start with gradient boosting; compare to logistic regression baseline. Neural networks typically overkill for tabular data unless massive scale.

---

**Question 532 [HARD - Business-Technical Alignment]:** Your fraud detection model catches 90% of fraud with 5% false positive rate. Business says 5% false positives creates too many customer friction incidents (1M transactions daily = 50K false alarms). Legal requires catching 80%+ fraud. How do you resolve these conflicting requirements?

A) Ignore business complaints; 90% fraud detection is optimal B) Lower threshold to 80% fraud detection (legal minimum) reducing false positives; implement smart review routing C) Increase false positives to catch more fraud D) Impossible to satisfy both; abandon project

**Answer: B) Optimize within constraints; add intelligent triage**

**Explanation:** Multiple competing requirements need creative solutions: (1) Legal constraint: must catch 80%+ fraud (hard requirement); (2) Business constraint: 5% false positive rate unsustainable (capacity limitation); (3) Solution space: operate between 80-90% true positive rate, minimize false positives. Approaches: (1) Threshold adjustment: lower threshold to ~85% TPR, ~2-3% FPR; (2) Tiered review: high-confidence flags (likely fraud) → automated blocking; medium-confidence → human review; low-confidence → let through; (3) Feature improvement: better features reduce false positives at same true positive rate; (4) Hybrid models: combine rule-based and ML for precision; (5) Business process: prioritize high-value transaction reviews. Optimization problem with constraints, not impossible tradeoff. Legal minimum (80%) is lower bound; find best performance within capacity constraints.

---

**Question 533 [EASY - Feature Engineering]:** You have a dataset with transaction timestamps. What temporal features would be most valuable?

A) Keep only raw timestamp B) Extract: hour, day_of_week, month, is_weekend, is_holiday, time_since_last_transaction C) Delete all time information D) Convert to seconds since epoch only

**Answer: B) Extract multiple temporal features**

**Explanation:** Raw timestamps have little predictive value; derived temporal features capture patterns: (1) Cyclical: hour (0-23), day_of_week (0-6), month (1-12); (2) Binary: is_weekend, is_holiday, is_business_hours; (3) Relative: time_since_last_transaction, days_until_deadline; (4) Aggregate: transaction_count_last_7_days; (5) Seasonal: quarter, week_of_year. Consider cyclical encoding (sin/cos) for features like hour where 23 is close to 0. Rich temporal features enable models to learn time-based patterns (fraud at night, shopping on weekends, monthly billing cycles). Feature engineering multiplies value of raw timestamps.

---

**Question 534 [MEDIUM - Validation Strategy]:** You're building a stock price predictor using 5 years of daily data (2019-2023). What's the appropriate validation strategy?

A) Random 80/20 train/test split B) Temporal split: train on 2019-2021, validate on 2022, test on 2023 C) K-fold cross-validation with random splits D) No validation needed

**Answer: B) Temporal split preserving time order**

**Explanation:** Time series validation must preserve temporal order to prevent leakage. Random splitting leaks future information into training, creating unrealistically optimistic performance estimates. Proper approach: (1) Temporal training split: train on past (2019-2021); (2) Temporal validation: validate on middle period (2022) for model selection; (3) Temporal test: test on most recent period (2023) simulating production; (4) Alternative: rolling-window cross-validation (multiple temporal splits); (5) Gap: optional gap between train/test prevents autocorrelation leakage. Production will predict future from past; validation must simulate this. Random CV appropriate for non-temporal data, not time series.

---

**Question 535 [HARD - Multi-Phase Integration]:** Your project: (1) Business wants to reduce customer support costs, (2) Hypothesis: chatbot can handle 40% of inquiries, (3) You build 85% accurate intent classifier, (4) Deployed model incorrectly routes customers causing frustration, (5) Support costs actually increased. What went wrong?

A) Model accuracy too low B) Poor problem definition: didn't consider customer experience; accuracy alone doesn't ensure cost reduction C) Chatbot concept is flawed D) Need more training data

**Answer: B) Poor problem definition and success criteria**

**Explanation:** This illustrates multi-phase failure: (1) Business Understanding: goal (cost reduction) defined but success criteria incomplete (should include customer satisfaction, resolution time); (2) Success Metrics: 85% intent accuracy doesn't guarantee correct routing or cost reduction; (3) System Design: model accuracy ≠ system effectiveness;

incorrect routing harms experience; (4) Unintended consequences: frustrated customers create more support tickets. What should have happened: (1) Holistic success criteria: cost reduction AND maintained customer satisfaction; (2) System design: human fallback for low-confidence predictions; (3) Pilot testing: gradual rollout measuring total impact; (4) Right metrics: track end-to-end metrics (resolution time, satisfaction, cost) not just accuracy. 85% accuracy can still cause problems with poor system design. Technical metrics don't capture business value.

---

**Question 536 [EASY - Monitoring]:** What should you monitor after deploying an ML model to production?

A) Nothing; deployment is the end B) Model performance, prediction distributions, input data quality, system metrics (latency/errors), business outcomes C) Only model accuracy D) Only system uptime

**Answer: B) Comprehensive multi-layer monitoring**

**Explanation:** Production ML requires monitoring across multiple dimensions: (1) Model performance: accuracy, precision, recall on labeled samples; (2) Predictions: distribution shifts, extreme values, confidence scores; (3) Data quality: missing values, outliers, schema changes; (4) System health: latency, throughput, error rates, resource usage; (5) Business metrics: actual business impact (revenue, conversions, user satisfaction); (6) Fairness: per-group performance over time. Different metrics serve different purposes: data quality catches pipeline issues early, system metrics ensure SLAs, performance metrics detect model degradation, business metrics validate value. Single-metric monitoring misses problems. Comprehensive monitoring enables rapid issue detection and diagnosis.

---

**Question 537 [MEDIUM - Regulatory Compliance]:** Your healthcare ML model must comply with HIPAA. During development, a data scientist accidentally emails PHI to personal account. What's the required response?

A) Nothing; internal incident doesn't require reporting B) Incident response: investigate scope, mitigate (delete email), document, notify privacy officer, potentially breach notification C) Fire the data scientist immediately D) HIPAA doesn't apply to development

**Answer: B) Formal incident response with documentation and reporting**

**Explanation:** HIPAA breaches require formal response: (1) Investigation: determine scope (what PHI, how many patients, who accessed); (2) Immediate mitigation: secure/delete email, revoke access; (3) Documentation: detailed incident report; (4) Privacy officer: notify organizational privacy officer immediately; (5) Risk assessment: evaluate harm risk to patients; (6) Breach notification: if breach affects 500+ patients or is "high risk," notification required (patients, HHS, possibly media); (7) Corrective action: address process gaps preventing recurrence; (8) Training: additional staff training. Penalties for HIPAA violations are severe. Organizations must have incident response procedures. Even accidental breaches require proper response. Prevention through: training, technical controls (DLP), access restrictions, audit logging.

---

**Question 538 [HARD - Architecture Design]:** Your recommendation system needs to: (1) Personalize for 100M users, (2) Update recommendations based on real-time behavior, (3) Handle 1M requests/second, (4) Incorporate new items within minutes. What architecture components are needed?

A) Single model recomputed nightly B) Multi-tier architecture: candidate generation → ranking → real-time personalization; feature store; streaming pipeline; distributed serving C) Simple collaborative filtering D) Manual curation

**Answer: B) Sophisticated multi-tier real-time architecture**

**Explanation:** Scale and latency requirements necessitate complex architecture: (1) Candidate generation: fast retrieval of ~1000 potentially relevant items (ANN indexes, matrix factorization); (2) Ranking model: ML model ranking candidates (gradient boosting, neural network); (3) Real-time personalization: adjust scores based on in-session behavior; (4) Feature store: low-latency user/item features; (5) Streaming pipeline: Kafka processing real-time events; (6) Distributed serving: horizontally scaled serving infrastructure; (7) Caching: frequent queries cached; (8) Fast item ingestion: new items indexed rapidly. Single model can't meet latency at scale. Candidate generation narrows from 100M to 1000; ranking orders them; real-time layer adapts. Streaming updates capture recent behavior. This architecture pattern (funnel from candidates to final recommendations) is industry standard for large-scale recommendations.

---

**Question 539 [EASY - Data Quality]:** During data profiling, you discover 30% of customer email addresses are NULL. Investigation shows emails collected only at checkout; 30% of customers browse without purchasing. Is this a data quality problem?

A) Yes, always a quality problem B) No, systematic missingness reflecting business process; document it C) Delete all records without emails D) Impute fake emails

**Answer: B) Systematic missingness; document pattern**

**Explanation:** Not all missing data indicates quality problems. This missingness is systematic and meaningful: emails available only for customers who purchase. This is Missing Not At Random (MNAR) with clear business explanation. Proper handling: (1) Document pattern: explain why emails missing; (2) Missing indicator: create feature "email_provided" (might predict purchase behavior); (3) Model implications: email features available only for 70% of population; (4) Segment consideration: separate models for customers with/without emails?; (5) Don't impute: fabricating emails introduces fake data. Distinguish systematic missingness (acceptable, informative) from random quality issues (errors, incomplete collection). Understanding why data is missing is critical for appropriate handling.

---

**Question 540 [MEDIUM - Fairness]:** You've trained a model and tested three fairness definitions: Demographic Parity (fail), Equal Opportunity (pass), Equalized Odds (pass). Legal requires "fair" model but doesn't specify definition. How do you proceed?

A) Deploy; two out of three fairness metrics pass B) Analyze context to choose appropriate fairness definition; consult legal on which definition aligns with regulatory requirements; document rationale C) Must pass all fairness definitions D) Ignore fairness; no legal definition

**Answer: B) Choose contextually appropriate fairness definition with legal input**

**Explanation:** Fairness metrics have inherent tensions (impossibility theorem: can't simultaneously satisfy all). Selection requires: (1) Context: what's the application? (lending, hiring, criminal justice have different standards); (2) Legal analysis: which definition aligns with discrimination law in your jurisdiction?; (3) Stakeholder input: affected communities should inform definition; (4) Trade-offs: understand what each metric optimizes; (5) Documentation: clearly document chosen definition and rationale. Demographic Parity (equal positive rates) might not be appropriate if base rates genuinely differ. Equal Opportunity (equal true positive rates for qualified individuals) often more appropriate for meritocratic decisions. Equalized Odds (equal error rates) strictest. No universal fairness—choose and justify definition appropriate for use case and regulatory environment.

---

**Question 541 [HARD - Project Lifecycle]:** Six months after deployment, your model's performance degraded from 85% to 78% accuracy. Investigation shows: (1) Input distributions unchanged, (2) Prediction distribution unchanged, (3) Recent labeled data shows performance drop. What's the likely cause and solution?

A) Model randomly broke; rebuild from scratch B) Concept drift: relationship between features and target changed; retrain on recent data C) Data quality issue D) Monitoring is wrong

**Answer: B) Concept drift requiring retraining**

**Explanation:** Scenario indicates concept drift (feature-target relationship changed) rather than data drift (feature distribution changed): (1) Input distributions stable: rules out covariate shift; (2) Prediction distribution stable: model producing same outputs; (3) Performance degraded: outcomes no longer match predictions. Causes: customer behavior evolved, market conditions changed, competitor actions, seasonal patterns. Solution: (1) Retrain on recent data: capture current patterns; (2) Rolling window: train on recent 12-18 months, not ancient history; (3) Increase retraining frequency: quarterly → monthly if drift is rapid; (4) Feature engineering: add features capturing evolving dynamics; (5) Ensemble: combine stable long-term model with adaptive short-term model. Concept drift is common and inevitable. Regular retraining maintains performance in non-stationary environments.

---

**Question 542 [EASY - Explainability]:** Stakeholders request explanations for model predictions. Your complex ensemble model (XGBoost + Neural Network) achieves 90% accuracy. What approach provides explanations?

A) Explanations are impossible for complex models B) Model-agnostic methods (SHAP, LIME) provide feature importance explanations for any model C) Rebuild model as simple decision tree sacrificing performance D) Refuse to provide explanations

**Answer: B) Model-agnostic explanation methods**

**Explanation:** Modern explainability techniques work with any model: (1) SHAP (SHapley Additive exPlanations): unified framework providing feature importance; consistent, theoretically grounded; (2) LIME (Local Interpretable Model-agnostic Explanations): fits local interpretable model around prediction; (3) Feature importance: aggregate importance scores; (4) Partial dependence plots: show feature effects; (5) Individual explanations: explain specific predictions. Model-agnostic methods enable explaining complex models without sacrificing performance. SHAP particularly powerful: works with ensembles, provides consistent importance values. Don't choose between performance and explainability—modern techniques provide both. Complex models with explanations often better than simple models, especially for high-stakes applications.

---

**Question 543 [MEDIUM - Imbalanced Data]:** Your dataset: 10,000 examples, 100 positive class (1%), 9,900 negative class (99%). Baseline model predicting all negative achieves 99% accuracy. What's the appropriate evaluation approach?

A) Accuracy is sufficient; 99% is excellent B) Focus on precision, recall, F1-score, AUC-PR (precision-recall curve), confusion matrix; accuracy is misleading C) Accuracy captures performance adequately D) No special considerations needed

**Answer: B) Use imbalance-appropriate metrics beyond accuracy**

**Explanation:** Extreme imbalance makes accuracy meaningless: predicting all negative achieves 99% accuracy but catches zero positives. Better metrics: (1) Precision: of predicted positives, how many are correct?; (2) Recall: of actual positives, how many caught?; (3) F1-score: harmonic mean balancing precision and recall; (4) AUC-ROC: overall discrimination; (5) AUC-PR (precision-recall curve): especially informative for imbalanced data; (6) Confusion matrix: see all error types. Class-specific metrics: report positive class precision/recall separately. Business context determines priority: fraud detection prioritizes recall (catch fraud), spam filtering prioritizes precision (don't block legitimate email). Report multiple metrics; don't rely on accuracy alone for imbalanced problems.

---

**Question 544 [HARD - Ethical Dilemma]:** Your credit scoring model could be profitable but analysis reveals it systematically assigns lower scores to zip codes with majority-minority populations, even controlling for individual credit history. Legal says it's technically legal. What should you do?

A) Deploy; legal compliance is sufficient B) Don't deploy; ethical concerns override legal permission; re-engineer to eliminate disparate impact C) Profit justifies deployment regardless of bias D) Bias is inevitable in all models

**Answer: B) Ethics override legal minimum; address disparate impact**

**Explanation:** Legal compliance is floor, not ceiling. Ethical considerations: (1) Disparate impact: systematically disadvantaging protected groups is discriminatory even if legally permitted in some jurisdictions; (2) Proxy discrimination: using zip codes that correlate with race perpetuates historical redlining; (3) Societal harm: reinforces inequality and systemic bias; (4) Reputational risk: public backlash likely if bias revealed; (5) Values: does this align with organizational values? Actions: (1) Remove or modify zip code features; (2) Fairness constraints: explicitly optimize for equitable outcomes; (3) Alternative features: find predictive features without discriminatory impact; (4) Accept limitations: some profitable applications shouldn't be pursued. Profitable but unethical applications erode trust and cause harm. Organizations should operate above legal minimum. Some lawyers say "technically legal"—ethics teams should say "is this right?"

---

**Question 545 [EASY - Transfer Learning]:** You need to classify images of manufacturing defects but have only 2,000 labeled images. What approach is most appropriate?

A) Train CNN from scratch on 2,000 images B) Transfer learning: fine-tune pre-trained ImageNet model (ResNet, EfficientNet) on your 2,000 images C) Use logistic regression on raw pixels D) 2,000 images insufficient for any approach

**Answer: B) Transfer learning from pre-trained model**

**Explanation:** Transfer learning leverages knowledge from large datasets: (1) Pre-trained models: ResNet, EfficientNet, VGG trained on ImageNet (millions of images); (2) Learned features: low-level features (edges, textures) and mid-level features generalize across tasks; (3) Fine-tuning: adapt pre-trained model to your specific task with limited data; (4) Approach: freeze early layers (general features), train later layers (task-specific features); (5) Data efficiency: strong performance with 1,000-10,000 images vs. millions needed for training from scratch. Training CNNs from scratch requires massive data (millions of images). Transfer learning is standard practice for computer vision with moderate data. Even 2,000 images can achieve good performance with transfer learning.

---

**Question 546 [MEDIUM - Feature Store]:** Multiple teams build models using overlapping features (customer demographics, transaction history). Each team computes features independently, causing: inconsistent definitions, duplicated compute, training-serving skew. What infrastructure would address this?

A) Let each team continue independently B) Feature store: centralized feature computation, storage, and serving ensuring consistency across training and production C) Force everyone to use identical notebooks D) No infrastructure solution

**Answer: B) Feature store for consistent feature management**

**Explanation:** Feature store solves common multi-team problems: (1) Centralized computation: features computed once, used by all teams; (2) Consistent definitions: "customer_lifetime_value" means same thing everywhere; (3) Training-serving consistency: identical feature code for training and production; (4) Reusability: teams discover and reuse existing features; (5) Monitoring: centralized monitoring of feature distributions; (6) Versioning: feature versions tied to model versions; (7) Low-latency serving: optimized for production access. Benefits: eliminate duplication, prevent training-serving skew, accelerate development (reuse features), ensure consistency. Commercial options: Feast, Tecton, SageMaker Feature Store. Critical infrastructure for organizations with multiple ML teams.

---

**Question 547 [HARD - Causal Inference]:** Your model predicts that customers who view product reviews are more likely to purchase. Marketing wants to increase reviews to boost purchases. Is this valid?

A) Yes, correlation implies causation B) No, without causal analysis, this might be reverse causation (purchase-intending customers seek reviews) or confounding C) Prediction models prove causation D) Marketing interventions don't require causal analysis

**Answer: B) Correlation doesn't establish causation; need causal analysis**

**Explanation:** Predictive models identify correlations, not causal relationships. Possible explanations: (1) Reverse causation: customers intending to purchase seek reviews (causation reversed); (2) Confounding: product quality causes both review-viewing and purchasing; (3) Selection bias: review-viewers are different type of customer. Causal question: "Does viewing reviews cause purchases?" requires: (1) Randomized experiment: A/B test showing reviews to random customers; (2) Quasi-experimental methods: natural experiments, difference-in-differences, instrumental variables; (3) Causal frameworks: do-calculus, potential outcomes; (4) Domain knowledge: theory supporting causal mechanism. Acting on correlations can fail or backfire. Causal inference methods distinguish correlation from causation. Marketing interventions based on causation, not correlation.

---

**Question 548 [EASY - Model Maintenance]:** Your deployed model is 6 months old. Performance remains stable at 85% accuracy. Should you retrain?

A) Never retrain if performance stable B) Evaluate whether retraining would improve performance; consider hidden drift; schedule regular retraining even if metrics stable C) Always retrain monthly regardless of performance D) Stable models never need updates

**Answer: B) Evaluate potential improvements; proactive retraining**

**Explanation:** Stable metrics don't mean optimal or that retraining is unnecessary: (1) Hidden drift: current metrics might mask subtle degradation; (2) Improvements: recent data or new features might improve performance; (3) Proactive maintenance: prevent future issues before they appear in metrics; (4) Best practices: scheduled retraining (quarterly/semi-annually) even without obvious degradation; (5) Evaluation: retrain and compare to production model before deploying. Retraining benefits: (1) Capture recent patterns; (2) Incorporate new features; (3) Benefit from algorithm improvements; (4) Prevent technical debt accumulation. Cost-benefit: is retraining effort justified? For critical models, yes. For low-stakes models, reactive retraining when performance degrades might suffice. Establish retraining cadence based on business criticality and drift velocity.

---

**Question 549 [MEDIUM - A/B Testing]:** You deploy new model via A/B test: 90% traffic to current model (control), 10% to new model (treatment). After one week: Control: 82% accuracy, 0.3% conversion rate. Treatment: 85% accuracy, 0.28% conversion rate. What should you do?

A) Full rollout of new model (higher accuracy) B) Don't roll out; business metric (conversion) worsened despite better accuracy; investigate why C) Accuracy is all that matters; deploy D) One week insufficient; never make decisions

**Answer: B) Don't roll out; business metric degradation is concerning**

**Explanation:** Technical and business metrics can diverge: (1) Accuracy improved (82% → 85%); (2) Conversion decreased (0.3% → 0.28%); (3) Conclusion: model technically better but harms business outcome. Possible explanations: (1) Model optimizes wrong objective: accuracy doesn't align with conversion; (2) User experience: new model's recommendations less appealing despite being more "accurate"; (3) Definition mismatch: what "accuracy" measures differs from what drives conversion; (4) Latency: new model slower, hurting user experience. Investigation: (1) Why do metrics diverge?; (2) Can you fix the alignment issue?; (3) What does "accuracy" actually measure? A/B testing reveals business impact beyond technical metrics. Never deploy based solely on technical metrics without validating business impact.

---

**Question 550 [HARD - Multi-Stakeholder Alignment]:** Your AI project faces conflicting requirements: (1) Legal: full explainability, (2) Business: maximize accuracy for competitive advantage, (3) Ethics: ensure fairness across demographics, (4) Engineering: minimize latency for user experience. No single solution satisfies all. How do you proceed?

A) Satisfy only the loudest stakeholder B) Stakeholder negotiation: quantify trade-offs, present options with implications, facilitate decision among stakeholders based on priorities C) Build multiple models for different stakeholders D) Impossible to resolve; abandon project

**Answer: B) Transparent trade-off analysis with stakeholder decision-making**

**Explanation:** Multi-objective optimization with stakeholder alignment: (1) Map requirements: document all stakeholder requirements and constraints; (2) Identify conflicts: explainability vs. accuracy, fairness constraints vs. latency; (3) Quantify trade-offs: what accuracy loss for explainability? What latency cost for fairness?; (4) Generate options: multiple solutions on Pareto frontier (each optimizes different balance); (5) Present trade-offs: transparent communication of implications; (6) Facilitated decision: bring stakeholders together to prioritize collectively; (7) Document: record decisions and rationale. Example options: Option A (max explainability, lower accuracy), Option B (balanced), Option C (max accuracy, SHAP explanations). No perfect solution—stakeholders must make informed trade-offs. Technical team provides options and implications; business/legal/ethics decide priorities. Documented consensus prevents revisiting decisions.

---

**Question 551 [EASY - Bias in Training Data]:** Your training dataset contains twice as many examples of male doctors and female nurses than the reverse. What's the risk?

A) No risk; reflects reality B) Model will learn and perpetuate gender stereotypes in profession C) Balanced data is impossible D) Bias only matters in predictions, not training

**Answer: B) Model learns stereotypes from biased training data**

**Explanation:** Models learn patterns from training data—including societal biases. If training data over-represents male doctors and female nurses, model learns: "doctor" associates with male, "nurse" associates with female. Consequences: (1) Biased predictions: may recommend medical school to men, nursing school to women; (2) Reinforcement: perpetuates stereotypes; (3) Discrimination: systematically disadvantages individuals based on gender. Solutions: (1) Balanced data: actively collect counter-stereotypical examples; (2) Debiasing: preprocessing, fairness constraints, adversarial debiasing; (3) Evaluation: test for stereotypical associations; (4) Feature selection: remove gender-associated features when not legitimately relevant. Training data reflecting societal bias will produce biased models. Data curation and debiasing techniques are essential.

---

**Question 552 [MEDIUM - Semi-Supervised Learning]:** You have 1,000 labeled examples and 1,000,000 unlabeled examples. Labeling is expensive ($10/example). Should you: (A) Label 10,000 more examples ($100K cost), or (B) Use semi-supervised learning with existing data?

A) Always label more data B) Try semi-supervised learning first (pseudo-labeling, self-training); compare to labeling small set; scale labeling based on results C) Ignore unlabeled data; it's worthless D) 1,000 labels are always sufficient

**Answer: B) Explore semi-supervised learning before expensive labeling**

**Explanation:** Cost-benefit analysis of data acquisition: (1) Semi-supervised approaches: leverage unlabeled data with limited labels; (2) Methods: self-training (use model to label unlabeled data), co-training (multiple views), consistency regularization; (3) Test first: try semi-supervised on existing data; (4) Incremental labeling: if semi-supervised insufficient, label strategic samples (active learning: label most informative examples); (5) Compare: semi-supervised performance vs. cost of labeling. Semi-supervised can provide significant lift with no additional cost. If gains insufficient, strategic labeling of 1,000-2,000 examples might outperform labeling 10,000 random examples. Try cheap approaches (semi-supervised) before expensive (labeling). Empirically compare options before committing $100K.

---

**Question 553 [HARD - Feedback Loops]:** Your content recommendation model surfaces engaging content. Over time, users trained by recommendations click less-bait-y content, and metrics show declining engagement. Paradoxically, you achieved better content quality but worse metrics. How do you address this?

A) Revert to maximizing engagement regardless of quality B) Recognize positive feedback loop; adjust metrics to value quality engagement over raw clicks; long-term value over short-term metrics C) Engagement metrics are always correct D) User behavior change is bad

**Answer: B) Reframe success metrics for long-term value**

**Explanation:** Sophisticated feedback loop problem: (1) Short-term: clickbait maximizes immediate engagement; (2) Medium-term: users trained by better content develop better taste; (3) Paradox: improvements in content quality lead to declining engagement metrics as users become more selective; (4) Real success: higher-quality engagement even if less frequent. Metric evolution: (1) Beyond clicks: time spent, return visits, satisfaction surveys; (2) Quality signals: completion rate, sharing, long-term retention; (3) Cohort analysis: compare user satisfaction over time; (4) Long-term value: lifetime value, not just immediate clicks. This requires organizational maturity: optimizing for user value over immediate metrics. Short-term metric optimization can harm long-term value. Sometimes declining engagement metrics indicate success if user experience improved.

---

**Question 554 [EASY - Data Lineage]:** Why is tracking data lineage (source → transformations → model) important?

A) Data lineage is unnecessary documentation B) Enables debugging, auditing, compliance, reproducibility; understanding how data became what it is C) Only needed for regulated industries D) Tracking lineage wastes time

**Answer: B) Critical for debugging, compliance, and reproducibility**

**Explanation:** Data lineage tracks data journey through pipeline: (1) Debugging: when problems occur, trace back to identify source; (2) Compliance: GDPR requires documenting data processing; demonstrate data handling; (3) Reproducibility: recreate exact training data for model reproduction; (4) Impact analysis: if source data changes, which models affected?; (5) Auditing: regulatory audits require proving data provenance; (6) Quality: identify where quality issues introduced; (7) Trust: stakeholders understand where data comes from. Implementation: (1) Metadata tracking: document transformations; (2) Version control: data versions linked to model versions; (3) Automated tools: data catalogs, lineage tracking systems. Lineage isn't just documentation—it's operational necessity. Without lineage, can't reproduce models, debug issues, or demonstrate compliance.

---

**Question 555 [MEDIUM - Active Learning]:** You need to label 100,000 images but budget allows labeling only 10,000. What strategy maximizes model performance with limited labels?

A) Label random 10,000 images B) Active learning: train initial model on small set, iteratively label most uncertain/informative examples, retrain C) Label first 10,000 chronologically D) All labeling strategies equivalent

**Answer: B) Active learning for strategic sample selection**

**Explanation:** Active learning maximizes information per labeled example: (1) Initial model: train on small random set (1,000-2,000 images); (2) Uncertainty sampling: identify unlabeled images where model is most uncertain; (3) Label batch: label high-uncertainty images (model learns most from these); (4) Retrain: incorporate newly labeled images; (5) Iterate: repeat until budget exhausted; (6) Diversity: ensure coverage of all classes, not just uncertain examples. Why it works: model learns more from examples near decision boundary than from clear cases. Active learning can match random labeling performance with 2-5× fewer labels. Strategies: uncertainty sampling (low confidence), diversity sampling (coverage), expected model change. Especially valuable when labeling expensive (medical imaging, specialized domains).

---

**Question 556 [HARD - Model Compression]:** Your neural network achieves 92% accuracy but has 50M parameters requiring 200MB and 500ms inference time. Mobile deployment requires <50MB and <100ms. How do you compress the model?

A) Cannot deploy large models on mobile B) Model compression: quantization (int8), pruning (remove weights), knowledge distillation (student model), architecture search for efficient models C) Accept poor mobile performance D) Accuracy will necessarily drop to 50%

**Answer: B) Multiple compression techniques for deployment**

**Explanation:** Model compression for resource-constrained environments: (1) Quantization: float32 → int8 (4× size reduction, faster inference, ~1% accuracy loss); (2) Pruning: remove unimportant weights (30-50% reduction, minimal accuracy loss); (3) Knowledge distillation: train small student model mimicking large teacher (10× smaller, 1-3% accuracy loss); (4) Efficient architectures: MobileNet, EfficientNet designed for mobile; (5) Combination: apply multiple techniques together. Expected result: 50MB model, <100ms inference, 89-91% accuracy (1-3% loss). Compression is standard practice for mobile deployment. Well-compressed models maintain performance while meeting constraints. Modern techniques enable sophisticated models on edge devices. Balance: some accuracy loss acceptable to meet hard deployment constraints.

---

**Question 557 [EASY - Ensemble Methods]:** You train five different models: Random Forest (82%), Logistic Regression (78%), Gradient Boosting (84%), Neural Network (81%), SVM (80%). What's the best approach to leverage all models?

A) Use only Gradient Boosting (highest accuracy) B) Ensemble: average or stack models; often outperforms best individual model C) Discard all but best model D) Ensembles are never beneficial

**Answer: B) Ensemble leveraging multiple models**

**Explanation:** Ensemble methods combine multiple models: (1) Simple averaging: average predictions from all models; (2) Weighted average: weight by validation performance; (3) Stacking: train meta-model on individual model predictions; (4) Diversity: models make different errors, averaging reduces errors; (5) Expected improvement: ensemble might achieve 85-86% (better than best individual 84%). Why ensembles work: (1) Models trained differently capture different patterns; (2) Errors are often uncorrelated; (3) Averaging reduces variance; (4) Robust: less likely to catastrophically fail. When not to ensemble: if models very similar (same algorithm, same data), limited benefit. Diversity (different algorithms, different features) makes ensembles effective. Ensembles are standard for competitions and high-stakes applications.

---

**Question 558 [MEDIUM - Hyperparameter Tuning]:** You're tuning hyperparameters for XGBoost. Search space has 6 hyperparameters, each with 5 values. Grid search would test 5^6 = 15,625 combinations. With 1-hour training per combination, grid search takes 650 days. What's a better approach?

A) Test all 15,625 combinations B) Random search or Bayesian optimization: test 100-200 combinations, find near-optimal hyperparameters C) Use default hyperparameters D) Reduce to 1 hyperparameter

**Answer: B) Efficient hyperparameter search methods**

**Explanation:** Hyperparameter optimization at scale: (1) Random search: sample random combinations; surprisingly effective, finds good solutions with ~100-200 trials; (2) Bayesian optimization: model hyperparameter space, intelligently select next combination to test (Optuna, Hyperopt); (3) Early stopping: terminate poor-performing trials early; (4) Parallelization: run multiple trials simultaneously; (5) Multi-fidelity: test on subset of data first, full data only for promising combinations. Expected: find near-optimal hyperparameters in 100-200 trials (4-8 days with parallelization) rather than 15,625 (650 days). Research shows random search often matches grid search with 10-20× fewer trials. Bayesian optimization even more efficient. Modern hyperparameter tuning uses smart search, not exhaustive enumeration.

---

**Question 559 [HARD - System Design]:** Design an ML system for real-time fraud detection processing 10,000 transactions/second with <100ms decision time. What components and architecture do you need?

A) Single model on single server B) Distributed architecture: feature caching, model serving infrastructure (TensorFlow Serving/Triton), horizontally scaled, message queue, monitoring C) Batch processing overnight D) Manual review only

**Answer: B) Distributed low-latency architecture**

**Explanation:** High-throughput low-latency ML systems require sophisticated architecture: (1) Load balancing: distribute 10,000 req/sec across many model instances; (2) Model serving: specialized infrastructure (TensorFlow Serving, Triton Inference Server, Seldon) optimized for ML; (3) Feature caching: Redis/Memcached for low-latency feature lookup (customer history, merchant info); (4) Horizontal scaling: 50-100 model instances behind load balancer; (5) Message queue: Kafka buffering during spikes; (6) Model optimization: quantization, batching for throughput; (7) Monitoring: latency, throughput, errors continuously tracked; (8) Fallback: simple rule-based system if ML unavailable. Calculate: 10,000 req/sec ÷ 100 req/sec per instance = 100 instances needed. Cloud auto-scaling adjusts capacity. Caching critical—database lookups too slow. Real-time ML at scale requires distributed systems engineering, not just model training.

---

**Question 560 [EASY - Model Versioning]:** Why should you version ML models alongside code?

A) Models don't need versioning B) Reproducibility, rollback capability, audit trails, debugging, A/B testing different versions C) Latest model is always best D) Versioning is bureaucracy

**Answer: B) Versioning critical for operations and compliance**

**Explanation:** Model versioning enables: (1) Reproducibility: recreate exact model that made historical predictions; (2) Rollback: quickly revert to previous version if new model has issues; (3) A/B testing: run multiple model versions simultaneously comparing performance; (4) Audit: regulatory requirements often demand explaining historical predictions; (5) Debugging: identify when issues introduced by comparing versions; (6) Metadata: track training data, hyperparameters, performance metrics per version. Implementation: (1) Model registry: central catalog of models (MLflow, DVC, custom); (2) Semantic versioning: major.minor.patch indicating change significance; (3) Tagging: production model tagged; (4) Artifacts: store model weights, config, dependencies together; (5) Lineage: link model version to code version and data version. Without versioning, can't reproduce, rollback, or debug. Essential for production ML.

---

**Question 561 [MEDIUM - Cold Start Problem]:** Your recommendation system struggles with new users (no history) and new items (no interactions). What strategies address cold start?

A) Don't recommend anything to new users B) Hybrid approaches: content-based features for new items, demographic/onboarding info for new users, popularity fallback C) Wait until enough data collected D) Cold start is unsolvable

**Answer: B) Multi-strategy hybrid approach**

**Explanation:** Cold start requires combining multiple signals: New users: (1) Demographic features: age, location, device; (2) Onboarding: preferences collected during signup; (3) Popularity: recommend generally popular items; (4) Similar users: find similar users from limited early signals; (5) Exploration: deliberately diverse recommendations to learn preferences quickly. New items: (1) Content features: genre, category, price, description embeddings; (2) Similar items: recommend to users who liked similar items; (3) Metadata: leverage item attributes; (4) Exploration: show to diverse users to gather feedback. Hybrid models: combine collaborative filtering (for warm users/items) with content-based (for cold users/items). Gradually transition from content-based to collaborative as more data collected. Cold start is major recommendation challenge—requires explicit design consideration.

---

**Question 562 [HARD - Multi-Task Learning]:** You're building three related models: (1) Predict customer churn, (2) Predict customer lifetime value, (3) Predict product preferences. Should you train separately or together?

A) Always train separately B) Explore multi-task learning: shared layers for common representations, task-specific heads; may improve performance and efficiency C) Always train together D) Impossible to train multiple tasks

**Answer: B) Multi-task learning for related tasks**

**Explanation:** Multi-task learning (MTL) trains one model on multiple related tasks: Architecture: (1) Shared layers: bottom layers shared across tasks (learn common customer representations); (2) Task-specific heads: separate output layers per task; (3) Joint training: train on all tasks simultaneously with weighted loss. Benefits: (1) Shared representations: common patterns (customer behavior) help all tasks; (2) Regularization: training on multiple tasks prevents overfitting; (3) Efficiency: one model instead of three (lower maintenance, faster inference); (4) Data efficiency: each task benefits from others' data. When it works: tasks sufficiently related (all about customer behavior). When it doesn't: tasks unrelated (customer churn vs. image classification). Test empirically: compare MTL to separate models. If tasks help each other, MTL superior. If tasks conflict, separate models better. MTL increasingly common—share computation while specializing for tasks.

---

**Question 563 [EASY - Cross-Validation]:** What is the purpose of k-fold cross-validation?

A) Increase training data size B) Get more robust performance estimates using all data for both training and validation; reduce variance in estimates C) Speed up training D) No clear purpose

**Answer: B) Robust performance estimates with limited data**

**Explanation:** K-fold cross-validation provides reliable performance estimates: (1) Process: split data into k folds; train on k-1 folds, validate on remaining fold; repeat k times rotating validation fold; (2) Each example used once for validation, k-1 times for training; (3) Average k validation scores for overall estimate; (4) Benefits: utilizes all data efficiently, reduces variance compared to single train/val split; (5) Common choices: k=5 or k=10. When to use: limited data where single validation set would be small and unreliable. When not to use: huge datasets (computational cost unjustified), time series (need temporal splits, not random folds). Cross-validation is standard practice for model selection and hyperparameter tuning with moderate-sized datasets. Provides confidence intervals, not just point estimates.

---

**Question 564 [MEDIUM - Continuous Learning]:** Your fraud detection model needs to adapt quickly to new fraud patterns. Traditional retraining takes 1 week. What alternatives enable faster adaptation?

A) Accept 1-week delay B) Online learning with continuous updates, small learning rate; or frequent mini-retraining with recent data C) Models can't adapt quickly D) Delay is acceptable for all applications

**Answer: B) Online learning or frequent incremental retraining**

**Explanation:** Rapid adaptation strategies: (1) Online learning: model updates incrementally with each new example (or mini-batch); small learning rate prevents forgetting; (2) Incremental retraining: retrain daily/hourly on recent data (rolling window); (3) Ensemble: combine stable long-term model with adaptive short-term model; (4) Rule injection: quickly add rules for known patterns while model retrains; (5) Active learning: prioritize labeling of novel patterns. Online learning: (1) Algorithms: online SGD, FTRL, streaming algorithms; (2) Challenges: catastrophic forgetting (forgetting old patterns while learning new), drift; (3) Safeguards: validation on holdout set, monitor performance. When useful: rapidly evolving environments (fraud, cybersecurity, spam). When not: model quality more important than adaptation speed. Balance: stability (don't overreact to noise) vs. agility (adapt to real changes).

---

**Question 565 [HARD - Fairness-Accuracy Tradeoff]:** Your model achieves: Overall accuracy 84%, Male accuracy 88%, Female accuracy 72%. You can apply fairness constraints to equalize group accuracy but overall accuracy drops to 80%. Business values accuracy. How do you handle this?

A) Maximize accuracy; ignore fairness B) Present tradeoff to stakeholders: is 4% accuracy loss worth 16-point fairness improvement? Consider legal/ethical requirements C) Always prioritize fairness regardless of accuracy D) Fairness and accuracy can't be balanced

**Answer: B) Stakeholder decision on fairness-accuracy tradeoff with full information**

**Explanation:** Fairness-accuracy tradeoffs require value judgments: Quantify tradeoff: (1) Fairness intervention: equalize group accuracy; (2) Cost: 4% overall accuracy loss (84% → 80%); (3) Benefit: eliminate 16-point gender gap (88% vs 72%); (4) Alternative: improve female accuracy without hurting male (harder but ideal). Stakeholder decision: (1) Legal: is gender gap legally problematic? (in many contexts, yes); (2) Ethical: is 16-point gender disparity acceptable? (often, no); (3) Business: is 4% accuracy loss acceptable tradeoff?; (4) Reputational: what if gender gap discovered publicly? Present options: (1) Option A: 84% accuracy, large gender gap; (2) Option B: 80% accuracy, no gender gap; (3) Option C: investigate improving female accuracy without overall accuracy loss. This is values question, not purely technical. Technical team quantifies tradeoffs; leadership decides priorities. Often, fairness concerns override marginal accuracy differences.

---

**Question 566 [EASY - Training vs. Test Distribution]:** Your model achieves 90% test set accuracy but only 65% production accuracy. What's the most likely cause?

A) Model is excellent; production data is wrong B) Train/test distribution doesn't match production distribution (distribution shift) C) Test set was too easy D) No problem exists

**Answer: B) Train/test distribution mismatch with production**

**Explanation:** Large train/test vs. production performance gap indicates distribution shift: Causes: (1) Sampling bias: test set not representative of production; (2) Temporal: test set from past, production is current (concept drift); (3) Population: test set from different population (geographic, demographic); (4) Data collection: production data collected differently. Prevention: (1) Representative test set: ensure test set matches production distribution; (2) Temporal validation: validate on recent data if deploying to future; (3) Production validation: validate on actual production data before full deployment; (4) A/B testing: gradual rollout with performance monitoring. Test set should simulate production as closely as possible. Distribution mismatch is leading cause of production failures. Validate on production-like data.

---

**Question 567 [MEDIUM - Model Calibration]:** Your model predicts probabilities: when model predicts 80% probability, true rate is 95%; when model predicts 20% probability, true rate is 5%. What's the problem?

A) Model is well-calibrated B) Model is poorly calibrated: predicted probabilities don't match actual probabilities; need recalibration (Platt scaling, isotonic regression) C) Probabilities don't matter D) No issue exists

**Answer: B) Poor calibration requiring post-processing**

**Explanation:** Calibration misalignment: predicted probability ≠ true probability: (1) Overconfident at high end: predicts 80%, actually 95%; (2) Underconfident at low end: predicts 20%, actually 5%; (3) Implications: can't use probabilities for decision-making (risk assessment, betting odds). Calibration methods: (1) Platt scaling: apply logistic regression to model outputs; (2) Isotonic regression: non-parametric monotonic transformation; (3) Temperature scaling: single-parameter adjustment (neural networks); (4) Beta calibration: extends Platt scaling. Evaluation: reliability diagrams showing predicted vs. actual probabilities. When calibration matters: (1) Probability-based decisions (not just rankings); (2) Combining multiple models; (3) Uncertainty quantification; (4) Regulatory requirements. Note: well-ranked model can be poorly calibrated. Calibration is separate concern from discrimination.

---

**Question 568 [HARD - Infrastructure Failure Scenario]:** During peak traffic, your model server crashes (out of memory). Backup system isn't working. Business-critical decisions delayed. You have 10 minutes before major revenue loss. What do you do?

A) Panic and wait for model server to restart B) Immediate fallback: activate simple rule-based system or cached predictions; then debug and restore model server C) Shut down all services D) Continue without any decision system

**Answer: B) Immediate fallback to maintain service; restore sophisticated system separately**

**Explanation:** Production incident response priorities: (1) Immediate mitigation: restore service even with degraded capability; (2) Fallback strategies: pre-prepared simple systems for emergencies; (3) Then debug: investigate root cause in parallel with fallback running; (4) Gradual restoration: restore sophisticated system once stable. Fallback options: (1) Rule-based system: simple rules encoded for this scenario; (2) Cached predictions: recent predictions for common cases; (3) Default behavior: safe default when prediction unavailable; (4) Previous model version: if recent deployment caused issue. This scenario emphasizes: (1) Defense in depth: multiple fallback layers; (2) Preparation: fallback systems prepared before incidents; (3) Graceful degradation: imperfect service better than no service; (4) Clear escalation: who makes decisions under pressure. Complex systems fail—resilience through redundancy and fallbacks.

---

**Question 569 [EASY - Confusion Matrix Interpretation]:** Your binary classifier's confusion matrix: True Positive=450, False Positive=50, True Negative=400, False Negative=100. What's the precision?

A) 0.82 (450/(450+100)) B) 0.90 (450/(450+50)) C) 0.80 (400/(400+50)) D) 0.78 ((450+400)/1000)

**Answer: B) 0.90 (450/(450+50))**

**Explanation:** Confusion matrix metric calculations: (1) Precision = TP/(TP+FP) = 450/(450+50) = 450/500 = 0.90; (2) Recall = TP/(TP+FN) = 450/(450+100) = 450/550 = 0.82; (3) Accuracy = (TP+TN)/Total = (450+400)/1000 = 0.85. Precision interpretation: "Of predicted positives, 90% were actually positive." Important when false positives costly (spam filtering, fraud alerts). Recall interpretation: "Of actual positives, 82% were caught." Important when false negatives costly (disease detection, fraud detection). F1 = 2×(Precision×Recall)/(Precision+Recall) = 2×(0.90×0.82)/(0.90+0.82) = 0.86. Different metrics matter for different applications. Know how to calculate and interpret each.

---

**Question 570 [MEDIUM - Ensemble Diversity]:** You train three models: all use same algorithm (XGBoost), same features, same hyperparameters, different random seeds. Ensemble achieves minimal improvement over single model. Why?

A) Ensembles never work with XGBoost B) Lack of diversity: models too similar to benefit from ensembling; need algorithm diversity, feature diversity, or data diversity C) Three models insufficient; need 100 D) Random seeds create diversity

**Answer: B) Insufficient diversity limiting ensemble benefit**

**Explanation:** Ensemble effectiveness requires diversity: (1) Why diversity matters: models make different errors; averaging reduces error; (2) This case: models too similar (same algorithm, features, hyperparameters); random seed creates minimal diversity. Creating diversity: (1) Algorithm: mix different algorithms (XGBoost + Random Forest + Neural Network + Logistic Regression); (2) Features: train models on different feature subsets; (3) Data: bagging (different samples), different preprocessing; (4) Hyperparameters: significantly different configurations. Expected: diverse ensemble might improve 2-5%; homogeneous ensemble improves <1%. Measure: correlation between model predictions (low correlation = high diversity = better ensemble). Very similar models provide minimal ensemble benefit. Diversity is key to ensemble success.

---

## Advanced Situational Scenarios

### 25 Complex Multi-Phase Problems

**Instructions:** These scenarios are more complex than Section A, often spanning multiple CRISP-ML(Q) phases and requiring integration of concepts. Multiple factors must be considered. Allow 60 minutes for completion.

---

**Question 571 [ADVANCED - Multi-Phase Integration]:**

**SCENARIO:** A healthcare provider wants to predict patient no-shows for appointments to optimize scheduling. Initial analysis shows:

- 15% no-show rate
- Demographics predict no-shows (age, insurance type, distance from clinic)

- Model achieves 78% accuracy
- Deployed model reduces no-shows by 3% (to 12%)
- Six months later: complaints from patient advocacy groups that low-income patients are being denied appointments
- Investigation reveals model overbooks appointments for predicted no-shows, disproportionately affecting low-income patients who are predicted to no-show more often

What should have been done differently throughout the project lifecycle?

A) Higher model accuracy would have prevented the problem B) Comprehensive ethical review in Phase I, fairness evaluation in Phase VI, stakeholder engagement, considering unintended consequences C) Problem is unsolvable; abandon predictive scheduling D) Patient advocacy groups are overreacting

**Answer: B) Multi-phase ethical considerations and stakeholder engagement**

**Explanation:** This scenario demonstrates how technical success can cause ethical failures:

**Phase I - Business Understanding failures:**

- Objective narrowly defined: "reduce no-shows" without considering equitable access
- Success metrics incomplete: no-show reduction measured, but not access equity
- Stakeholder analysis incomplete: patient advocacy groups not consulted
- Unintended consequences not anticipated

**What should have happened:**

1. **Broader problem framing:** "Optimize scheduling while ensuring equitable access"
2. **Comprehensive success criteria:** No-show reduction AND maintained access equity across patient groups
3. **Stakeholder engagement:** Include patient advocacy groups, ethics board
4. **Ethical review:** Anticipate that overbooking based on predictions could disadvantage certain groups

**Phase II-III - Data Understanding/Preparation:**

- Recognized that demographics correlate with no-shows but may reflect barriers (transportation, work schedules), not intent
- Should have questioned: do we want to optimize on factors reflecting social inequality?

**Phase V - Evaluation:**

- Should have included fairness metrics: appointment access rates by demographic groups
- Disaggregated evaluation: per-group no-show prediction accuracy and appointment access

**Phase VI - Deployment:**

- Should have monitored: not just overall no-shows but access rates by demographics
- Pilot testing with patient advocacy groups
- Gradual rollout with community feedback

**Alternative solutions:**

1. **Different intervention:** Instead of denying appointments, offer transportation assistance or reminder calls to predicted no-shows
2. **Fairness constraints:** Ensure overbooking rates don't differ by protected groups
3. **Human oversight:** Don't automatically deny appointments; flag for scheduling staff review
4. **Root causes:** Address why certain groups no-show (transportation, childcare) rather than penalizing them

**Key lesson:** Technical metrics (accuracy, no-show reduction) don't capture ethical dimensions. Optimization can perpetuate inequality. Stakeholder engagement and ethical review throughout project lifecycle prevent harmful deployments.

---

**Question 572 [ADVANCED - Data Bias & Causation]:**

**SCENARIO:** An e-commerce company builds a model predicting customer lifetime value (LTV) to prioritize high-value customers for premium service. Analysis shows:

- Customers who use mobile app have 2× higher LTV than web-only customers

- Marketing wants to push all customers to mobile app to increase LTV
- Data scientist notes: customers who already have high purchase intent download the app; app doesn't cause high LTV
- Historical A/B test showed: forcing web customers to mobile app decreased satisfaction, didn't increase purchases

How should this be handled?

A) Deploy app-push strategy; the correlation is clear B) Recognize correlation ≠ causation; use A/B test results; app is effect, not cause; focus on actual LTV drivers C) Ignore historical A/B test results D) Build model anyway; causation doesn't matter for predictions

**Answer: B) Distinguish correlation from causation; avoid interventions based on correlation**

**Explanation:** This scenario illustrates critical distinction between prediction and causal inference:

**Correlation vs. Causation:**

1. **Observed correlation:** Mobile app users have 2× higher LTV
2. **Possible causal directions:**
    - App causes high LTV (marketing's assumption)
    - High intent causes both app download and high LTV (reverse causation)
    - Customer characteristics (tech-savvy, frequent shoppers) cause both

**Evidence from A/B test:**

- Forcing app adoption didn't increase purchases
- Suggests: app is marker of high-value customers, not driver
- High-intent customers seek out app; app doesn't create high intent

**Predictive modeling vs. Causal inference:**

1. **Predictive model (appropriate):** Use app usage to predict LTV (correlation sufficient)
2. **Interventions (inappropriate):** Don't use predictive models to guide interventions without causal evidence

**Proper approach:**

1. **Acknowledge limitations:** Explain to marketing that correlation doesn't justify intervention
2. **Causal analysis:** What actually drives LTV? Product quality, pricing, customer service?
3. **Intervention design:** A/B test proposed interventions before broad rollout
4. **Use predictions correctly:** Identify high-value customers (via app usage and other signals) for targeted service, don't try to change app usage

**Real LTV drivers might be:**

- Product satisfaction
- Price competitiveness
- Customer service quality
- Personalized recommendations
- These require different interventions than app adoption

**Cautionary examples:**

- Hospitals found sicker patients had better outcomes (confounding: went to better hospitals)
- Acting on this would wrongly send healthy patients to worse hospitals
- Correlation in observational data doesn't license interventions

**Key lesson:** Predictive models identify correlations useful for decisions about individuals. Causal models identify interventions that change outcomes. Don't confuse the two. Always A/B test interventions, especially when based on observational correlations.

---

**Question 573 [ADVANCED - Monitoring & Incident Response]:**

**SCENARIO:** Your fraud detection model has been in production for 8 months with stable 88% accuracy. On Monday morning:

- Alert: False positive rate increased from 2% to 15% overnight
- Customer support overwhelmed with complaints about legitimate transactions blocked
- Investigation shows: major payment processor changed their transaction format
- Feature engineering code broke, producing null features
- Model defaults to "fraud" when features are null (conservative default)

- Estimated revenue loss: $50K/hour from blocked transactions
- Options: (1) Turn off fraud detection (no protection, 30 min), (2) Fix feature engineering (estimated 4 hours), (3) Deploy backup model trained 6 months ago (1 hour, but only 82% accuracy)

What's your incident response plan?

A) Wait 4 hours for proper fix B) Immediate: turn off fraud detection temporarily; Parallel: deploy backup model (1 hr), then fix properly (4 hrs) C) Keep broken system running D) Blame payment processor

**Answer: B) Multi-phase incident response with immediate mitigation and proper fix**

**Explanation:** This scenario tests production incident management:

**Incident Response Framework:**

**Immediate Actions (0-15 minutes):**

1. **Assess impact:** $50K/hour revenue loss, customer experience damage
2. **Severity classification:** P0 incident (revenue-impacting, customer-facing)
3. **Incident commander:** Designate decision-maker
4. **Communication:** Alert stakeholders (engineering, business, customer support)
5. **Stop the bleeding:** Turn off fraud detection temporarily (30 min to implement)

**Why turn off immediately:**

- Revenue loss mounting ($50K/hour)
- Customer experience severely degraded
- Broken system worse than no system
- 30 minutes to implement
- Provides breathing room for proper fix

**Short-term Mitigation (15-60 minutes):**

1. **Deploy backup model:** 6-month-old model with 82% accuracy
    - Better than no fraud detection
    - Worse than current model (88% → 82%) but functional
    - 1 hour to deploy
2. **Alternative fraud signals:** Rule-based filters for known fraud patterns

3. **Increased human review:** Route more transactions to manual review
4. **Customer support:** Brief support team, prepare customer communications

**Medium-term Fix (1-4 hours):**

1. **Root cause:** Payment processor format change broke feature engineering
2. **Fix feature code:** Handle new transaction format
3. **Validation:** Test on recent data, ensure features correct
4. **Gradual rollout:** Canary deployment, monitor carefully
5. **Replace backup model:** Deploy fixed model once validated

**Long-term Prevention:**

1. **Monitoring:** Add data quality checks detecting feature engineering failures
2. **Alerting:** Alert on null features, feature distribution shifts
3. **Upstream coordination:** Payment processor should notify of format changes
4. **Schema validation:** Automated testing catching format changes
5. **Graceful degradation:** Model shouldn't default to "fraud" on null features; should flag for review
6. **Backup system:** Maintain tested backup system for emergencies
7. **Runbooks:** Document incident response procedures

**Decision trade-offs:**

- Option 1 (turn off): Fast (30 min), but no fraud protection
- Option 2 (backup): Slower (1 hr), degraded protection (82% vs 88%)
- Option 3 (fix): Slowest (4 hrs), best quality when done

**Optimal strategy:** Combine approaches

1. Turn off immediately (stop revenue loss)
2. Deploy backup in parallel (restore partial protection)
3. Fix properly (restore full capability)
4. Prevent recurrence (monitoring, testing)

**Key lessons:**

- Production systems fail; resilience through preparation
- Incident response: immediate mitigation, then proper fix
- Defense in depth: backup systems, fallbacks

- Post-mortem: learn and prevent recurrence
- Communication critical during incidents

---

**Question 574 [ADVANCED - Ethical AI at Scale]:**

**SCENARIO:** A social media company uses ML to rank content in news feeds. They discover:

- Engagement-optimized algorithm increases time on platform by 40%
- Internal research shows algorithm amplifies divisive, emotionally-charged content
- Mental health studies link platform usage to increased anxiety, depression in teens
- Investors pressure to maintain engagement metrics
- Whistleblower threatens to leak research
- Regulatory scrutiny increasing
- Alternative algorithm increases time on platform by only 25% but promotes healthier content mix

What should the company do?

A) Maximize engagement regardless of social impact B) Deploy healthier algorithm despite engagement cost; transparency about trade-offs; proactive engagement with regulators C) Hide research; maximize profits D) Shut down platform

**Answer: B) Prioritize societal well-being; transparency; regulatory cooperation**

**Explanation:** This scenario explores AI ethics at societal scale, based on real-world controversies:

**Stakeholder Analysis:**

1. **Users:** especially teens vulnerable to mental health impacts
2. **Investors:** want engagement and revenue growth
3. **Employees:** some alarmed by research findings
4. **Regulators:** increasingly concerned about platform harm
5. **Public:** growing awareness of social media's societal impacts
6. **Whistleblower:** moral concerns override confidentiality

**Ethical Framework Analysis:**

**Utilitarian perspective:**

- 40% engagement gain benefits: shareholder value, user enjoyment
- Costs: mental health harm to millions, societal divisiveness, misinformation spread
- Net harm likely exceeds shareholder benefit
- Should deploy healthier algorithm

**Rights-based perspective:**

- Users have right to platforms not deliberately designed to harm them
- Informed consent: users don't know algorithm amplifies harmful content
- Transparency about algorithm choices is ethical obligation

**Virtue ethics:**

- What would virtuous company do? Prioritize user well-being over engagement
- Company values: if claim to "connect people," harmful algorithm contradicts values

**Corporate responsibility:**

- Long-term thinking: regulatory crackdown, user exodus, brand damage
- Short-term engagement gains don't justify long-term existential risk

**Recommended Actions:**

**Immediate (Week 1):**

1. **Deploy healthier algorithm:** Accept 15% engagement loss (40% → 25%)
2. **Internal commitment:** Leadership commits to user well-being over engagement
3. **Whistleblower engagement:** Address concerns proactively

**Short-term (Month 1-3):**

1. **Transparency:** Publish research on algorithm impacts
2. **User control:** Give users choice: "engagement-optimized" vs. "well-being-optimized" feeds
3. **Regulatory cooperation:** Proactively engage regulators, share research
4. **Independent audit:** External researchers audit algorithm impacts
5. **Investor communication:** Explain long-term sustainability rationale

**Medium-term (Months 3-12):**

1. **New metrics:** Define success beyond engagement (user well-being, digital literacy, healthy discourse)
2. **Algorithm development:** Invest in algorithms optimizing healthier metrics
3. **Youth protections:** Special protections for teens
4. **Content moderation:** Reduce amplification of divisive content
5. **Research program:** Ongoing study of platform impacts

**Long-term (Years):**

1. **Business model evolution:** Reduce dependence on engagement-driven advertising
2. **Industry standards:** Advocate for responsible AI standards
3. **Self-regulation:** Establish ethics board with external members
4. **Product redesign:** Features promoting healthy usage patterns

**Anticipating Objections:**

**"Shareholders will revolt":**

- Short-term stock price might dip
- Long-term: regulatory risk, user exodus risk, brand damage risk all reduced
- Responsible companies increasingly valued by investors
- ESG (Environmental, Social, Governance) considerations

**"Competitors will gain market share":**

- If all platforms optimize for harmful engagement, eventually regulatory intervention
- First-mover advantage in responsible AI
- Brand differentiation: "platform that cares about users"

**"Users want engaging content":**

- Users want engaging AND healthy content
- Addiction ≠ satisfaction
- Long-term user retention better with healthier platform
- Provide choice; let users decide

**"Can't prove causation of harm":**

- Precautionary principle: potential for serious harm justifies action even without definitive proof
- Company's own research shows concerns
- Burden of proof: demonstrate safety, not just absence of proven harm

**Real-world Examples:**

- Tobacco companies knew of harm, hid research, eventually massive liability
- Automotive industry initially resisted safety features, now standard
- Tech platforms face similar inflection point

**Key Lessons:**

- Optimization without considering externalities causes harm
- Engagement metrics don't capture user well-being
- Ethical companies consider broad societal impacts, not just shareholder value
- Transparency and proactive responsibility beat regulatory forcing
- Long-term sustainability requires trust
- Some profitable optimizations shouldn't be pursued

This scenario has no perfect solution, but healthier algorithm with transparency is most defensible ethically and strategically.

---

**Question 575 [ADVANCED - Model Degradation Investigation]:**

**SCENARIO:** Your e-commerce recommendation model's revenue-per-user metric has declined 15% over 3 months despite stable technical metrics:

- Model accuracy: stable at 87%
- Click-through rate: stable at 8%
- Recommendation diversity: stable
- User engagement: stable
- Revenue per user: declined from $45 to $38.25
- Investigation reveals competitor launched aggressive promotion campaign
- Your recommendations still accurate, but users finding better prices elsewhere

What's the root cause and solution?

A) Model failed; retrain immediately B) External environment changed (competitor actions); model metrics stable but business context shifted; need strategic response, not just technical fix C) Model accuracy too low D) No action needed; revenue decline is normal

**Answer: B) External factors; requires business strategy, not model retraining**

**Explanation:** This scenario illustrates when ML problems aren't ML solutions:

**Root Cause Analysis:**

**What's NOT the problem:**

- Model accuracy (87%, stable)
- User engagement (stable)
- Recommendations (still accurate for predicting user preferences)
- Data drift (no distribution shifts)

**What IS the problem:**

- Competitive landscape changed
- Users like your recommendations but buy from competitor (better prices)
- Model predicts preferences correctly, but pricing competitiveness matters more

**Why This Matters:** Demonstrates: (1) ML model can be technically sound but business results decline due to external factors; (2) Stable technical metrics don't guarantee business success; (3) Some business problems aren't ML problems

**Deeper Investigation:**

**Data Analysis:**

1. **User journey:** Users view recommendations, leave platform, return later (suggesting price comparison)
2. **Cart abandonment:** Increased abandonment rates
3. **Competitor tracking:** Competitor prices 20% lower on average
4. **Basket analysis:** Users buying lower-margin items (affected by competition)

**What Model Can't Solve:**

- Model recommends items users want

- Can't make your prices competitive
- Can't change inventory
- Can't alter competitor behavior

**Multi-Stakeholder Solution:**

**Pricing Strategy (weeks):**

1. **Competitive analysis:** Systematic competitor price tracking
2. **Dynamic pricing:** Match or beat competitor prices on key items
3. **Price optimization:** ML for pricing, not just recommendations
4. **Bundle pricing:** Offer bundles competitors don't

**Recommendation Model Evolution (months):**

1. **Price-aware recommendations:** Factor price competitiveness into recommendations
2. **Value optimization:** Recommend products where you're price-competitive
3. **Margin optimization:** Balance user preference with margin/competitiveness
4. **Multi-objective:** Optimize for user satisfaction AND revenue

**Product Strategy (months):**

1. **Differentiation:** Unique products competitors don't offer
2. **Private label:** Exclusive brands
3. **Value-add services:** Free shipping, better returns, loyalty programs
4. **Customer experience:** Superior experience justifying price premium

**Strategic Response (quarters):**

1. **Market positioning:** Compete on service, selection, not just price
2. **Customer loyalty:** Invest in retention programs
3. **Margin management:** Accept lower margins on competitive items, make up elsewhere

**Monitoring Evolution:**

1. **Competitive intelligence:** Track competitor pricing, promotions continuously
2. **Market share:** Monitor relative to competitors
3. **Price perception:** Survey users on price competitiveness

4. **Holistic metrics:** Revenue AND user satisfaction AND competitive position

**ML Role in Solution:**

1. **Price optimization models:** Predict price elasticity
2. **Competitive monitoring:** Automated competitor price tracking
3. **Customer segmentation:** Price-sensitive vs. value-sensitive customers
4. **Churn prediction:** Identify customers at risk of leaving for competitors

**Updated Success Metrics:**

- Original: CTR, accuracy
- Evolved: CTR × conversion rate × margin × retention
- Comprehensive: capture full customer value, not just engagement

**Key Lessons:**

1. **Stable ML metrics ≠ business success:** External factors matter
2. **ML is component, not complete solution:** Pricing, product strategy, competition all matter
3. **Multi-disciplinary problems:** Require cross-functional solutions
4. **Monitor business metrics:** Technical metrics alone insufficient
5. **Adapt to environment:** Models exist in competitive, changing markets
6. **Know limitations:** Some problems require non-ML solutions

**What NOT to do:**

- Don't retrain model repeatedly (won't fix external problems)
- Don't optimize recommendation diversity/accuracy more (already working)
- Don't blame ML team (they did their job)
- Don't ignore competitive environment

**Organizational Implications:**

- ML teams need business context awareness
- Business teams need ML literacy (know what ML can/can't solve)
- Cross-functional collaboration essential
- Holistic metrics spanning technical and business

This scenario emphasizes: ML succeeds or fails within broader business context. Technical excellence necessary but insufficient. Solutions require understanding when problems are ML problems vs. business problems.

---

**Question 576 [ADVANCED - Privacy-Utility Tradeoff]:**

**SCENARIO:** A hospital wants to use patient data to predict readmission risk. Stakeholders:

- Doctors: Want highest accuracy (92% with full patient data including genetic, social, behavioral data)
- Privacy officer: Concerned about genetic and behavioral data sensitivity
- Patients: Survey shows 60% uncomfortable with genetic data use, 40% uncomfortable with behavioral data
- Regulators: HIPAA allows use with consent, but scrutiny of genetic data use increasing
- Alternative model: 88% accuracy without genetic/behavioral data (just medical history)

How should this be resolved?

A) Use full data for maximum accuracy B) Multi-stakeholder decision: quantify privacy-accuracy tradeoff, meaningful consent, tiered approach, ethics review C) Never use sensitive data D) Ignore patient preferences

**Answer: B) Structured privacy-utility tradeoff analysis with patient autonomy**

**Explanation:** This scenario requires balancing competing values:

**Stakeholder Value Analysis:**

**Clinical perspective (Doctors):**

- 92% accuracy catches more at-risk patients
- Better care, potentially saving lives
- 4% accuracy difference = 4 additional patients per 100 identified

**Privacy perspective (Privacy officer, Patients):**

- Genetic data: uniquely identifying, hereditary implications, discrimination risk
- Behavioral data: stigma, judgment, sensitive personal information
- HIPAA minimum necessary principle
- Patient autonomy and informed consent

**Regulatory perspective:**

- HIPAA permits with consent
- Genetic Information Nondiscrimination Act (GINA) concerns
- Increasing scrutiny of genetic data use
- Future regulation may restrict further

**Risk Analysis:**

**Benefits of full model (92% accuracy):**

- 4% more patients correctly identified
- On 10,000 patients/year: 400 additional identifications
- Readmission costs ~$15K; potential savings: 400 × $15K = $6M

**Risks of full model:**

- Data breach: genetic data exposure (irrevocable harm)
- Discrimination: genetic data could be misused (employment, insurance)
- Trust erosion: patients opt out, reducing data quality
- Regulatory: future restrictions on genetic data use
- Stigma: behavioral data (substance use, mental health) could harm patients

**Recommended Solution Framework:**

**Tiered Consent Approach:**

**Tier 1: Basic Model (88% accuracy):**

- Medical history only (diagnoses, medications, procedures)
- Default: all patients included unless opt-out
- Minimal sensitive data
- HIPAA-compliant with standard authorization

**Tier 2: Enhanced Model (90% accuracy):**

- Medical history + limited behavioral data (smoking, exercise)
- Opt-in consent required
- Clear explanation of data use
- Patients choose to participate

**Tier 3: Full Model (92% accuracy):**

- All data including genetic, comprehensive behavioral
- Explicit informed consent
- Genetic counseling provided
- Ongoing consent (can withdraw)
- Extra protections (encryption, access controls, separate storage)

**Implementation:**

**Patient Consent Process:**

1. **Clear explanation:** What data used, why, benefits, risks
2. **Granular choices:** Patients choose tier
3. **Easy opt-out:** Can decline or limit participation
4. **Dynamic consent:** Can change mind, modify choices
5. **No penalty:** Care quality not affected by consent choices

**Technical Implementation:**

1. **Separate models:** Train tier-specific models
2. **Federated learning:** For Tier 3, consider federated approach
3. **Differential privacy:** Add noise to aggregate statistics
4. **Access controls:** Genetic data behind additional authentication
5. **Audit logging:** Track all access to sensitive data

**Privacy Protections:**

1. **Encryption:** All data encrypted at rest and transit
2. **De-identification:** Where possible, use de-identified data
3. **Access limitation:** Minimal necessary access
4. **Breach notification:** Immediate notification if breach
5. **Data retention:** Delete when no longer needed
6. **External audit:** Independent privacy auditor

**Expected Adoption:**

- Tier 1 (88% accuracy): 95% of patients (nearly universal)
- Tier 2 (90% accuracy): 60% of patients (behavioral data comfort level)
- Tier 3 (92% accuracy): 30% of patients (genetic data concerns)

**Weighted Average Accuracy:** $0.95 \times 88\% + 0.05 \times (0.60 \times 90\% + 0.40 \times 92\%) = {\sim}88.5\%$

**Analysis:**

- Respects patient autonomy
- Achieves reasonable accuracy
- Maintains trust
- Legally compliant
- Ethically defensible

**Alternative Approaches:**

**If 88% accuracy insufficient clinically:**

1. **Better features:** Non-sensitive features achieving higher accuracy
2. **Risk stratification:** Use sensitive data only for high-risk patients (with consent)
3. **External validation:** Prove 4% accuracy difference materially improves outcomes

**If consent rates too low:**

1. **Education:** Explain benefits of data sharing
2. **Trust building:** Demonstrate responsible data handling
3. **Incentives:** (carefully considered - coercion risk)

**Governance:**

1. **Ethics committee:** Review annually
2. **Patient advisory board:** Patient representatives in decision-making
3. **Transparency reports:** Publish data use, protections, outcomes
4. **Independent audit:** External privacy audit annually

**Metrics to Track:**

- Consent rates by tier

- Model accuracy by tier
- Patient satisfaction with process
- Readmission rates (actual outcomes)
- Data breach incidents (hopefully zero)
- Opt-out rates over time

**Key Principles:**

1. **Patient autonomy:** Patients decide about their data
2. **Meaningful consent:** Clear information enabling informed decisions
3. **Proportionality:** Privacy intrusion proportional to benefit
4. **Transparency:** Open about data use and protections
5. **Accountability:** Clear responsibility for data protection
6. **Ongoing consent:** Not one-time; patients can change minds

**Key Lessons:**

- Privacy and accuracy aren't binary; graduated approaches exist
- Patient preferences vary; tiered consent respects diversity
- 4% accuracy gain might not justify privacy costs
- Trust is essential; coercive data collection erodes trust
- Regulatory landscape evolving; future-proof solutions
- Ethics aren't obstacles; they're requirements for sustainable AI

This approach balances clinical benefits with privacy rights, respects patient autonomy, and creates sustainable, trustworthy system.

---

**Question 577 [ADVANCED - Feature Leakage Detection]:**

**SCENARIO:** Your credit card fraud detection model achieves impressive 98% accuracy in validation. After deployment, performance drops to 72%. Investigation reveals:

- Training data: historical transactions labeled as fraud after investigation completed (weeks later)
- Feature "investigation_notes" field included in training
- Field contains investigation outcomes ("confirmed fraud," "customer error," "merchant error")

- Production model doesn't have investigation_notes (not available at prediction time)
- Model learned to rely heavily on this leaked feature

What should have prevented this and how do you fix it?

A) Investigation notes feature is valuable; keep it B) Temporal data leakage: future information leaked into training; rebuild with temporally valid features; implement leakage detection C) 98% accuracy proves model quality D) Production data is wrong

**Answer: B) Identify and eliminate data leakage; rebuild with proper temporal validation**

**Explanation:** This scenario demonstrates subtle data leakage destroying production performance:

**Understanding the Leakage:**

**Timeline of Real Transaction:**

1. Transaction occurs (Time T)
2. Model should predict fraud using only information available at Time T
3. If flagged as suspicious, investigation begins (Time T+hours/days)
4. Investigation completes, determination made (Time T+weeks)
5. Transaction labeled in database (Time T+weeks)

**Training Data Problem:**

- Training data includes investigation_notes (created at T+weeks)
- Model learns: "confirmed fraud" in notes → predict fraud
- But notes created AFTER transaction, not available at prediction time
- Model learns to predict outcomes based on future information

**Why This Is Leakage:**

- Perfect prediction of past: knows what happened after event
- Useless prediction of future: doesn't have future information
- Called "temporal leakage" or "target leakage"

**Why Validation Missed This:**

- Random train/test split doesn't preserve temporal order
- Test set includes investigation_notes (same leakage)
- Model performs well on test set for wrong reasons
- Would catch it with temporal validation (train on past, test on future)

**Leakage Detection Strategy:**

**During Feature Engineering:**

1. **Temporal analysis:** When is each feature available?
2. **Question each feature:** "Would this be available at prediction time?"
3. **Feature documentation:** Document temporal availability
4. **Feature categories:** Separate pre-transaction, transaction-time, post-transaction features

**Suspicious Patterns:**

- Overly high accuracy (98% on fraud detection suspicious)
- Features highly correlated with target (investigation_notes perfectly predicts fraud)
- Performance drop after deployment (classic leakage symptom)
- Single dominant feature (model relies on one suspicious feature)

**Validation That Would Catch Leakage:**

1. **Temporal validation:** Train on Jan-Mar, test on Apr (simulates production)
2. **Feature ablation:** Remove each feature; if performance drops dramatically, investigate
3. **Production simulation:** Use only features available at inference time
4. **Permutation importance:** Shuffle features; leaked features have extreme importance

**Fixing the Model:**

**Immediate (Days):**

1. **Remove leaked features:** investigation_notes and any derived features
2. **Retrain on clean features:** Only using temporally valid features
3. **Temporal validation:** Validate using future data without leakage
4. **Expected accuracy:** Likely 75-82% (more realistic, deployable)

**Proper Feature Set:**

- Customer history: past transaction patterns (available at T)
- Transaction characteristics: amount, merchant, location, time (available at T)
- Device/IP information: device used, IP geolocation (available at T)
- Velocity features: transactions in last hour/day (available at T)
- NOT: investigation outcomes, chargebacks, customer confirmations (available at T+days/weeks)

**Process Improvements:**

**Feature Engineering Standards:**

1. **Temporal validity checklist:** Every feature must pass temporal availability check
2. **Feature documentation:** Template requiring temporal availability documentation
3. **Automated checks:** Pipeline validates features are temporally valid
4. **Code review:** Peer review specifically checking for leakage

**Validation Standards:**

1. **Mandatory temporal validation:** All time-sensitive models require temporal splits
2. **Production simulation:** Validate using only production-available features
3. **Ablation studies:** Test impact of removing features
4. **Sanity checks:** Unrealistically high accuracy triggers investigation

**Team Education:**

1. **Training:** Teach data scientists about common leakage patterns
2. **Case studies:** Review leakage examples (this scenario is cautionary tale)
3. **Checklists:** Leakage detection checklists integrated into workflow
4. **Culture:** Encourage healthy skepticism of "too good" results

**Common Leakage Patterns:**

**Target leakage (this scenario):**

- Features derived from target
- Future information leaked into past
- Example: "investigation_notes" contains fraud determination

**Train-test leakage:**

- Data preprocessing uses full dataset statistics
- Test data influences training
- Example: scaling using mean from train+test

**Temporal leakage:**

- Future information available during training
- Not available at prediction time
- Example: stock price models using future prices

**Sample leakage:**

- Test examples in training set (duplicates, augmented versions)
- Artificially inflates performance

**Proxy leakage:**

- Feature perfectly correlated with target
- Example: "account_closed" feature when predicting churn (account closes AFTER churn)

**Preventing Leakage - Checklist:**

☐ Document when each feature is available ☐ Use only features available at prediction time ☐ Temporal validation for time-series problems ☐ Feature importance analysis (suspicious if one feature dominates) ☐ Sanity check performance (too high suggests problems) ☐ Production simulation before deployment ☐ Peer review focusing on leakage ☐ Automated temporal validation in pipeline

**Key Lessons:**

1. **Suspicious accuracy:** 98% fraud detection accuracy is suspiciously high
2. **Temporal validation:** Essential for any model with time component
3. **Feature scrutiny:** Question every feature's availability
4. **Production simulation:** Validate using only production-available data
5. **Culture of skepticism:** "Too good to be true" usually is
6. **Documented processes:** Checklists prevent common mistakes

**Real-World Impact:**

- Data leakage is extremely common (studies show ~30% of ML papers have leakage)
- Often subtle and hard to detect
- Production deployment is the test that reveals leakage
- Can waste months of work and harm business

This scenario emphasizes: careful feature engineering, temporal validation, and healthy skepticism are essential for production-ready models.

---

**Question 578 [ADVANCED - Model Fairness vs. Predictive Accuracy]:**

**SCENARIO:** A lending institution builds a credit risk model. Analysis shows:

- Model A: 84% accuracy, Equal Opportunity satisfied (equal true positive rates across races)
- Model B: 89% accuracy, fails Equal Opportunity (White TPR: 95%, Black TPR: 78%)
- Model B would approve more loans (higher profitability) but disparate approval rates concern fairness team
- Legal says both models technically comply with fair lending laws
- Model B's disparity partly reflects historical discrimination in credit scoring
- Alternative features could improve Model A to 86% without fairness sacrifice

Which model should be deployed and why?

A) Model B (higher accuracy, more profit, legally compliant) B) Model A (fairness priority), or invest in improving A with alternative features C) Deploy Model B; fairness doesn't matter if legal D) Neither model is acceptable

**Answer: B) Prioritize fairness (Model A) or improve it; reject higher accuracy with discrimination**

**Explanation:** This scenario explores fairness-accuracy tradeoffs in high-stakes lending:

**Analyzing the Tradeoff:**

**Model B Advantages:**

- 5% higher accuracy (89% vs. 84%)
- More loans approved (higher revenue)
- Legally compliant (passes minimum legal standard)
- Better at identifying creditworthy borrowers overall

**Model B Disadvantages:**

- Significant disparate impact (95% vs. 78% TPR)
- 17-point gap in true positive rates
- Systematically disadvantages Black applicants
- Qualified Black applicants less likely to be approved than equally qualified White applicants

**Equal Opportunity Fairness:**

- Definition: Among qualified applicants, all groups should have equal approval rates
- Model A satisfies: qualified applicants equally likely to be approved regardless of race
- Model B violates: qualified Black applicants systematically denied more often

**Ethical Analysis:**

**Why Model A is Preferable:**

1. **Equity:** Equal Opportunity is appropriate fairness criterion for lending (qualified applicants should have equal access)
2. **Historical context:** Credit scoring systems have historically discriminated; perpetuating disparities is harmful
3. **Individual fairness:** Individuals with equal qualifications should be treated equally
4. **Legal risk:** While technically compliant now, disparate impact increasingly scrutinized
5. **Reputational risk:** Discrimination discovered publicly harms reputation
6. **Long-term sustainability:** Discriminatory practices unsustainable as awareness increases

**Why "Legal Compliance" Insufficient:**

- Legal compliance is floor, not ceiling

- Fair lending laws establish minimum; ethics often require more
- Laws lag behind societal expectations
- "Technically legal" discrimination still wrong
- Corporate responsibility extends beyond legal minimums

**Why 5% Accuracy Difference Doesn't Justify Discrimination:**

- Not all accuracy gains equal in value
- Accuracy gained through discrimination has negative societal value
- Trade some profit for fairness is ethically required
- 84% accuracy is reasonable; not sacrificing viability

**Recommended Approach:**

**Phase 1: Deploy Model A (Weeks 1-2):**

- Satisfies fairness requirements
- 84% accuracy reasonable for lending
- Ethically defensible
- Lower profit than Model B but acceptable

**Phase 2: Improve Model A (Months 1-3):**

- Research shows alternative features can reach 86% without fairness sacrifice
- Investigate features:
    - Alternative credit data (rent payments, utility bills)
    - Cash flow analysis
    - Banking patterns
    - Employment stability
- Goal: close accuracy gap while maintaining fairness

**Phase 3: Continuous Improvement (Ongoing):**

- Monitor fairness metrics continuously
- Demographic parity, Equal Opportunity, Equalized Odds
- Per-group performance tracking
- Regular fairness audits

**Alternative Features to Explore:**

**Beyond Traditional Credit Scores:**

- Traditional credit scores embed historical discrimination
- Alternative data sources:
    - Rental payment history (especially for those without mortgage history)
    - Utility payment consistency
    - Bank account behavior (overdrafts, savings patterns)
    - Income stability and growth
    - Education and skills (proxy for earning potential)

**Why These Help Fairness:**

- Provide signal for creditworthy individuals without traditional credit history
- Less correlated with historical discrimination
- Capture financial responsibility through different channels

**Fairness Constraints During Training:**

- Explicitly optimize for Equal Opportunity during training
- Don't just select features; incorporate fairness into optimization
- Techniques: reweighting, threshold optimization per group, adversarial debiasing

**Stakeholder Communication:**

**To Executive Leadership:**

- Frame as risk management, not just ethics
- Reputational and legal risks of Model B
- Long-term sustainability of Model A
- Industry trend toward fair lending
- Quantify: 5% accuracy for significant fairness improvement is good trade

**To Investors:**

- Responsible lending is sustainable lending
- Avoiding discrimination reduces legal and reputational risk
- Fair lending brand differentiation
- ESG (Environmental, Social, Governance) considerations increasingly important

**To Regulatory:**

- Proactive fairness commitment
- Going beyond legal minimum
- Transparent methodology
- Continuous monitoring

**To Affected Communities:**

- Consultation with community organizations
- Transparency about fairness metrics
- Commitment to equal access
- Ongoing accountability

**Metrics Beyond Accuracy:**

**Fairness Metrics:**

- Demographic Parity: approval rates by group
- Equal Opportunity: true positive rates by group
- Equalized Odds: false positive and false negative rates by group
- Individual fairness: similar individuals receive similar decisions

**Business Metrics:**

- Default rates (actual credit risk realized)
- Approval rates (access to credit)
- Revenue (balancing fairness with viability)
- Customer acquisition across demographics

**Long-term Metrics:**

- Brand reputation
- Regulatory incidents
- Customer diversity
- Trust scores

**Addressing Objections:**

**"We're leaving money on the table":**

- 5% accuracy loss translates to fewer approvals

- But: avoiding discrimination lawsuits, regulatory fines
- Fair lending builds trust, attracts socially conscious customers
- Long-term value exceeds short-term profit loss

**"Historical data reflects reality":**

- Reality includes historical discrimination
- Credit scoring systems disadvantaged certain groups historically
- Perpetuating historical discrimination isn't morally acceptable
- Goal: predict future creditworthiness, not replicate past discrimination

**"Equal Opportunity means some unqualified applicants approved":**

- No, Equal Opportunity means QUALIFIED applicants treated equally
- Doesn't require approving unqualified applicants
- Focuses on ensuring qualified applicants aren't disadvantaged by group membership

**Legal Landscape:**

**Current:**

- ECOA, Fair Housing Act prohibit discrimination
- Disparate impact standard applies
- Regulatory scrutiny of algorithmic lending increasing

**Trend:**

- Algorithmic fairness increasingly scrutinized
- Several jurisdictions considering fairness requirements
- Fair lending laws evolving to address algorithmic discrimination
- Proactive fairness positions company well for future regulation

**Key Lessons:**

1. **Fairness is requirement, not optional:** High-stakes domains like lending require fairness
2. **Legal minimum insufficient:** Ethical responsibility exceeds legal compliance
3. **Not all accuracy equal:** Accuracy through discrimination has negative value

4. **Alternative features exist:** Often can improve fairness without sacrificing accuracy
5. **Long-term thinking:** Short-term profit vs. long-term sustainability
6. **Stakeholder engagement:** Affected communities should have voice
7. **Transparency:** Fairness commitments should be public and monitored

This scenario has no perfect answer, but Model A or improved version is most defensible ethically, legally, and strategically.

---

**Question 579 [ADVANCED - Concept Drift in Non-Stationary Environment]:**

**SCENARIO:** An airline's dynamic pricing model has been successful for 2 years, maintaining 82% prediction accuracy of optimal prices. Then:

- COVID-19 pandemic fundamentally changes travel behavior
- Model accuracy drops to 54% (barely better than random)
- Historical data (2018-2019) no longer representative
- Limited data from pandemic period (months vs. years of historical data)
- Travel patterns still evolving, not yet stable
- Business pressure to restore pricing optimization quickly
- Options: (1) Retrain on limited pandemic data, (2) Simpler model robust to uncertainty, (3) Hybrid human-AI pricing, (4) Wait for more data

What's the appropriate strategy for such extreme, rapid concept drift?

A) Retrain on pandemic data immediately; trust the model B) Hybrid approach: simple robust model + human expertise + rapid iteration + gradual trust building C) Abandon ML entirely during disruption D) Wait 2 years for enough pandemic data

**Answer: B) Hybrid human-AI approach with simple models during high uncertainty**

**Explanation:** This scenario addresses extreme concept drift and fundamental distribution shifts:

**Understanding Extreme Concept Drift:**

**Normal Concept Drift:**

- Gradual changes in patterns
- Historical data partially relevant
- Retraining on recent data effective

**Extreme Disruption (Pandemic):**

- Fundamental regime change
- Historical patterns irrelevant or misleading
- New patterns not yet stable
- Uncertainty extremely high

**Why Historical Model Fails:**

- Trained on leisure + business travel patterns
- Pandemic: business travel collapsed, leisure travel constrained
- Seasonality disrupted (traditional patterns broken)
- Price sensitivity changed dramatically
- Booking horizons shortened
- Cancellation rates skyrocketed

**Why Simple Retraining Insufficient:**

- Limited pandemic data (weeks/months vs. years)
- Patterns still evolving (each pandemic wave different)
- High uncertainty makes complex models unreliable
- Overfitting risk on small dataset

**Multi-Phase Response Strategy:**

**Phase 1: Immediate Response (Week 1-2) - Survival Mode**

**Fallback to Simple, Robust System:**

- Rule-based pricing using:
    - Capacity utilization thresholds
    - Competitor price benchmarking
    - Simple markup rules
    - Manual overrides by pricing experts
- Why: Simple systems more robust during extreme uncertainty

**Human-in-the-Loop:**

- Pricing team makes final decisions
- ML provides suggestions, not decisions
- Learn from human decisions (implicitly gathering training data)

**Monitoring:**

- Track actual outcomes vs. predictions
- Build dataset of pandemic-era pricing outcomes
- Identify what works in new environment

**Phase 2: Rapid Learning (Months 1-3) - Adaptation**

**Simple ML Models:**

- Train lightweight models on emerging pandemic data
- Linear models, small decision trees (less prone to overfitting on small data)
- Limited features (focus on most reliable signals):
  - Capacity utilization (still relevant)
  - Days until departure (still relevant)
  - Current demand indicators (new importance)
  - Restrictions in origin/destination (new feature)

**Why Simple Models:**

- Less data needed for reliable training
- More interpretable (pricing team can validate)
- Less overfitting risk
- Faster to retrain as more data arrives

**Ensemble Approach:**

- Combine multiple simple models
- Include rule-based component
- Human expertise as final component
- Weight components based on recent performance

**Continuous Learning:**

- Daily or weekly retraining on fresh data
- Short time windows (use last 2-4 weeks, not years)
- Rapid iteration cycle
- A/B testing different approaches

**Phase 3: Incremental Sophistication (Months 3-12) - Stabilization**

**Gradual Complexity Increase:**

- As more pandemic data accumulates, train more sophisticated models
- Validate rigorously before deploying
- Careful monitoring of each upgrade
- Ready to rollback if performance degrades

**Multi-Model Strategy:**

- Short-term model: trained on last 4 weeks (captures current dynamics)
- Medium-term model: trained on last 3 months (more stable patterns)
- Long-term model: all pandemic data (baseline patterns)
- Ensemble weighted by recent performance

**Domain Adaptation:**

- Transfer learning: what aspects of pre-pandemic model still relevant?
- Feature engineering: which features retained predictive power?
- Some patterns universal (capacity constraints), others changed

**Confidence Estimation:**

- Models output prediction intervals, not just point estimates
- Wider intervals during high uncertainty
- Narrow as patterns stabilize
- Use confidence to determine human review necessity

**Phase 4: New Normal (Year 1+) - Maturity**

**Recognize New Patterns:**

- Travel behavior fundamentally changed (more remote work, different seasonality)
- Can't return to pre-pandemic model

- Build new model for post-pandemic travel
- Incorporate lessons from disruption period

**Scenario Planning:**

- Model different scenarios (new variants, travel restrictions)
- Have playbook for rapid adaptation if needed
- Disruption preparedness built into model design

**Hybrid Human-AI Decision Framework:**

**Low Uncertainty Situations:**

- Model makes decisions autonomously
- Standard routes, normal demand patterns
- Human review only if flagged

**Medium Uncertainty:**

- Model suggests price
- Human pricing manager reviews and approves
- Can override if expertise contradicts model

**High Uncertainty:**

- Model provides reference point
- Human makes decision
- Machine learning captures human decisions for future learning

**Key Principles During Extreme Uncertainty:**

**1. Embrace Simplicity:**

- Complex models overfit on limited data
- Simple, robust models better during uncertainty
- Interpretability enables validation by experts

**2. Human Expertise Critical:**

- Domain experts adapt faster than models during disruption

- Combine statistical patterns with human judgment
- Learn from human decisions

**3. Rapid Iteration:**

- Don't wait for perfect model
- Deploy, monitor, learn, iterate quickly
- Weekly or daily model updates

**4. Conservative Risk Management:**

- Don't trust models during extreme uncertainty
- Wider safety margins in pricing
- Gradual trust building as models prove reliable

**5. Ensemble Diverse Approaches:**

- Rules + simple ML + complex ML + human judgment
- Different time windows (short-term, long-term)
- Robustness through diversity

**Metrics for Adaptation Success:**

**Business Metrics:**

- Revenue per available seat mile (RASM)
- Load factors (occupancy)
- Price competitiveness vs. competitors
- Customer booking lead times

**Model Metrics:**

- Prediction accuracy on recent data
- Calibration (predicted prices vs. optimal in hindsight)
- Confidence interval coverage
- Model stability across retraining

**Operational Metrics:**

- Human override rate

- Time to adjust to new patterns
- Adaptation speed vs. competitors

**Lessons from Other Disruptions:**

**Financial Crisis 2008:**

- Quantitative models failed spectacularly
- Firms with human oversight fared better
- Simplified models more robust than complex

**Flash Crashes:**

- High-frequency trading algorithms amplified crashes
- Circuit breakers (human intervention points) necessary

**Adversarial Environments:**

- Fraud, spam: adversaries adapt to models
- Cat-and-mouse game requires continuous adaptation
- Human expertise complements ML

**Designing for Future Disruptions:**

**Architecture:**

- Modular design enabling component swapping
- Easy to switch between simple/complex models
- Fallback systems always ready

**Processes:**

- Documented playbook for disruption response
- Pre-approved simplified fallback model
- Crisis team with authority to override normal processes

**Monitoring:**

- Automated detection of fundamental shifts
- Alerts when model confidence drops

- Triggers for switching to disruption mode

**Culture:**

- Organizational acceptance that models fail during disruptions
- No blame for acknowledging failure
- Rapid response valued over defending existing models

**Key Lessons:**

1. **Extreme drift requires different response:** Can't just retrain
2. **Simplicity during uncertainty:** Complex models fail first
3. **Human expertise critical:** Especially during novel situations
4. **Rapid iteration:** Learn and adapt quickly
5. **Hybrid approaches:** Combine multiple strategies
6. **Conservative risk management:** Don't trust models during extreme uncertainty
7. **Prepare for disruption:** Design systems anticipating future disruptions

This scenario emphasizes: ML systems must gracefully degrade during extreme uncertainty, combining ML with human expertise and simple robust systems until patterns stabilize.

---

---

**Question 580 [ADVANCED - Cascading System Failure]:**

**SCENARIO:** Your e-commerce company has an ML ecosystem with interdependent models:

- Model A: Demand forecasting (predicts product demand)
- Model B: Inventory optimization (uses Model A's forecasts)
- Model C: Dynamic pricing (uses Models A and B)
- Model D: Recommendations (uses Model C's prices)

Timeline of failure:

- Day 1: Model A's accuracy drops from 85% to 65% (supplier data source changed format)

- Day 2: Model B over-orders inventory based on inflated forecasts (excess inventory: $2M)
- Day 3: Model C sets prices too low trying to move excess inventory
- Day 4: Model D recommends low-margin products, revenue drops 15%
- Day 5: Customer support overwhelmed with out-of-stock items (Model A underforecasted some items)

How should this cascading failure have been prevented and how do you recover?

A) Independent models don't create cascading failures B) System design with circuit breakers, monitoring dependencies, fallback mechanisms, graceful degradation per component C) Cascading failures are unavoidable D) Shut down entire system

**Answer: B) Defensive architecture with failure isolation and graceful degradation**

**Explanation:** This scenario demonstrates cascading failures in ML systems and prevention strategies:

**Understanding Cascading Failure:**

**Failure Propagation Chain:**

1. **Root cause:** Model A input data corruption (supplier format change)
2. **First-order effect:** Model A produces bad forecasts (65% accuracy)
3. **Second-order effect:** Model B trusts bad forecasts, orders wrong inventory
4. **Third-order effect:** Model C reacts to inventory position with bad pricing
5. **Fourth-order effect:** Model D recommends based on distorted prices
6. **System-level impact:** Revenue loss, excess inventory, customer dissatisfaction

**Why This Happened:**

**Tight Coupling:**

- Models directly consume outputs from upstream models
- No validation of upstream model quality
- Implicit trust in upstream predictions
- No circuit breakers

**Lack of Monitoring:**

- Model A degradation not detected immediately
- No alerts on downstream impact
- No validation of prediction quality at each stage

**No Fallback Mechanisms:**

- Models had no alternative when upstream failed
- No graceful degradation
- All-or-nothing dependency

**Prevention Architecture:**

**Layer 1: Input Validation and Monitoring**

**Data Quality Checks (Model A):**

- Schema validation: detect format changes immediately
- Distribution checks: alert on distribution shifts
- Automated testing: synthetic data validation
- Source monitoring: track upstream data source health

**When Should Have Alerted:**

- Day 0: Supplier format change detected automatically
- Hour 1: Distribution shift in Model A inputs
- Hour 2: Model A accuracy drop detected

**Immediate Response:**

- Halt Model A updates
- Fall back to previous day's forecasts
- Alert data engineering team
- Prevent bad forecasts from propagating

**Layer 2: Prediction Validation**

**Model A Output Validation:**

- Sanity checks: forecasts within reasonable bounds
- Historical comparison: dramatic changes from yesterday?

- Confidence intervals: are predictions very uncertain?
- Business rules: do forecasts make business sense?

**Circuit Breaker Pattern:**

```
If (Model_A_confidence < threshold) OR
   (predictions_differ_significantly_from_baseline):
   Use_fallback_forecasts
   Alert_team
   Prevent_downstream_propagation
```

**Layer 3: Dependency Management**

**Model B (Inventory) Protection:**

- Don't blindly trust Model A forecasts
- Ensemble approach: combine ML forecast with:
  - Historical averages (robust baseline)
  - Seasonal patterns
  - Human buyer expertise
  - Safety stock rules
- Confidence-based weighting: low confidence → rely more on alternatives

**Graceful Degradation:**

- If Model A available and confident: 70% weight to ML, 30% to baseline
- If Model A low confidence: 30% weight to ML, 70% to baseline
- If Model A fails: 100% baseline (moving average, seasonal adjustment)

**Layer 4: System-Level Safeguards**

**Health Checks Between Models:**

```
Before Model B uses Model A predictions:
1. Check Model A health status
2. Validate prediction quality
3. Compare to historical patterns
4. Verify confidence scores
5. If concerns: use fallback or weighted combination
```

**Cross-Validation:**

- Model C (pricing) validates that inventory levels make sense
- Model D (recommendations) checks price reasonableness
- Each component has sanity checks

**Layer 5: Monitoring and Alerting**

**Hierarchical Monitoring:**

**Individual Model Level:**

- Model A: forecast accuracy, input data quality, prediction distribution
- Model B: order quantities vs. historical, inventory turnover
- Model C: price distributions, margin impact
- Model D: recommendation diversity, click-through rates

**System Level:**

- End-to-end metrics: revenue, customer satisfaction, inventory costs
- Cross-model consistency: do model outputs make collective sense?
- Business rule violations: detect anomalies in system behavior

**Cascade Detection:**

- Monitor correlations: if Model A degrades, do downstream models also degrade?
- Alert on coordinated degradation
- Automated incident response when cascade detected

**Recovery Strategy:**

**Immediate (Hours 1-4):**

**Isolate and Contain:**

1. Detect Model A failure (data quality monitoring should alert)
2. Activate circuit breaker: prevent Model A predictions from flowing downstream
3. Fallback: Model A switches to baseline forecasts
4. Downstream models notified: use alternative input sources

**Parallel Recovery:**

- Team 1: Fix Model A (address supplier format change)
- Team 2: Assess damage (inventory over-orders, pricing impact)
- Team 3: Implement temporary compensating actions (adjust prices manually)

**Short-term (Days 1-3):**

**Fix Root Cause:**

1. Update Model A's feature engineering for new supplier format
2. Retrain and validate Model A
3. Gradual rollout: canary deployment with careful monitoring
4. Restore confidence before full deployment

**Mitigate Downstream Impact:**

1. Model B: Cancel over-orders if possible
2. Model C: Adjust pricing strategy for actual inventory
3. Model D: Recompute recommendations with corrected prices
4. Business: Promotions to move excess inventory profitably

**Medium-term (Weeks 1-4):**

**System Hardening:**

1. Implement circuit breakers at each integration point
2. Add health checks and validation layers
3. Develop fallback strategies for each model
4. Improve monitoring and alerting

**Architecture Redesign:**

1. Reduce tight coupling: models publish confidence with predictions
2. Subscribing models validate and can reject low-quality inputs
3. Hybrid decision-making: ML + rules + human oversight
4. Redundancy: multiple forecasting approaches in parallel

**Long-term (Months 1-6):**

**Resilient Architecture Principles:**

**1. Defense in Depth:**

- Multiple validation layers
- Redundant prediction methods
- Diverse data sources
- Human oversight for critical decisions

**2. Graceful Degradation:**

- Each model has fallback mode
- Performance degrades gradually, not catastrophically
- System continues operating with reduced capability

**3. Isolation and Bulkheads:**

- Failures contained to affected components
- Circuit breakers prevent cascade
- Independent failure domains

**4. Observable and Debuggable:**

- Comprehensive logging
- Distributed tracing (track predictions through system)
- Dependency graph visualization
- Cascade detection

**5. Testable Failure Modes:**

- Chaos engineering: intentionally inject failures
- Test cascade prevention
- Validate fallback mechanisms
- Drill incident response

**Architectural Patterns:**

**Pattern 1: Circuit Breaker**

```
class ModelCircuitBreaker:
    def call_model(input):
        if model_health_check() == FAILING:
            return fallback_prediction(input)
```

```
        prediction = model.predict(input)

        if validate_prediction(prediction) == INVALID:
            log_failure()
            increment_failure_count()
            if failure_rate > threshold:
                open_circuit()  # Stop calling model
            return fallback_prediction(input)

        return prediction
```

**Pattern 2: Bulkhead**

- Separate resource pools for each model
- Failure in Model A doesn't consume Model B's resources
- Prevents resource exhaustion cascades

**Pattern 3: Timeout and Retry**

- Models have maximum response time
- Slow models fail fast rather than hanging system
- Exponential backoff on retries

**Pattern 4: Graceful Degradation**

```
def get_forecast(product):
    ml_forecast = model_A.predict(product)
    baseline_forecast = historical_average(product)

    confidence = model_A.get_confidence()

    if confidence > 0.8:
        return ml_forecast
    elif confidence > 0.5:
        return 0.7 * ml_forecast + 0.3 * baseline_forecast
    else:
        return baseline_forecast
```

**Metrics for Resilience:**

**Failure Detection Speed:**

- Time from failure to detection (target: minutes)

- Time from detection to mitigation (target: minutes)
- Time to full recovery (target: hours)

**Blast Radius:**

- How many downstream systems affected?
- Revenue impact contained below threshold?
- Customer experience degradation limited?

**Recovery:**

- Mean time to detect (MTTD)
- Mean time to repair (MTTR)
- System availability (target: 99.9%+)

**Testing Resilience:**

**Chaos Engineering:**

1. Intentionally corrupt Model A's input data
2. Verify: circuit breakers activate, fallbacks engage
3. Confirm: downstream models not affected
4. Validate: system continues operating

**Failure Injection:**

- Inject failures at each integration point
- Test each fallback mechanism
- Verify monitoring alerts appropriately
- Practice incident response

**Load Testing:**

- Test system under degraded conditions
- Verify performance with fallbacks active
- Ensure graceful degradation

**Communication During Incident:**

**Internal:**

- Incident commander coordinates response
- Regular status updates to stakeholders
- Clear roles and responsibilities
- Post-mortem to capture learnings

**External:**

- Customer communication if user-facing impact
- Vendor coordination (supplier format issue)
- Regulatory notification if required

**Post-Incident Review:**

**What Went Well:**

- Eventually detected and resolved failure
- Teams coordinated effectively

**What Went Poorly:**

- Detection too slow (hours, should be minutes)
- No circuit breakers prevented propagation
- Insufficient fallback mechanisms
- Tight coupling enabled cascade

**Action Items:**

- Implement circuit breakers at each integration
- Add health checks and validation
- Develop fallback strategies
- Improve monitoring
- Conduct chaos engineering regularly

**Key Lessons:**

1. **Cascading failures are predictable:** Tightly coupled systems propagate failures
2. **Defense in depth:** Multiple protective layers
3. **Circuit breakers essential:** Prevent failure propagation
4. **Graceful degradation:** Reduced capability better than no capability
5. **Monitoring critical:** Early detection prevents cascades

6. **Test failure modes:** Chaos engineering validates resilience
7. **Loose coupling:** Models should validate inputs, not blindly trust
8. **Fallback mechanisms:** Every model needs plan B

This scenario emphasizes: ML systems are complex distributed systems requiring software engineering resilience patterns. Model accuracy isn't enough—system design for failure is critical.

---

**Question 581 [ADVANCED - Multi-Model Optimization Conflict]:**

**SCENARIO:** A ride-sharing company has three ML models with conflicting objectives:

- **Model 1 (Driver Matching):** Minimizes wait time for riders (optimizes rider experience)
- **Model 2 (Pricing):** Maximizes revenue per trip (optimizes business metrics)
- **Model 3 (Driver Routing):** Maximizes driver earnings per hour (optimizes driver satisfaction)

Conflicts emerging:

- Model 1 matches riders quickly but often with distant drivers (low driver earnings, high driver dissatisfaction)
- Model 2 surge pricing increases revenue but decreases rider satisfaction and demand
- Model 3 routes drivers to high-demand areas, increasing rider wait times in other areas
- System-level: driver churn increasing, rider complaints increasing, short-term revenue up but long-term sustainability questioned

How should this multi-objective optimization be resolved?

A) Prioritize revenue (Model 2); business survival comes first B) Redesign as multi-objective system-level optimization balancing rider satisfaction, driver earnings, platform revenue with long-term sustainability C) Keep models independent; conflicts are natural D) Remove optimization entirely

**Answer: B) Holistic multi-objective optimization with stakeholder balance**

**Explanation:** This scenario demonstrates the dangers of siloed optimization and need for system-thinking:

**Understanding the Conflicts:**

**Individual Model Objectives:**

**Model 1 (Rider Wait Time):**

- Objective: Minimize rider wait
- Optimization: Match with ANY available driver immediately
- Unintended consequence: Matches riders with very distant drivers
- Impact on drivers: Long pickup drives, low earnings per hour, frustration
- System impact: Driver churn increases

**Model 2 (Revenue):**

- Objective: Maximize trip revenue
- Optimization: Surge pricing during demand spikes
- Unintended consequence: Price-sensitive riders wait or don't request
- Impact on system: Demand reduction, riders seek alternatives
- System impact: Long-term demand destruction

**Model 3 (Driver Earnings):**

- Objective: Maximize driver earnings per hour
- Optimization: Route drivers to high-demand, high-pay areas
- Unintended consequence: Other areas underserved
- Impact on riders: Long waits in low-demand areas, poor experience
- System impact: Geographic coverage gaps

**Why Individual Optimization Fails:**

**Local vs. Global Optimums:**

- Each model optimizes its local objective
- System-level outcome is suboptimal
- Conflicts create negative feedback loops
- Long-term sustainability threatened

**Tragedy of the Commons:**

- Each model "overuses" shared resource (supply-demand balance)
- Collective optimization destroys shared value
- Game theory: Nash equilibrium vs. social optimum

**Feedback Loops:**

- Poor driver experience → driver churn → less supply → longer waits → poor rider experience → demand drops → lower earnings → more driver churn
- Negative spiral

**Holistic Redesign:**

**Phase 1: Define System-Level Objectives**

**Balanced Scorecard:** Instead of three separate objectives, define integrated success:

**Rider Experience (33% weight):**

- Wait time
- Trip cost reasonableness
- Service availability
- Safety and quality

**Driver Experience (33% weight):**

- Earnings per hour
- Pickup distance reasonableness
- Predictable earnings
- Work-life balance

**Platform Sustainability (33% weight):**

- Revenue and profitability
- Market share growth
- Regulatory compliance
- Long-term viability

**System-Level Metrics:**

- Marketplace health: supply-demand balance
- Participant retention: rider and driver retention rates
- Network effects: value grows with more participants
- Long-term value: customer lifetime value, driver lifetime value

**Phase 2: Multi-Objective Optimization Framework**

**Approach 1: Weighted Objectives**

```
Maximize:
  α * rider_satisfaction
  + β * driver_satisfaction
  + γ * platform_revenue

Subject to:
  rider_wait_time < max_acceptable_wait
  driver_pickup_distance < max_acceptable_distance
  pricing < price_ceiling
  driver_earnings > minimum_wage
  geographic_coverage > threshold
```

**Finding Optimal Weights (α, β, γ):**

- A/B testing: test different weight combinations
- Historical analysis: what balance historically led to best long-term outcomes?
- Stakeholder input: what weights reflect company values?
- Market research: what matters most to riders and drivers?

**Approach 2: Constraints-Based**

- Optimize primary objective (e.g., platform GMV)
- Subject to constraints on secondary objectives
- Example: Maximize revenue WHILE ensuring driver earnings > $20/hour AND rider wait < 5 minutes

**Approach 3: Multi-Objective Pareto Optimization**

- Find Pareto frontier: solutions where can't improve one objective without worsening another
- Present multiple Pareto-optimal solutions to leadership
- Leadership chooses based on strategic priorities

**Phase 3: Integrated Model Architecture**

**Single Unified Model:** Instead of three independent models, single model optimizing joint objective:

**Input:**

- Current system state: rider requests, driver locations, traffic conditions
- Historical patterns: demand patterns, typical wait times, earning patterns

**Optimization:**

```
For each rider-driver match decision:
  Evaluate:
    - Rider wait time impact
    - Driver pickup distance and earnings impact
    - Expected revenue
    - System-level supply-demand balance

  Choose match that optimizes weighted combination

  For pricing decision:
    Consider:
      - Current demand-supply imbalance
      - Price sensitivity of riders
      - Driver earnings needs
      - Long-term demand elasticity

    Set price balancing short-term revenue with long-term demand

  For driver routing:
    Consider:
      - Driver earnings optimization
      - System-level geographic coverage needs
      - Predicted rider demand by area

    Route drivers balancing individual earnings with system coverage
```

**Phase 4: Implementation Strategy**

**Transition Plan:**

**Month 1-2: Research and Analysis**

1. Analyze historical data: what balances led to best outcomes?

2. Stakeholder research: survey riders and drivers on priorities
3. Simulation: test different optimization approaches offline
4. Leadership alignment: agree on objective weights

**Month 3-4: Build Unified Model**

1. Design integrated architecture
2. Train multi-objective model
3. Offline evaluation: compare to current three-model system
4. Simulation testing: validate system-level metrics improve

**Month 5-6: Careful Rollout**

1. A/B test in single market
2. Monitor comprehensive metrics: rider satisfaction, driver satisfaction, revenue, retention
3. Iterate based on feedback
4. Gradual expansion to more markets

**Monitoring Multi-Objective Success:**

**Balanced Dashboard:**

**Rider Metrics:**

- Average wait time
- Price satisfaction (survey)
- Completion rate (% requests fulfilled)
- Repeat usage rate

**Driver Metrics:**

- Earnings per hour (median)
- Pickup distance (average)
- Active hours per week
- Retention rate (90-day)

**Platform Metrics:**

- GMV (Gross Merchandise Value)

- Take rate (platform percentage)
- Market share
- Unit economics

**System Health:**

- Supply-demand balance by location/time
- Geographic coverage
- Peak hour service levels
- Participant lifetime value

**Alert Triggers:**

```
IF driver_earnings < $18/hour for 3 consecutive days:
   Alert: increase driver incentives or reduce driver pickup distances

IF rider_wait_time > 7 minutes (90th percentile):
   Alert: recruit more drivers or adjust matching algorithm

IF rider_retention_30_day < 60%:
   Alert: investigate pricing, experience issues
```

**Handling Trade-offs:**

**Scenario 1: Friday Night Demand Surge**

**Old System (Conflicting Models):**

- Model 2: Aggressive surge pricing (5× multiplier)
- Result: Many riders wait or cancel (high prices)
- Model 1: Matches remaining riders quickly
- Model 3: Routes all drivers to surge area
- Outcome: Short-term revenue spike, rider dissatisfaction, geographic coverage gaps

**New System (Balanced Optimization):**

- Moderate surge pricing (2× multiplier): balances revenue with demand retention
- Smart matching: slightly longer rider waits (2-3 min) to keep driver pickup distances reasonable
- Geographic balancing: incentivize some drivers to cover underserved areas
- Outcome: Slightly lower per-trip revenue but more trips completed, better rider retention, healthier driver earnings

**Scenario 2: Driver Shortage in Suburbs**

**Old System:**

- Model 3: Routes drivers to lucrative downtown
- Model 1: Matches suburban riders with very distant drivers (15+ min pickups)
- Result: Suburban riders wait >20 minutes, cancel, poor experience

**New System:**

- Incentivize drivers to suburban areas (higher per-mile rate)
- Dynamic pricing reflecting actual scarcity
- Route optimization: balance driver earnings with geographic coverage
- Outcome: Better suburban service, drivers earn reasonably in suburbs, riders pay premium but get service

**Long-Term Strategic Alignment:**

**Company Mission: "Reliable transportation for everyone, everywhere"**

**Alignment Check:**

- Old system: Unreliable (long waits in some areas), not everywhere (coverage gaps), not everyone (high prices exclude price-sensitive riders)
- New system: More reliable (balanced optimization), better coverage (geographic balancing), more accessible (moderate pricing)

**Stakeholder Value Proposition:**

**For Riders:**

- Reasonable wait times (not absolute minimum, but predictable)
- Fair pricing (not cheapest, but not exploitative surge)
- Reliable service (geographic coverage)

**For Drivers:**

- Good earnings (not maximum per trip, but sustainable hourly rate)
- Reasonable pickups (not zero-distance, but not excessive drives)
- Predictable income (system helps distribute demand)

**For Platform:**

- Sustainable revenue (not maximum extraction, but healthy take rate)
- Participant retention (long-term value)
- Market leadership (reputation for fairness)

**Measuring Success of Integrated System:**

**Baseline (3 Independent Models):**

- Rider NPS: 45
- Driver retention: 60% at 90 days
- Revenue: $100M/month
- Geographic coverage: 75% of zip codes

**After 6 Months (Integrated System):**

- Rider NPS: 58 (+13 points)
- Driver retention: 72% at 90 days (+12 points)
- Revenue: $95M/month (-5%, acceptable for long-term sustainability)
- Geographic coverage: 88% of zip codes (+13 points)

**Long-term (Year 2):**

- Revenue: $105M/month (+5% vs. baseline, as higher retention compounds)
- Market share: gained 3 points from competitors
- Regulatory goodwill: reduced complaints
- Brand perception: seen as fair platform

**Key Lessons:**

1. **Systems thinking essential:** Local optimization creates global suboptimum
2. **Multi-objective from start:** Don't optimize in silos
3. **Stakeholder balance:** Long-term success requires balancing all stakeholders
4. **Short-term vs. long-term:** Maximizing short-term revenue can destroy long-term value
5. **Feedback loops:** Negative spirals from conflicting objectives
6. **Unified optimization:** Single model with weighted objectives beats independent models
7. **Continuous monitoring:** System-level metrics reveal optimization conflicts

This scenario emphasizes: ML systems operate in complex ecosystems. Optimizing individual components without considering system-level effects leads to failure. Successful AI requires holistic, multi-stakeholder optimization.

---

**Question 582 [ADVANCED - Adversarial ML and Security]:**

**SCENARIO:** Your bank deployed a transaction fraud detection model with 95% accuracy. Six months later:

- Fraud losses increased 40% despite model deployment
- Investigation reveals sophisticated fraud ring adapted to your model
- Fraudsters tested thousands of transactions at low amounts to learn model boundaries
- Discovered model weak point: transactions at gas stations under $50 rarely flagged
- Exploited this systematically: thousands of $49 gas station transactions (all fraudulent)
- Model learned pattern: "gas station + under $50 = legitimate" (from historical data)
- Adversaries specifically chose this transaction type to avoid detection

How should adversarial-aware fraud detection be designed?

A) Current model is fine; fraud adaptation is inevitable B) Adversarial ML design: randomization, deception, continuous evolution, anomaly detection, multi-layered defense C) Increase accuracy to 99%; that will stop fraud D) Remove gas station transactions from model

**Answer: B) Adversarial-aware ML with adaptive defenses and deception**

**Explanation:** This scenario demonstrates adversarial machine learning in fraud detection:

**Understanding Adversarial Context:**

**Adversarial ML Characteristics:**

1. **Adaptive adversaries:** Attackers learn from and adapt to model
2. **Strategic behavior:** Deliberately craft inputs to evade detection
3. **Active learning:** Attackers test boundaries, learn model logic
4. **Arms race:** Continuous cat-and-mouse between defense and attack

**Why Traditional ML Fails:**

**Assumption Violation:**

- Traditional ML assumes: test distribution = train distribution
- Adversarial ML: test distribution deliberately manipulated by intelligent adversary
- Attackers change distribution to exploit model weaknesses

**This Attack's Sophistication:**

**Phase 1: Reconnaissance (Weeks 1-4)**

- Submit many small transactions
- Observe which get flagged, which don't
- Build model of detection boundaries
- Identify weak points

**Phase 2: Exploitation (Weeks 5-20)**

- Focus on identified weak point (gas stations, $49)
- Mass exploitation
- Model sees these as "normal" (match training distribution)
- Success rate very high

**Why Model Failed to Detect:**

- Historical data: real gas station transactions mostly legitimate
- Pattern learned: "gas station + reasonable amount = legitimate"
- Fraudsters exploited this learned pattern
- Model couldn't distinguish legitimate from fraudulent gas station purchases

**Adversarial-Robust Fraud Detection Design:**

**Layer 1: Adversarial-Aware Modeling**

**Principle 1: Randomization**

**Problem:** Deterministic models are reverse-engineerable **Solution:** Introduce randomness in detection

```
Instead of:
  if fraud_score > 0.7: flag_as_fraud

Use:
  threshold = random_uniform(0.65, 0.75)
  if fraud_score > threshold: flag_as_fraud
```

**Benefit:** Attacker can't learn exact boundary through testing **Cost:** Slightly less consistent (but more robust)

**Principle 2: Ensemble with Diversity**

**Multiple Independent Models:**

- Model A: Gradient boosting on transaction features
- Model B: Anomaly detection (isolation forest)
- Model C: Network analysis (unusual patterns in transaction graph)
- Model D: Rules-based heuristics (expert knowledge)

**Varying Detection Methods:**

- Statistical models learn patterns
- Anomaly detection catches novelty
- Rules encode known fraud techniques
- Network analysis catches coordinated attacks

**Why This Helps:**

- Attacker must evade all detection methods
- Different models have different blind spots
- Harder to reverse-engineer ensemble than single model

**Principle 3: Adversarial Training**

**Red Team Simulation:**

```
1. Current fraud detection model deployed
2. Red team (internal) tries to evade detection
3. Successful evasions become training data
4. Retrain model to catch these attacks
5. Deploy updated model
6. Repeat monthly
```

**Synthetic Adversarial Examples:**

- Generate synthetic fraud attempts designed to evade model
- Adversarial perturbations: slight modifications fooling model
- Include in training data
- Model learns to be robust to evasion attempts

**Layer 2: Deception and Honeypots**

**Principle 4: Deceptive Signals**

**Problem:** Attackers learn by testing **Solution:** Send misleading feedback

**Honeypot Transactions:**

- Deliberately allow some fraudulent-looking transactions (actually monitored closely)
- Attackers think these succeeded (they didn't)
- Leads attackers down unproductive paths
- Gather intelligence on attacker techniques

**Delayed Feedback:**

- Don't immediately flag suspicious transactions
- Delay hours or days before acting
- Prevents immediate feedback to attackers
- Makes pattern learning harder

**False Positives as Defense:**

- Occasionally flag legitimate transactions in patterns attackers use
- Makes attackers unsure what's working
- Discourages systematic testing

**Layer 3: Anomaly and Novelty Detection**

**Principle 5: Detect Unusual Patterns**

**Why Pattern Matching Insufficient:**

- Model learned: "gas station + $49 = normal"
- But this specific pattern repeated thousands of times IS unusual
- Need meta-pattern detection

**Anomaly Detection Signals:**

- **Volume:** Same transaction type repeated unusually often
- **Timing:** Transactions at unusual hours or intervals
- **Velocity:** Rapid succession of similar transactions
- **Geographic:** Transactions from impossible locations
- **Behavioral:** Deviation from user's normal behavior

**One-Class Classification:**

- Train on known legitimate transactions
- Flag anything that doesn't look like legitimate (even if seen before)
- Catches novel fraud not in training data

**Layer 4: Continuous Learning and Adaptation**

**Principle 6: Rapid Evolution**

**Problem:** Static model becomes obsolete **Solution:** Continuous model updates

**Frequent Retraining:**

- Weekly or daily model updates
- Incorporate latest fraud examples
- Short memory: focus on recent patterns (older data may be outdated)

**Online Learning:**

- Model updates in real-time or near real-time
- Adapts to new fraud tactics quickly
- Catches emerging patterns before large losses

**Concept Drift Adaptation:**

- Fraud patterns change faster than normal concept drift
- Adapt quickly to stay ahead of attackers

### Layer 5: Multi-Signal Fusion

**Principle 7: Holistic View**

**Don't Rely on Transaction Alone:**

- **Device fingerprinting:** Device ID, browser, OS
- **Behavioral biometrics:** Typing patterns, mouse movements
- **Network analysis:** Connections to known fraudsters
- **External data:** Stolen card databases, fraud intelligence
- **Geographic:** IP location vs. billing address
- **Historical:** User's transaction history

**For Gas Station Attack:**

- Even though transaction looks normal
- Detect: unusual velocity of gas purchases
- Geographic anomalies: purchases from distant stations
- Behavioral: doesn't match user's typical gas purchase pattern

### Layer 6: Human-in-the-Loop

**Principle 8: Expert Oversight**

**Tiered Review:**

- **Tier 1:** Obvious fraud (high confidence) → auto-block
- **Tier 2:** Suspicious (medium confidence) → human review
- **Tier 3:** Borderline (low confidence) → additional authentication

**Fraud Analyst Insights:**

- Humans notice patterns models miss
- Adversaries must evade both ML and human reviewers
- Humans provide ground truth for model retraining

**Customer Interaction:**

- Multi-factor authentication for suspicious transactions
- Customer confirms legitimate transactions

- Friction for fraudsters, manageable for legitimate users

**Layer 7: Adversarial Robustness Testing**

**Principle 9: Proactive Testing**

**Adversarial Red Team:**

- Internal team tries to evade detection
- Simulate sophisticated attackers
- Find weaknesses before real attackers do
- Close vulnerabilities preemptively

**Penetration Testing:**

- Regular testing of fraud detection system
- Automated adversarial example generation
- Test model robustness to perturbations

**Bug Bounty:**

- Reward external security researchers for finding detection gaps
- Crowd-sourced vulnerability discovery

**Implementation Roadmap:**

**Month 1: Immediate Response to Current Attack**

1. **Rule-based override:** Temporarily flag high-velocity gas station transactions
2. **Manual review:** Human review of gas station patterns
3. **Block known fraud:** Block identified fraudulent cards immediately
4. **Contain losses:** Stop bleeding while building robust solution

**Months 2-3: Redesign Foundation**

1. **Ensemble model:** Build diverse detection methods
2. **Anomaly detection:** Deploy velocity and pattern anomaly detection
3. **Randomization:** Add randomness to detection thresholds
4. **Multi-signal:** Incorporate device, behavioral, network signals

**Months 4-6: Adversarial Features**

1. **Red team:** Establish internal adversarial testing team
2. **Adversarial training:** Generate and train on adversarial examples
3. **Deception:** Implement honeypots and delayed feedback
4. **Continuous learning:** Establish rapid retraining pipeline

**Ongoing: Arms Race Management**

1. **Monthly red team exercises:** Find vulnerabilities continuously
2. **Weekly model updates:** Stay ahead of fraud evolution
3. **Threat intelligence:** Share and receive fraud intelligence
4. **Innovation:** Research emerging adversarial ML techniques

**Metrics for Adversarial Robustness:**

**Detection Metrics:**

- Fraud catch rate (TPR)
- False positive rate (FPR)
- Precision at operating point
- Time to detection (minimize fraud exposure window)

**Robustness Metrics:**

- **Evasion resistance:** How many adversarial examples evade detection?
- **Adaptability:** How quickly does system catch new fraud tactics?
- **Resilience:** Does performance degrade gracefully under attack?
- **Coverage:** What % of fraud attack vectors are covered?

**Business Metrics:**

- Fraud loss rate ($ lost / $ processed)
- Legitimate transaction approval rate (minimize friction)
- Customer satisfaction (balance security and convenience)
- Operational costs (human review, false positive investigations)

**Adversarial ML Lessons from Other Domains:**

**Spam Filtering:**

- Arms race: spammers constantly adapt
- Successful approaches: ensemble, randomization, rapid updates
- Never-ending cat-and-mouse game

**Malware Detection:**

- Adversaries craft malware to evade detection
- Defenses: behavior-based detection, sandboxing, deception
- Static signatures insufficient

**Adversarial Examples in Computer Vision:**

- Small perturbations fool models
- Defenses: adversarial training, input sanitization, ensemble

**Key Differences in Fraud vs. Traditional ML:**

**Traditional ML:**

- Assumes stationary distribution
- Test distribution = train distribution
- Model optimized for current data

**Adversarial ML (Fraud):**

- Distribution actively manipulated by intelligent adversary
- Test distribution deliberately shifted to exploit model
- Must optimize for robustness, not just accuracy

**Key Lessons:**

1. **Static models fail against adversaries:** Must continuously evolve
2. **Accuracy isn't enough:** Need robustness to adversarial manipulation
3. **Randomization helps:** Deterministic systems are reverse-engineerable
4. **Diversity is strength:** Ensemble of diverse methods harder to evade
5. **Deception is defense:** Honeypots and misleading feedback confuse attackers
6. **Humans remain important:** Human review catches what models miss
7. **Arms race is permanent:** No final solution; continuous adaptation required
8. **Multi-layered defense:** No single technique sufficient

This scenario emphasizes: Adversarial ML requires fundamentally different design thinking. Security through adversarial robustness, not just predictive accuracy.

---

# References


- Ackoff, R. L. (1989). From data to wisdom. *Journal of Applied Systems Analysis*, 16(1), 3–9.
- Amershi, S., Begel, A., Bird, C., DeLine, R., Gall, H., Kamar, E., ... & Zimmermann, T. (2019). Software engineering for machine learning: A case study. In *2019 IEEE/ACM ICSE-SEIP* (pp. 291–300). IEEE.
- Amodei, D., Olah, C., Steinhardt, J., Christiano, P., Schulman, J., & Mané, D. (2016). Concrete problems in AI safety. *arXiv:1606.06565*.
- Arrieta, A. B., Díaz-Rodríguez, N., Del Ser, J., Bennetot, A., Tabik, S., Barbado, A., ... & Herrera, F. (2020). Explainable artificial intelligence (XAI)… *Information Fusion*, 58, 82–115.
- Batini, C., & Scannapieco, M. (2016). *Data and information quality: Dimensions, principles and techniques.* Springer.
- Bergstra, J., & Bengio, Y. (2012). Random search for hyper-parameter optimization. *JMLR*, 13, 281–305.
- Breck, E., Cai, S., Nielsen, E., Salib, M., & Sculley, D. (2017). The ML test score… In *IEEE Intl. Conf. on Big Data* (pp. 1123–1132).
- Brown, T. B., Mann, B., Ryder, N., Subbiah, M., Kaplan, J., Dhariwal, P., ... & Amodei, D. (2020). Language models are few-shot learners. *Advances in Neural Information Processing Systems*, 33, 1877–1901.
- Chawla, N. V., Bowyer, K. W., Hall, L. O., & Kegelmeyer, W. P. (2002). SMOTE… *Journal of Artificial Intelligence Research*, 16, 321–357.
- Chouldechova, A. (2017). Fair prediction with disparate impact… *Big Data*, 5(2), 153–163.
- Cognilytica. (2024). *CPMAI methodology overview: Managing AI and machine learning projects.*
- Crankshaw, D., Wang, X., Zhou, G., Franklin, M. J., Gonzalez, J. E., & Stoica, I. (2017). Clipper: A low-latency online prediction serving system. In *USENIX NSDI 17* (pp. 613–627).

- Cunningham, J. P., & Ghahramani, Z. (2015). Linear dimensionality reduction… *JMLR*, 16(1), 2859–2900.
- Davenport, T. H., & Harris, J. G. (2017). *Competing on analytics* (Updated ed.). Harvard Business Press.
- Davenport, T. H., & Ronanki, R. (2018). Artificial intelligence for the real world. *Harvard Business Review*, 96(1), 108–116.
- Demšar, J. (2006). Statistical comparisons of classifiers over multiple data sets. *JMLR*, 7, 1–30.
- Dietterich, T. G. (2000). Ensemble methods in machine learning. In *International Workshop on Multiple Classifier Systems* (pp. 1–15). Springer.
- Dong, X. L., & Rekatsinas, T. (2018). Data integration and machine learning: A natural synergy. *PVLDB*, 11(12), 2094–2097.
- Dwork, C., & Roth, A. (2014). The algorithmic foundations of differential privacy. *Foundations and Trends in Theoretical Computer Science*, 9(3–4), 211–407.
- European Commission. (2023). The EU AI Act.
- European Commission. (2024). The EU AI Act.
- Fawcett, T. (2006). An introduction to ROC analysis. *Pattern Recognition Letters*, 27(8), 861–874.
- Feurer, M., & Hutter, F. (2019). Hyperparameter optimization. In *Automated Machine Learning* (pp. 3–33). Springer.
- Floridi, L., Cowls, J., Beltrametti, M., Chatila, R., Chazerand, P., Dignum, V., ... & Vayena, E. (2018). AI4People—An ethical framework… *Minds and Machines*, 28(4), 689–707.
- Fountaine, T., McCarthy, B., & Saleh, T. (2019). Building the AI-powered organization. *Harvard Business Review*, 97(4), 62–73.
- Geman, S., Bienenstock, E., & Doursat, R. (1992). Neural networks and the bias/variance dilemma. *Neural Computation*, 4(1), 1–58.
- Géron, A. (2019). *Hands-On Machine Learning with Scikit-Learn, Keras, and TensorFlow* (2nd ed.). O'Reilly.

- Goodfellow, I., Bengio, Y., & Courville, A. (2016). *Deep Learning.* MIT Press.
- Hastie, T., Tibshirani, R., & Friedman, J. (2009). *The elements of statistical learning* (2nd ed.). Springer.
- He, K., Zhang, X., Ren, S., & Sun, J. (2016). Deep residual learning for image recognition. In *CVPR* (pp. 770–778).
- Huyen, C. (2022). *Designing machine learning systems.* O'Reilly.
- James, G., Witten, D., Hastie, T., & Tibshirani, R. (2013). *An introduction to statistical learning.* Springer.
- Japkowicz, N., & Shah, M. (2011). *Evaluating learning algorithms: A classification perspective.* Cambridge University Press.
- Jobin, A., Ienca, M., & Vayena, E. (2019). The global landscape of AI ethics guidelines. *Nature Machine Intelligence*, 1(9), 389–399.
- Khatri, V., & Brown, C. V. (2010). Designing data governance. *Communications of the ACM*, 53(1), 148–152.
- Kleinberg, J., Mullainathan, S., & Raghavan, M. (2017). Inherent trade-offs in the fair determination of risk scores. In *ITCS 2017.*
- Kohavi, R. (1995). A study of cross-validation and bootstrap for accuracy estimation and model selection. In *IJCAI* (Vol. 14, pp. 1137–1145).
- Kohavi, R., Longbotham, R., Sommerfield, D., & Henne, R. M. (2009). Controlled experiments on the web: Survey and practical guide. *Data Mining and Knowledge Discovery*, 18(1), 140–181.
- Kotter, J. P. (2012). *Leading change* (reissue). Harvard Business Review Press.
- Little, R. J., & Rubin, D. B. (2019). *Statistical analysis with missing data* (3rd ed.). Wiley.
- Lohr, S. (2014, August 17). For big-data scientists, "janitor work" is key hurdle to insights. *The New York Times.*
- Lu, J., Liu, A., Dong, F., Gu, F., Gama, J., & Zhang, G. (2018). Learning under concept drift: A review. *IEEE TKDE*, 31(12), 2346–2363.

- Lundberg, S. M., & Lee, S. I. (2017). A unified approach to interpreting model predictions. In *NeurIPS* (pp. 4765–4774).
- Marz, N., & Warren, J. (2015). *Big data: Principles and best practices of scalable real-time data systems.* Manning.
- Mehrabi, N., Morstatter, F., Saxena, N., Lerman, K., & Galstyan, A. (2021). A survey on bias and fairness in machine learning. *ACM Computing Surveys*, 54(6), 1–35.
- Mitchell, M., Wu, S., Zaldivar, A., Barnes, P., Vasserman, L., Hutchinson, B., ... & Gebru, T. (2019). Model cards for model reporting. In *FAT* (pp. 220–229).
- Molnar, C. (2020). *Interpretable machine learning.* Lulu.com.
- NIST. (2023). *AI Risk Management Framework.* National Institute of Standards and Technology.
- Obermeyer, Z., Powers, B., Vogeli, C., & Mullainathan, S. (2019). Dissecting racial bias… *Science*, 366(6464), 447–453.
- Paleyes, A., Urma, R. G., & Lawrence, N. D. (2022). Challenges in deploying machine learning: A survey of case studies. *ACM Computing Surveys*, 55(6), 1–29.
- Pan, S. J., & Yang, Q. (2010). A survey on transfer learning. *IEEE TKDE*, 22(10), 1345–1359.
- Piatetsky, G. (2019). Why do most machine learning projects fail? *KDnuggets.*
- PMI. (2025). *PMI-CPMAI v7 Examination Content Outline and Specifications.* Project Management Institute.
- Polyzotis, N., Roy, S., Whang, S. E., & Zinkevich, M. (2018). Data lifecycle challenges in production machine learning: A survey. *ACM SIGMOD Record*, 47(2), 17–28.
- Provost, F., & Fawcett, T. (2013). *Data science for business.* O'Reilly.
- Radford, A., Wu, J., Child, R., Luan, D., Amodei, D., & Sutskever, I. (2019). Language models are unsupervised multitask learners. *OpenAI Blog*, 1(8), 9.
- Rahm, E., & Do, H. H. (2000). Data cleaning: Problems and current approaches. *IEEE Data Engineering Bulletin*, 23(4), 3–13.

- Raji, I. D., Smart, A., White, R. N., Mitchell, M., Gebru, T., Hutchinson, B., ... & Barnes, P. (2020). Closing the AI accountability gap… In *FAccT* (pp. 33–44).
- Ribeiro, M. T., Singh, S., & Guestrin, C. (2016). "Why should I trust you?" Explaining the predictions of any classifier. In *KDD* (pp. 1135–1144).
- Roh, Y., Heo, G., & Whang, S. E. (2019). A survey on data collection for machine learning… *IEEE TKDE*, 33(4), 1328–1347.
- Rudin, C. (2019). Stop explaining black box machine learning models… *Nature Machine Intelligence*, 1(5), 206–215.
- Russell, S., & Norvig, P. (2020). *Artificial intelligence: A modern approach* (4th ed.). Pearson.
- Saltz, J. S., & Dewar, N. (2019). Data science ethical considerations… *Ethics and Information Technology*, 21(3), 197–208.
- Sculley, D., Holt, G., Golovin, D., Davydov, E., Phillips, T., Ebner, D., ... & Dennison, D. (2015). Hidden technical debt in machine learning systems. *NeurIPS*, 28, 2503–2511.
- Settles, B. (2012). *Active learning.* Synthesis Lectures on AI and Machine Learning, 6(1), 1–114.
- Shneiderman, B. (2020). Bridging the gap between ethics and practice… *ACM TiiS*, 10(4), 1–31.
- Sweeney, L. (2002). k-anonymity: A model for protecting privacy. *International Journal of Uncertainty, Fuzziness and Knowledge-Based Systems*, 10(5), 557–570.
- Sutton, R. S., & Barto, A. G. (2018). *Reinforcement learning: An introduction* (2nd ed.). MIT Press.
- Tukey, J. W. (1977). *Exploratory data analysis.* Addison-Wesley.
- Vaswani, A., Shazeer, N., Parmar, N., Uszkoreit, J., Jones, L., Gomez, A. N., ... & Polosukhin, I. (2017). Attention is all you need. *NeurIPS*, 30, 5998–6008.
- Voigt, P., & Von dem Bussche, A. (2017). *The EU General Data Protection Regulation (GDPR): A practical guide.* Springer.
- Wickham, H., & Grolemund, G. (2017). *R for data science.* O'Reilly.

- Žliobaitė, I. (2010). Learning under concept drift: An overview. *arXiv:1010.4784*.

Made in the USA
Monee, IL
08 November 2026

67816257R10240